NEW TECHNIQUES IN FOOD AND BEVERAGE MICROBIOLOGY

A complete list of titles in the Society for Applied Bacteriology Technical Series appears at the end of this volume

THE SOCIETY FOR APPLIED BACTERIOLOGY
TECHNICAL SERIES NO 31

New Techniques in Food and Beverage Microbiology

Edited by

R.G.KROLL
Torry Research Station
Aberdeen

A.GILMOUR
Agriculture and Food Science Centre
The Queen's University of Belfast

M.SUSSMAN
Department of Microbiology
Medical School
University of Newcastle upon Tyne

OXFORD
BLACKWELL SCIENTIFIC PUBLICATIONS
LONDON EDINBURGH BOSTON
MELBOURNE PARIS BERLIN VIENNA

and published for them by
Blackwell Scientific Publications
Editorial Offices:
Osney Mead, Oxford OX2 0EL
25 John Street, London WC1N 2BL
23 Ainslie Place, Edinburgh EH3 6AJ
238 Main Street, Cambridge
Massachusetts 02142, USA
54 University Street, Carlton
Victoria 3053, Australia

Other Editorial Offices:
Librairie Arnette SA
2, rue Casimir-Delavigne
75006 Paris
France

Blackwell Wissenschafts-Verlag GmbH
Düsseldorfer St. 38
D-10707 Berlin
Germany

Blackwell MZV
Feldgasse 13
A-1238 Wien
Austria

First published 1993

Set by Excel Typesetters Company, Hong Kong
Printed and bound in Great Britain
at The University Press, Cambridge

DISTRIBUTORS

Marston Book Services Ltd
PO Box 87
Oxford OX2 0DT
(*Orders*: Tel: 0865 791155
Fax: 0865 791927
Telex: 837515)

USA
Blackwell Scientific Publications, Inc.
238 Main Street
Cambridge, MA 02142
(*Orders*: Tel: 800 759-6102
617 876-7000)

Canada
Oxford University Press
70 Wynford Drive
Don Mills
Ontario M3C 1J9
(*Orders*: Tel: 416 441-2941)

Australia
Blackwell Scientific Publications Pty Ltd
54 University Street
Carlton, Victoria 3053
(*Orders*: Tel: 03 347-5552)

A catalogue record for this title
is available from the British Library

ISBN 0-632-03755-5

Library of Congress
Cataloging-in-Publication Data

New techniques in food and beverage microbiology/
[edited by] R.G. Kroll, A. Gilmour, M. Sussman.
p. cm. — (Technical series; no. 31)
Includes bibliographical references and index.
ISBN 0-632-03755-5
1. Food—Microbiology—Methodology.
2. Beverages—Microbiology—Methodology.
I. Kroll, R. G. II. Gilmour, A. III. Sussman,
Max. IV. Series: Technical series (Society for
Applied Bacteriology); no. 31.
QR115.N45 1993
576'.163—dc20

Contents

Contributors

H.J.Ball, *Veterinary Sciences Division, Department of Agriculture for Northern Ireland, Stoney Road, Stormont, Belfast BT4 3SD, UK*

R.P.Betts, *Campden Food and Drink Research Association, Chipping Campden, Gloucestershire GL55 6LD, UK*

A.J.Bjourson, *Food Microbiology Division, Department of Agriculture for Northern Ireland, Newforge Lane, Belfast BT9 5PX, UK*

T.H.Birkbeck, *Department of Microbiology, University of Glasgow, Glasgow G12 8QQ, UK*

M.Blackburn, *Ministry of Agriculture, Fisheries and Food, Fisheries Laboratory, Remembrance Avenue, Burnham-on-Crouch, Essex CM0 8HA, UK*

J.A.Bobbitt, *Campden Food and Drink Research Association, Chipping Campden, Gloucestershire GL55 6LD, UK*

M.A.Brailsford, *Chemunex UK, Opas Centre, St John's Innovation Park, Cowley Road, Cambridge CB4 4WS, UK*

M.Byrne, *Thames Water Utilities, Spencer House Laboratory, Manor Farm Road, Reading RG2 0JN, UK*

Shona Campbell, *Department of Microbiology, AFRC Institute of Food Research, Reading Laboratory, Earley Gate, Whiteknights Road, Reading RG6 2EF, UK*

P.E.Carter, *Department of Medical Microbiology, University of Aberdeen Medical School, Foresterhill, Aberdeen AB9 2ZD, UK*

D.J.Clarke, *Sensor Development Group, Centre for Applied Microbiology and Research, Public Health Laboratory Service, Porton Down, Salisbury, Wiltshire SP4 0JG, UK*

Rosemary G.Clarke, *AFRC Institute of Food Research, Norwich Research Park, Colney, Norwich, Norfolk NR4 7UA, UK*

J.E.Cooper, *Agriculture and Food Science Centre, Queen's University of Belfast, Newforge Lane, Belfast BT9 5PX, UK*

K.S.Cudjoe, *Norwegian College of Veterinary Medicine, Department of Food Hygiene, PO Box 8146 Dep. N-0033, Oslo, Norway*

J.P.Diaper, *Department of Genetics and Microbiology, Life Sciences Building, University of Liverpool, Liverpool L69 3BX, UK*

Joanne H.Dickinson, *AFRC Institute of Food Research, Reading Laboratory, Earley Gate, Whiteknights Road, Reading RG6 2EF, UK*

P.Druggan, *Life Technologies Ltd, PO Box 35, Trident House, Renfrew Road, Paisley PA3 4EF, UK*

C.Edwards, *Department of Genetics and Microbiology, Life Sciences Building, University of Liverpool, Liverpool L69 3BX, UK*

KAREN EVANS, *Ministry of Agriculture, Fisheries and Food, Fisheries Laboratory, Remembrance Avenue, Burnham-on-Crouch, Essex CM0 8HA, UK*

S.J.FORSYTHE, *Department of Life Sciences, Nottingham Polytechnic, Clifton Lane, Nottingham NG11 8NS, UK*

R.FREEMAN, *Department of Microbiology, Medical School, Newcastle upon Tyne NE2 4HH, UK*

C.R.FRICKER, *Thames Water Utilities, Spencer House Laboratory, Manor Farm Road, Reading RG2 0JN, UK*

SUSAN GALLACHER, *Scottish Office Agriculture and Fisheries Department, Marine Laboratory, Victoria Road, Aberdeen AB9 8DB, UK*

S.GATLEY, *Chemunex SA, 41 Rue du 11 Novembre, 94700 Maisons Alfort, Paris, France*

A.GILMOUR, *Agriculture and Food Science Centre, The Queen's University of Belfast, Newforge Lane, Belfast BT9 5PX, UK*

KATHLEEN A.GRANT, *AFRC Institute of Food Research, Reading Laboratory, Earley Gate, Whiteknights Road, Reading RG6 2EF, UK*

S.D.HAINES, *bioMérieux UK Ltd, Grafton House, Grafton Way, Basingstoke RG22 6HY, UK*

J.HERNANDEZ, *Departamento de Biotecnologia, Laboratorio de Microbiologia, Universidad Politecnica de Valencia, Camino de Vera 14, 46022 Valencia, Spain*

C.J.HEUVELMAN, *Laboratory for Water and Food Microbiology, National Institute of Public Health and Environmental Protection, PO Box 1, NL-3720 BA Bilthoven, The Netherlands*

J.L.JONES, *Campden Food and Drink Research Association, Chipping Campden, Gloucestershire GL55 6LD, UK*

S.KERR, *Veterinary Sciences Division, Department of Agriculture for Northern Ireland, Stoney Road, Stormont, Belfast BT4 3SD, UK*

R.G.KROLL, *MAFF, Torry Research Station, PO Box 31, 135 Abbey Road, Aberdeen AB9 8DG, UK*

N.F.LIGHTFOOT, *Regional Public Health Laboratory, Newcastle General Hospital, Westgate Road, Newcastle upon Tyne NE4 6BE, UK*

D.P.MACKIE, *Veterinary Sciences Division, Department of Agriculture for Northern Ireland, Stoney Road, Stormont, Belfast BT4 3SD, UK*

P.G.MURPHY, *Bacteriology Department and Northern Ireland Public Health Laboratory, Belfast City Hospital, Lisburn Road, Belfast BT9 7AD, UK*

E.OLSEN, *Dynal A/S, PO Box 158, Skøyen, N-0212 Oslo, Norway*

Ø.OLSVIK, *Enteric Diseases Branch, MSCO3, Center for Disease Control, Atlanta, Georgia 30333, USA*

R.J.OWEN, *National Collection of Type Cultures, Central Public Health Laboratory, London NW9 5HT, UK*

R.W.OWEN, *Sensor Development Group, Centre for Applied Microbiology and Research, Public Health Laboratory Service, Porton Down, Salisbury, Wiltshire SP4 0JG, UK*

P.D.PATEL, *Leatherhead Food Research Association, Randalls Road, Leatherhead, Surrey KT22 7RY, UK*

M.J.PAYNE, *Department of Microbiology, AFRC Institute of Food Research, Reading Laboratory, Earley Gate, Whiteknights Road, Reading RG6 2EF, UK*

T.H.PENNINGTON, *Department of Medical Microbiology, University of Aberdeen Medical School, Foresterhill, Aberdeen AB9 2ZD, UK*

A.C.PINDER, *AFRC Institute of Food Research, Norwich Research Park, Colney, Norwich, Norfolk NR4 7UA, UK*

SANDY A.G.POULTER, *AFRC Institute of Food Research, Norwich Research Park, Colney, Norwich, Norfolk NR4 7UA, UK*

CAROLE L.REID, *Department of Chemical Engineering, Loughborough University of Technology, Loughborough, Leicestershire LE12 3TU, UK*

ANNE M.RIDLEY, *Laboratory of Microbiological Reagents, Central Public Health Laboratory, 61 Colindale Avenue, London NW9 5HT, UK*

A.H.RITCHIE, *Torry Research Station, PO Box 31, 135 Abbey Road, Aberdeen AB9 8DG, UK*

N.A.SAUNDERS, *Laboratory of Microbiological Reagents, Central Public Health Laboratory, 61 Colindale Avenue, London NW9 5HT, UK*

K.SHEPHERD, *Thames Water Utilities, Spencer House Laboratory, Manor Farm Road, Reading RG2 0JN, UK*

P.SILLEY, *Don Whitley Scientific Ltd, 14 Otley Road, Shipley, West Yorkshire BD17 7SE, UK*

PENELOPE R.SISSON, *Regional Public Health Laboratory, Newcastle General Hospital, Westgate Road, Newcastle upon Tyne NE4 6BE, UK*

E.SKJERVE, *Norwegian College of Veterinary Medicine, Department of Food Hygiene, PO Box 8146 Dep, N-0033 Oslo, Norway*

J.S.SLADE, *Thames Water Utilities, Spencer House Laboratory, Manor Farm Road, Reading RG2 0JN, UK*

J.P.STEER, *Campden Food and Drink Research Association, Chipping Campden, Gloucestershire GL55 6LD, UK*

M.SUSSMAN, *Department of Microbiology, Medical School, Newcastle upon Tyne NE2 4HH, UK*

G.S.A.B.STEWART, *Department of Applied Biochemistry and Food Science, University of Nottingham, Sutton Bonington, Loughborough, Leicestershire LE12 5RD, UK*

FIONA M.THOMSON-CARTER, *Department of Medical Microbiology, University of Aberdeen Medical School, Foresterhill, Aberdeen AB9 2ZD, UK*

G.VESEY, *Thames Water Utilities, Spencer House Laboratory, Manor Farm Road, Reading RG2 0JN, UK*

W.M.WAITES, *Department of Applied Biochemistry and Food Science, University of Nottingham, Sutton Bonington, Loughborough, Leicestershire LE12 5RD, UK*

M.WALDOCK, *Ministry of Agriculture, Fisheries and Food, Fisheries Laboratory, Remembrance Avenue, Burnham-on-Crouch, Essex CM0 8HA, UK*

K.WERNARS, *Laboratory for Water and Food Microbiology, National Institute of Public Health and Environmental Protection, PO Box 1, NL-3720 BA Bilthoven, The Netherlands*

D.W.WILLIAMS, *Leatherhead Food Research Association, Randalls Road, Leatherhead, Surrey KT22 7RY, UK*

I.G.WILSON, *Public Health Laboratory, Belfast City Hospital, Lisburn Road, Belfast BT9 7AD, UK*

A.F.ZAMANI, *Sensor Development Group, Centre for Applied Microbiology and Research, Public Health Laboratory Service, Porton Down, Salisbury, Wiltshire SP4 0JG, UK*

Preface

Throughout the Western world there is an increasing public demand for safer foods of higher quality. Microorganisms play a highly significant role in both the theoretical and the practical aspects of food and beverage collection, processing, retailing and distribution. Current methods for detecting and identifying foodborne organisms have proved inadequate in speed, precision and accuracy, and their improvement has been a concern of the Society for Applied Bacteriology for many years. It is also a subject of interest to producers, manufacturers and retailers, as well as regulatory and enforcement bodies. Since many of the existing methods have been established for many years, there is a need for new approaches and concepts. This is, therefore, a scientifically and technically challenging subject for investigation.

Over the last two decades progress in the analytical microbiology of foods and beverages has been the subject of several books, for example M.D.Pierson & N.J.Stern (eds), *Foodborne Microorganisms and their Toxins*; Developing Methodology (New York: Marcel Dekker, 1986), C.J.Stannard, S.B.Petit & F.A.Skinner (eds), *Rapid Microbiological Methods for Foods, Beverages and Pharmaceuticals*, Society for Applied Bacteriology Technical Series No. 25 (Oxford: Blackwell Scientific Publications, 1989) and M.R.Adams & C.F.A.Hope (eds), *Rapid Methods in Food Microbiology*, Progress in Industrial Microbiology Vol. 26 (Amsterdam: Elsevier, 1989). Rather than attempt to update these volumes, we have assembled contributions that reflect some recent technical developments. To this end, and reflecting the emphasis of the Society for Applied Bacteriology Technical Series, on practical matters, there are several contributions on similar novel approaches from members of different laboratories. The varied emphasis of these chapters reflects the authors' different technical and scientific experience.

The contributions to this book fall into three main groups: novel detection methods (including flow cytometry, impedance microbiology, polymerase chain reactions, oligonucleotide probes and algal toxin detection by tissue culture and physicochemical methods), separation methods (including antibody and lectin bioaffinity and novel physical methods) and new typing methods (including random PCR priming, restriction fragment length polymorphism, enzyme electrophoresis, pyrolysis mass spectrometry and pulsed-field gel electrophoresis). Though none of the methods described in this volume has so far found widespread routine application, we hope that these pages will contribute to improving the safety and quality of foods and beverages.

This book, Volume 31 in the Society for Applied Bacteriology Technical Series, is based on a Demonstration Meeting held at the King's Road Campus of King's College, London, in September 1992. The editors wish to thank the contributors for their presentations at the meeting and for the skilful preparation of the chapters that constitute this book. Their efforts were greatly enhanced on the day by Dr Sally Bloomfield, the Local Organizer of the meeting, and by Mr Phil Masters, the SAB Meetings Secretary.

R.G.Kroll
A.Gilmour
M.Sussman

1: Extraction and Concentration of Foodborne Pathogenic Bacteria by Lectin Magnetic Separation

M.J.Payne,[1] Shona Campbell[1] and R.G.Kroll[2]
[1]*Department of Microbiology, AFRC Institute of Food Research, Reading Laboratory, Earley Gate, Whiteknights Road, Reading RG6 2EF, UK;*
and [2]MAFF, Torry Research Station, PO Box 31, 135 Abbey Road, Aberdeen AB9 8DG, UK

Many different rapid methods of detecting pathogenic bacteria in foods and clinical samples have been developed over recent years. The sensitivity of many of these can, however, be greatly impaired by background interference due to the presence of sample components. Furthermore, pathogens and some spoilage organisms have to be detected when present in low numbers, often against a background of a comparatively large number of saprophytic microorganisms. The separation of bacteria from samples before detection should reduce background interference and has the potential for enhancing the sensitivity of many rapid methods (Payne & Kroll 1991). Separation methods, in addition to increasing the concentration of cells to levels necessary for detection, can concentrate bacteria, so further lowering detection thresholds. This is important, because it is commonly found that 'rapid' methods can only be applied after enrichment for many hours to increase the numbers of target organisms. Thus, separation methods may make possible the direct application of rapid methods to samples and facilitate the truly rapid, sensitive and specific detection of pathogens.

Separation Processes for Isolating Bacteria

Physical and chemical methods have been used to concentrate bacteria from foods and clinical samples. The major physical separation techniques are filtration and centrifugation. There are problems with both because the cell size or density of bacteria is similar to many components of the food matrix.

New Techniques in
Food and Beverage Microbiology

0-632-03755-5

Prefiltration steps can remove larger particulate matter (Entis 1990), but can cause a loss of bacteria at each filtration step in addition to prolonging the time taken to concentrate the organisms.

Separation techniques that depend on the specific recognition of various cell-surface characteristics of microorganisms have been developed, and these afford more selectivity than simple physical methods for the isolation of bacteria from food matrices. These techniques exploit hydrophobic/hydrophilic interactions (Entis 1990; Wibawan *et al.* 1992), charge effects (Wood & Gibbs 1982), antibody specificity (Doyle & Schoeni 1987; Skjerve *et al.* 1990; Cudjoe *et al.* 1991; Skjerve & Olsvik 1991), lectins (Payne *et al.* 1992, 1993a) and other proteins that bind bacteria (Payne *et al.* 1993b). Although these methods involve chemical interactions, physical removal of adsorbed bacteria remains necessary. The ligands have, therefore, to be bound to an inert immobilization support. This must be chosen carefully because, even if the affinity ligand used is highly specific and able to recognize target organism cell surfaces in complex media, the lack of an efficient immobilization support will severely affect performance (Payne *et al.* 1992). Hence, the support must be able to bind to the affinity ligand so that activity of the ligand is maximized and it must be easily removable from food samples.

The ability of immobilized bacterial binding components to adsorb and extract bacteria from buffered systems has been studied with a wide range of immobilization supports, including sepharose, agarose gels, polystyrene beads and membranes, all of which possess active groups. The effectiveness of these techniques, as monitored by conventional plating regimes, can be variable (Payne *et al.* 1992). Sepharose and agarose gels (45–200 μm diameter beads) proved difficult to remove from many foods after bacteria had been adsorbed to their ligands. Conversely, polystyrene beads are easily recovered from food samples but, because of their large size (7 mm diameter), they present a smaller surface area than a similar volume of gel beads, and so have fewer binding sites. Membranes to which bacterial binding proteins are linked are ineffective, because large pieces of food matrix inhibit adsorption of bacteria to the membrane, which easily becomes blocked, allowing only small samples of food to be filtered. Membranes may also be suspended in food samples, rather than being used as a filter pad. However, we have observed variable binding of bacteria to lectin-liganded membranes when used in this way (Payne *et al.* 1992). Magnetic microspheres (2.8–40 μm in diameter) provide a more effective method of extracting bacteria, partly because removal of microspheres from food relies on a magnetic field, to which no components of the food respond, and partly because of their potential to bind to large numbers of bacteria because of their small size and large surface area. Microspheres are readily recovered after incubation with contaminated samples, without the need for centrifugation or lengthy filtration steps (see also Chapter 2).

Adsorbed bacteria are then removed to another environment. The time necessary to extract the bacteria is therefore minimized and loss of target organisms is reduced, because only one separation step is required.

Lectin magnetic separation

Lectin magnetic separation involves the immobilization of a lectin(s) on the surface of magnetic microspheres, the attachment of this lectin to a specific sugar or aminosugar present in the bacterial cell envelope and the physical extraction of the adsorbed cells with a magnetic field. Lectins are generally described as non-enzymic proteins that possess at least two sugar-binding sites. Their ability to bind specifically to different cell types is, therefore, often determined by simple cell agglutination assays. Lectins are ubiquitous throughout the plant and animal kingdoms, but little has been learned about their endogenous cellular functions (Barondes 1988). Lectins are used in eukaryotic cell separation (Etzler 1985) and for labelling cell-surface components for tissue typing (Lis & Sharon 1986). Their ability to recognize and bind bacteria in invertebrates implies that they may form part of a primitive defence or 'pseudo-immune' system analogous to antibodies in higher organisms. Unfortunately, most commercially available lectins have been characterized by means of haemagglutination assays. Hence, the range of binding specificities is limited. There must be many more lectins with a higher specificity for the surface structural carbohydrates of pathogenic, and other, bacteria. Nevertheless, several commercially available lectins have potential in the isolation of bacteria from food samples. In addition to sufficient specificity to separate bacteria from pure cultures and complex food matrices (Patchett *et al.* 1991; Payne *et al.* 1992, 1993a), some of these lectins are capable of agglutinating many strains of microorganism and they therefore afford advantages over the use of specific antibodies.

Determination of bacterial lectin affinity

Cell agglutination assays

These can be performed by incubating bacteria that have been washed in buffer and resuspended to a concentration of 10^8 cfu/ml, with serial dilutions of lectin, for example from 1 mg/ml downwards, for 30 min, in dual-welled microscope slides with a maximum sample size capacity of 0.1 ml. Typical lectin and bacterial cell suspension cell volumes are 0.025 ml each. Samples are checked for cell agglutination visually, or with a light microscope. Binocular stereomicroscopes are particularly useful. The dual-welled slides allow direct comparison of the test sample with a control in which the lectin has been omitted. Such controls are essential because of occasional non-lectin-mediated cell agglutination. If, under these conditions, the cells have on their surface only one type of sugar-binding site, at least two sugar-binding sites on the lectin must be of the same specificity for cell agglutination to occur. Agglutination assays should be carried out at several temperatures, because some lectins, such as Concanavalin A, are unstable at 4 °C and so fail to agglutinate as effectively as at room temperature. It is also important to adjust the temperature of the initial immobilization process, which normally occurs at 4 °C. In addition, the buffer ionic strength is important. High ionic strength weakens the adsorption of lectins to bacteria and inhibits cell agglutination. A buffer of low ionic strength, such as quarter-strength Ringer's solution, must therefore be used. Many lectins retain their ability to agglutinate cells over a wide range of pH and some lectins are dependent on manganese, magnesium and calcium ions, which should be included in all the buffers used, at 0.1 mmol/l.

Lectins that agglutinate a wide range of strains of a particular bacterial species, or a broad range of bacterial species, are of obvious interest, as are those that specifically agglutinate pathogenic strains of a particular bacterium. The specificity of a lectin for a particular sugar substrate present in a bacterial cell wall is checked by performing agglutination assays in the presence of specific competing sugar substrates. If agglutination is non-specific, release of cells after lectin magnetic separation can be difficult. This may not be important if the subsequent rapid detection method employed is based on identification of cell components that can be extracted from the immobilized bacteria, such as cell-surface antigens or nucleic acids, but with other techniques, such as flow cytometry, this can be a problem. However, agglutination assays alone are purely qualitative and are not sufficiently stringent to conclude that either no agglutination or total agglutination of cells has occurred.

Immobilization of lectins on magnetic microspheres

Various sizes and activated forms of magnetic microspheres are available. Tosyl-activated 2.8 μm diameter Dynabeads® (Dynal UK, Wirral, UK) are routinely used to bind a range of lectins. The process of immobilization is simple, as illustrated in Fig. 1.1. Larger BioMag® microspheres (Metachem Diagnostics Ltd, Northampton, UK), which are 10–40 μm in diameter, have also been used. However, the effect of bead surface area on bacterial adsorption cannot be accurately quantified, because the larger BioMag® particles are liganded with lectin by a method different from that used for Dynabeads®. This probably results in different numbers of active lectin molecules associated, per unit surface area, with each type of bead. BioMag® particles are supplied preliganded with lectins from *Canavalia ensiformis*, *Arachis hypogea*, *Phaseolus*

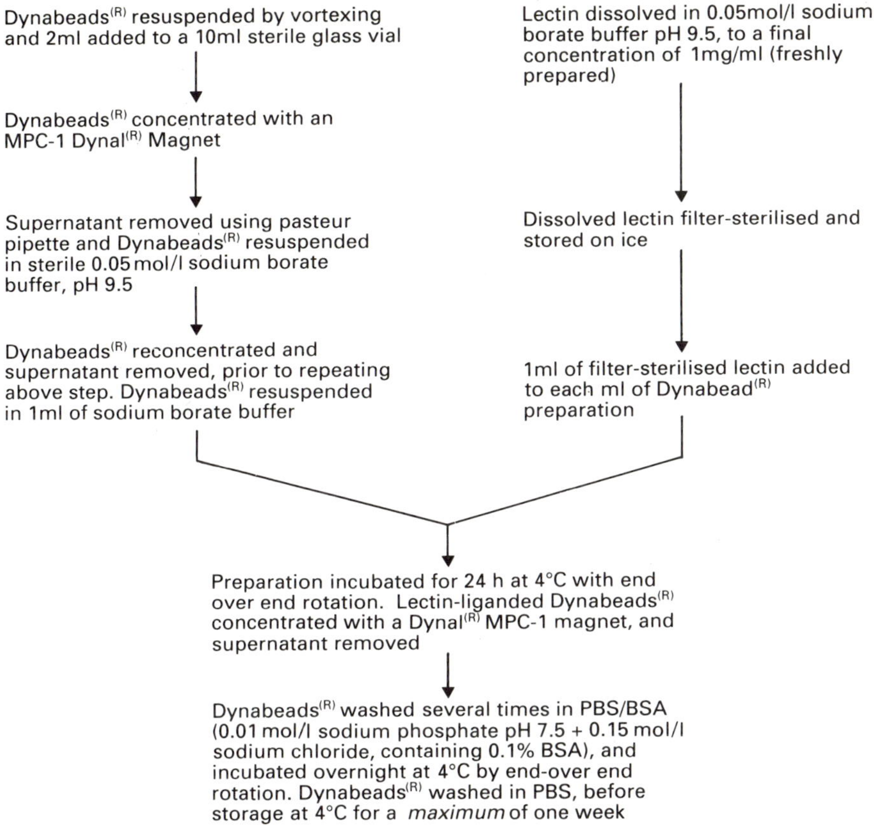

FIG. 1.1. Immobilization of lectins with tosyl-activated M-280 Dynabeads®. All manipulations are performed at room temperature. PBS, Phosphate buffered saline; BSA, Bovine serum albumen.

vulgaris or *Triticum vulgaris*. An advantage of these larger microspheres is that they appear to be more readily magnetized, and hence concentrated more quickly from suspension. This is not usually a problem in buffered systems, but in milk samples, for example, small 2.8 μm diameter beads take much longer to separate out (typically 20 min per 100 ml sample). This may be due partly to larger particles settling more readily under gravity and becoming localized around the magnet at the bottom of the glass container. It is necessary to use higher concentrations of magnetic particles (e.g. 10^5 particles/bacterium) than recommended by the manufacturer (e.g. 10^1 particles/bacterium) to recover bacteria from pure cultures and foods (Patchett *et al.* 1991; Payne *et al.* 1992, 1993a). This is an obvious disadvantage in terms of cost.

Once bacteria have been concentrated, their removal from the magnetic microspheres involves a 15–30 min incubation in the presence of the competing sugar substrate. Sometimes quite high concentrations of 100–200 mg/ml of competing sugar may be necessary, and physical shear by vortexing for several seconds is also recommended, to optimize release. Lower ionic strengths of assay solution require higher concentrations of competing substrate. The removal of bacteria from the microspheres can be an important step. When determining the effectiveness of lectin magnetic separation (LMS) by conventional plating, for example, it is often necessary to remove bacteria from the microspheres, because they can be difficult to spread, particularly if residual food particulate matter is present, which may cause clumping. However, clumping tends only to become a major problem in fatty foods, such as milk, pâté or cheese, or in foods of high viscosity, such as raw egg.

Inoculation of food samples

Samples of food are weighed into stomacher bags and inoculated with low numbers of the required target organism, usually 1 ml of a 10^3 cfu/ml inoculum per 10 g or 25 g sample and at least five 1-ml serial dilutions of this inoculum. The inoculum is mulched into the sample to ensure direct contact of bacteria with the solid food. Bacteria should be heat-stressed before inoculation, in order to create a situation similar to that found in processed food. For instance, for *Salmonella enteritidis* this can be achieved by heating resuspended cells in quarter-strength Ringer's solution at 51 °C for 5 min and immediately cooling on ice, before inoculation of food samples. Cells treated in this way will grow on yeast–dextrose agar but fail to grow on *Salmonella–Shigella* selective agar (SSSA). A 6-h preenrichment in YDB at 37 °C is necessary to promote cell recovery, followed by a 12-h selective enrichment in RVS medium at 43 °C, in order to remove competing organisms.

Examination of foods for bacteria

Food samples of 10 g and 25 g are homogenized with 90 ml and 225 ml respectively of either buffer or non-selective or selective growth medium. Non-selective enrichment is used where a direct lectin magnetic separation assay can be used. If the inoculated organism is not detectable by direct plate count and must be isolated from a food source, in which no competing microorganisms are normally present (e.g. *Salmonella enteritidis* from raw egg), selective enrichment should be unnecessary. In that case incubation in a non-selective enrichment medium, such as yeast–dextrose broth (YDB) at 37 °C, will result in faster cell growth. This should enhance the rate of detection, because selective enrichment, which often inhibits the growth of stressed target organisms, is unnecessary. Where high numbers of background microflora with low numbers of target organism are present, food samples must be selectively enriched to promote the growth of the target organism and inhibit the growth of other organisms. It is not normally advisable to perform a lectin magnetic separation directly with processed foods, because the target organisms may have been stressed by heat, freezing, salt, etc. It will, therefore, be necessary to allow cell recovery under milder conditions, by use of non-selective, followed by selective, growth media. We have not assessed the effect that stress has on cell adsorption by LMS. By performing enrichment regimes before the separation step, this potential problem is avoided.

To promote growth of damaged or stressed *S. enteritidis*, it is necessary to preenrich in yeast–dextrose broth (YDB) for 6 h at 37 °C, followed by selective enrichment in Rappaport–Vassiliadis medium (RVS) for 12 h at 43 °C. At this stage the LMS assay is employed. *Triticum vulgaris* lectin-liganded BioMag® particles resuspended in quarter-strength Ringer's solution (1 ml containing 5×10^6 particles), are added to 100 ml of sample in RVS, and incubated at 30 °C for 2 h in conical flasks with shaking. After concentration of BioMag® particles under magnetic influence, the supernatant is removed and adsorbed bacteria are released from the magnetic microspheres by incubation with 120 mg/ml *N*-acetyl glucosamine for 15 min at room temperature. Samples are vortexed before reconcentration of the beads, and the supernatant containing concentrated released bacteria is serially diluted in quarter-strength Ringer's solution and plated on YDA and SSSA. Agar plates are incubated for 24 h at 37 °C, before colonies are counted.

Staphylococcus aureus may be enriched selectively from food by incubation at 30 °C for 12 h in Giolitti Cantoni broth supplemented with 3.5% potassium tellurite and layered with liquid paraffin. After incubation, 20-ml samples are diluted in 80 ml of quarter-strength Ringer's solution containing 1 ml of *T. vulgaris* lectin-liganded BioMag® particles. The diluted sample is incubated for 2 h at 30 °C, before extraction of the magnetic particles as described above. *Staph. aureus* is routinely plated on YDA and Baird–Parker agar before incubation at 37 °C for 24 h.

Listeria monocytogenes can be enriched selectively in Trypticase Soy Broth (Lab M, Bury, UK) containing 0.6% yeast extract (TSB-YE), supplemented with 40 mg/l nalidixic acid, 50 mg/l cyclohexamide and 20 mg/l acriflavin, and incubated at 30 °C for 12 h, before removing 20-ml samples and adding to 80 ml of quarter-strength Ringer's solution containing either 1 ml of BioMag® particles (5×10^6 particles/ml final concentration) or 0.4 ml of *Agaricus bisporus* lectin-liganded Dynabeads® (2.6×10^6 beads/ml final concentration). To remove adsorbed bacteria from the beads in the latter case, 1 ml of 120 mg/ml fetuin is used as the competing sugar substrate. Samples are plated on coryneform agar and *Listeria* selective agar (LSA) and incubated at 30 °C for 24 and 48 h respectively before colonies are counted.

Adsorption of Bacteria to Magnetic Particles

Parameters that affect adsorption

Our repeated experience has been that increased sample volume, in the presence of defined numbers of magnetic particles, causes increased adsorption of bacteria. The reason is not known. It may be expected that an effective reduction in the concentration of magnetic particles would result in decreased adsorption of the bacteria. A possible explanation is that, with low sample volumes, the initial adsorption of bacteria from the medium results in a decrease in the residual concentration of unattached bacteria below the threshold concentration necessary for bacteria and beads to interact. Clearly, if a larger volume is present, the same number of magnetic particles will be able to adsorb more bacteria before the concentration of target organisms present falls below this threshold.

The kinetics of lectin magnetic separation are complex and worthy of further investigation. Results have indicated that, at low cell concentrations, a small proportion of cells present are actually adsorbed by a fixed concentration of magnetic particles (Payne *et al.* 1992). If cell concentrations are raised above 10^3-10^4 cfu/ml, the proportion of cfu bound and released from the lectin-liganded magnetic particles becomes more stable. This may be because the collision frequency between magnetic particles and bacteria is increased. In addition, it is possible that there is some positive cooperation between binding sites on the lectin molecules, i.e. the attachment of a bacterium to one binding site on the lectin increases the affinity of another binding site for another bacterium in spite of the possibility of steric hindrance. If this were the case, it would be necessary for a bacterium to bind to the lectin, before its affinity for other bacteria is enhanced. This would depend on the concentration of bacteria in the medium.

The lectins from *T. vulgaris* and *A. bisporus* successfully extracted all 13 strains of *Listeria* examined, including nine *L. monocytogenes* strains, from pure culture (Payne *et al.* 1992). It may be possible to exploit the lack of strain specificity exhibited by some lectins, to extract a range of bacterial species from food samples.

Interaction of bacteria with pseudo-immune isolates from invertebrates

The lower affinity of the lectins used in our studies for Gram-negative organisms has warranted further investigation. The pseudo-immune systems of invertebrates were chosen for further exploration because they must possess antibacterial mechanisms that do not involve antibodies and that are capable of recognizing and attaching to both Gram-positive and Gram-negative pathogens. If the components of the haemolymph or coelomic fluid respon-

TABLE 1.1. *Adsorption and release of bacteria using immobilized purified haemolymph components from* Limulus polyphemus

Bacterial strain	Initial colony count (cfu/ml)	Proportion of cfu adsorbed (%)	Proportion of adsorbed cells released after substrate* addition (%)
Listeria monocytogenes C200	2.7×10^4	53	53
Salmonella enteritidis 37782	3.3×10^4	15	27
Staphylococcus aureus NCDO 949	5.2×10^4	92	62
Escherichia coli E4936/76	8.7×10^3	19	100
Escherichia coli W2-2	5.2×10^4	65	58

* A cocktail of *N*-acetyl glucosamine, *N*-acetyl galactosamine, glucose, mannose and fetuin was used as substrate, each at a final concentration of 37 mg/ml. Haemolymph (Sigma, UK) was subjected to ammonium sulphate precipitation with 20, 40, 60 and 80% saturated ammonium sulphate in phosphate-buffered saline, pH 7.2, followed by centrifugation and dialysis of resuspended pellet. (Reproduced with permission from Payne *et al.* 1993b.)

sible for such bacterial interactions could be isolated and immobilized, they may provide an alternative, and highly specific, method for extracting bacteria from foods. We have studied the ability of components isolated from the haemolymph of horseshoe crab (*Limulus polyphemus*), to isolate bacteria (Payne *et al.* 1993b). These components were immobilized on M280-Dynabeads® as described (Fig. 1.1), and used to extract *L. monocytogenes*, *S. enteritidis*, *Staph. aureus* and *Escherichia coli* from pure cultures (Table 1.1). *Staphylococcus aureus* appeared to have the highest affinity for the purified haemolymph component, and *S. enteritidis* the lowest affinity and specificity. Binding of *E. coli* was variable, and strain E4936/76 had a lower affinity than strain W2-2, which bound to a greater extent than *L. monocytogenes* C200. However, even though only 19% of *E. coli* E4936/76 bound to the Dynabeads®, attachment appeared to be of high specificity, because all bound bacteria were released on addition of a cocktail of sugars/aminosugars (Table 1.1). This indicated that components in the purified haemolymph interacted with the bacteria in a lectin-like manner, by binding to sugar to aminosugar residues on the cell walls.

Lectin magnetic separation of Gram-positive bacteria

Lectins such as those isolated from *Helix pomatia*, *Triticum vulgaris*, Concanavalin A, *Agaricus bisporus*, *Cicer arietinum* and *Phaseolus coccineus* have proved particularly useful in the extraction of pathogenic bacteria. All appear to show a preference for binding to Gram-positive organisms. In general, Gram-negative bacteria compete poorly with Gram-positive organisms for

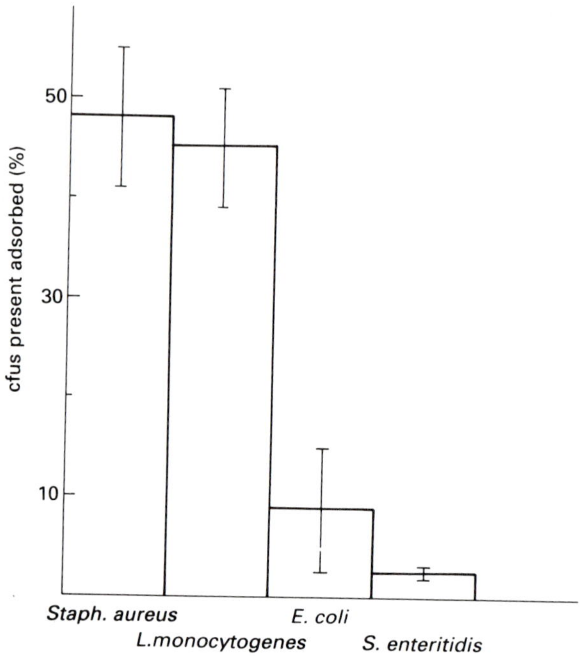

FIG. 1.2. Competitive binding of Gram-negative and Gram-positive bacteria to Dynabeads® liganded with lectins from *Canavalia ensiformis*, *Phaseolus coccineus*, *Triticum vulgaris* and *Helix pomatia*. *Staphylococcus aureus*, *Listeria monocytogenes*, *Escherichia coli* and *Salmonella enteritidis* were separated from mixed culture, each at an initial concentration of 10^3 cfu/ml (Dynabead® concentration approximately 6×10^6 particles/ml).

similar sugar-binding sites on lectins. Lectin magnetic separation has clearly indicated that the above lectins possess a higher affinity for *Staph. aureus* and *L. monocytogenes* than for *S. enteritidis* and *E. coli*. When cocktails of these lectins, except Concanavalin A, were liganded to magnetic particles and incubated with a mixed culture of four separate strains of each of the above four bacterial species, at approximately 10^3 cfu/ml, the Gram-positive organisms were extracted from the total culture in significantly greater numbers than the Gram-negative organisms (Fig. 1.2). This clearly indicates the differences in the composition, or accessibility, of the cell-surface sugars, aminosugars and oligosaccharides of each type of microorganism.

Isolation of Bacteria from Foods by LMS

When bacteria are isolated directly from a food sample, various parameters can be manipulated to maximize the adsorption of bacteria to the magnetic

TABLE 1.2. *Removal of pathogenic bacteria from foods by lectin magnetic separation*

Lectin	Organism	Isolated from:	Enrichment	Fold concentration
Agaricus bisporus	*Listeria monocytogenes*	Braising steak	Selective	1–4
Triticum vulgaris	*Listeria monocytogenes*	Braising steak	Selective	1–5
Triticum vulgaris	*Listeria monocytogenes*	Camembert cheese	Selective	1–10
Triticum vulgaris	*Listeria monocytogenes*	Full-fat milk (pasteurized)	None	0.3
Triticum vulgaris	*Staphylococcus aureus*	Braising steak	Selective	1
Triticum vulgaris	*Staphylococcus aureus*	Full-fat milk (pasteurized)	None	0.6–0.7
Triticum vulgaris	*Salmonella enteritidis*	Raw egg	Non-selective	0.2–0.5
Triticum vulgaris	*Salmonella enteritidis*	Braising steak	Non-selective + selective	10–100
Triticum vulgaris	*Salmonella enteritidis*	Braising steak	Selective	100–1000

particles. For instance, adsorption of bacteria can often be increased by diluting the food sample. Extraction of bacteria from full-fat pasteurized milk, for example, was intensified by 1:4 dilution in quarter-strength Ringer's solution (Payne *et al.* 1992). As there was a reduction in ionic strength, this may have been enough to significantly increase bacterial adsorption. Similar results were obtained with raw egg samples, although the increase in adsorption of bacteria can partly be attributed to the reduction in viscosity, which results in an increased mobility of both bacteria and magnetic beads during the shaking incubation.

The potential ability of lectin magnetic separation to isolate various bacteria from food samples is shown in Table 1.2. *Triticum vulgaris* lectin exhibits a strong affinity for the Gram-positive organisms (e.g. *L. monocytogenes* and *Staph. aureus*), making possible rapid (2 h) isolation, from braising steak or full-fat milk. *Salmonella enteritidis* may also be extracted from braising steak and raw egg, provided that competitive microflora are minimized by selective enrichment. Successful isolation of *L. monocytogenes* after selective enrichment from braising steak was achieved with either *A. bisporus* lectin or *T. vulgaris* lectin (Table 1.2). Although the bacteria are not always concentrated from the food by LMS, they are successfully removed, which should increase the sensitivity of rapid and conventional microbiological methods, by reducing background interference. Any additional concentration effect observed will, therefore, further intensify the detection. Lectin magnetic separation has already been observed to intensify conventional plating techniques for the isolation of *S. enteritidis* (Payne *et al.* 1993a) and has made possible the rapid detection of *Brochothrix* spp. in foods (Grant *et al.* 1993) with the polymerase chain reaction with 16S rRNA sequences.

Sugar specificity of lectin-binding sites

Some of the specific affinities associated with the lectins which have been used to extract bacteria from food samples are shown in Table 1.3 (Payne *et al.* 1992, 1993a). *A. bisporus*, *C. arietinum* and *P. coccineus* lectins possess some affinity for fetuin, which competitively inhibits the attachment of lectin to the bacterial cell wall. Fetuin has been used successfully to release bacteria from immobilized lectin in our studies. We doubt whether the affinity of these lectins for fetuin is specific, because it is a protein that has many terminal saccharide groups, including *N*-acetyl neuraminidase, *N*-acetyl glucosamine, *N*-acetyl galactosamine and galactosamine residues (Table 1.3). It is likely that the presence of these sugars and aminosugars, which are presented on the outer surface and throughout the bacterial cell wall, in the teichoic acid and peptidoglycan layer of the Gram-positive organisms, is significant in the attachment of lectins to the cell surface. Although oligosaccharide units associated with the lipopolysaccharide of the cell envelope of Gram-negative bacteria possess similar sugars and aminosugars, fewer binding sites for lectins may be present.

Selecting for Gram-negative bacteria

The lectins that possess an affinity for Gram-negative organisms that we have investigated have an even higher affinity for Gram-positive organisms. If

TABLE 1.3. *Lectin specificity for sugars and aminosugars*

Lectin	Sugar specificity
Agaricus bisporus	Fetuin*
Cicer arietinum	Fetuin
Concanavalin A	α-D-Mannose/α-D-Glucose
Helix pomatia	*N*-acetyl galactosamine
Phaseolus coccineus	Oligosaccharide
Triticum vulgaris	*N*-acetyl glucosamine

* Two NeuAcα2-3Galβ-1-3GalNAc residues, one NeuAcα2-3Galβ-1-3GalNAc(6-2αNeuAc) residue and nine NeuAcα2-3Galβ-1-4GlcNAc residues are present as terminal saccharide parts on each fetuin molecule.

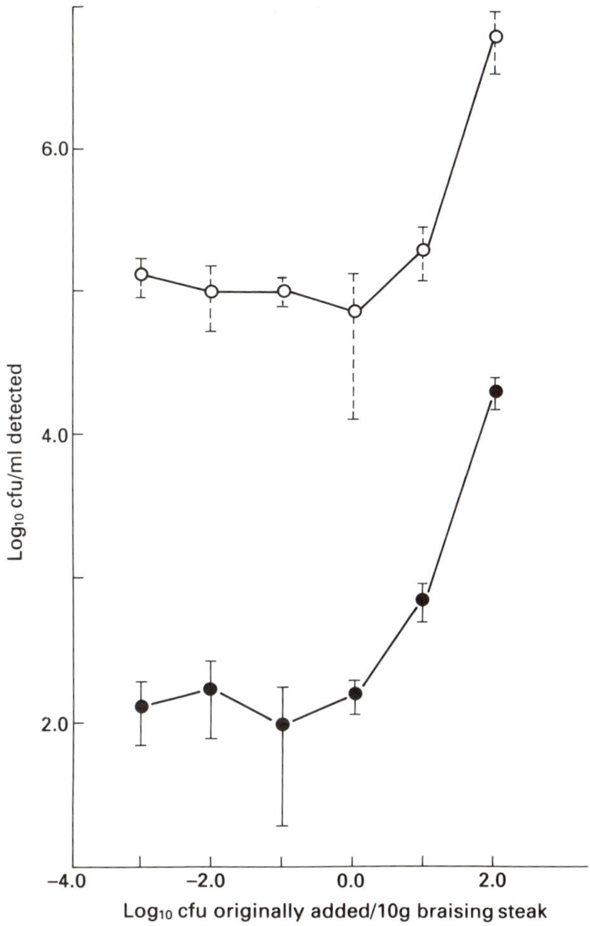

FIG. 1.3. Isolation of *Salmonella enteritidis* from braising steak with *Triticum vulgaris* lectin liganded BioMag® particles after selective enrichment. After inoculating 10 g of braising steak with 10^2 cfu, or 1 ml of five serial dilutions of this cell concentration, samples were incubated for 12 h at 43 °C in Rappaport–Vassiliadis medium, before lectin magnetic separation for 2 h at 30 °C. ○, 12-h incubation at 43 °C followed by LMS for 2 h, before spread-plating of substrate-released *S. enteritidis*; ●, 14-h incubation at 43 °C without LMS, before spread-plating. (Reproduced with permission from Payne *et al.* 1993a.)

the target organism is Gram-negative, selective enrichment, before lectin magnetic separation, will reduce this problem, and this approach has been effective in the isolation of *S. enteritidis* from meat samples (Payne *et al.* 1993a; Fig. 1.3). When 10-g samples of braising steak were inoculated with either 1 ml of a culture containing 200 cfu/ml or 1 ml of a 1:9 serial dilution of this culture (diluted five times in total), and incubated in RVS medium for 12 h at 43 °C, immobilized *T. vulgaris* lectin was able to concentrate the bacteria to detection levels two orders of magnitude higher than those detectable by conventional selective enrichment (Fig. 1.3). This is a highly significant concentrating effect, which raised the concentration of *S. enteritidis* to numbers above 10^4 cfu/ml, a level detectable by many rapid methods. Because of the lack of specificity of adsorption by the above lectins, selective enrichment regimes have also been necessary for the isolation of *Staph. aureus* from beef, and *L. monocytogenes* from beef and Camembert cheese (Payne *et al.* 1993a). In each case lectin magnetic separation is applied at the end of the selective enrichment procedure. In spite of the poorer affinity of Gramnegative organisms for lectins, LMS is more efficient at concentrating *S. enteritidis* after selective enrichment in RVS than it is at concentrating the Gram-positive organisms. This is probably because the enrichment regime for *Salmonella* is much more selective than that used for *Staph. aureus* or *L. monocytogenes* and because the RVS, used for *Salmonella* isolation, does not to any significant extent inhibit the interaction of lectin with bacteria. In an environment in which there is less competition for lectin-binding sites, more *S. enteritidis* will bind to the magnetic particles. In addition, when extracting *S. enteritidis*, there is no need to dilute samples of raw braising steak homogenized in selective RVS. However, samples of *Staph. aureus* in Giolitti Cantoni broth, or *L. monocytogenes* in supplemented TSB-YE, are extracted more efficiently if diluted 1:4 in quarter-strength Ringer's solution. Therefore, the concentrations of target bacteria are diluted. Clearly, parameters need to be optimized for each selective medium and food system involved.

Conclusion

LMS, used in our study, and immunomagnetic separation (IMS), used elsewhere, have considerable potential for increasing the speed and sensitivity of detection of foodborne pathogens by both routine conventional and rapid methods. We have successfully used LMS to isolate pathogenic bacteria directly from food samples and after non-selective or selective enrichment. LMS therefore has the potential to present cleaner samples of bacteria to rapid detection systems, which are at present inhibited by background interference from particulate food matter. In some cases, bacteria present in food samples in low numbers are significantly concentrated to levels above those normally detectable by rapid methods. However, it should be emphasized that the introduction of additional steps in the detection and isolation of pathogens is costly in materials and manpower. Further refinements to these separation techniques are necessary, if they are to provide real benefits after incorporation into routine methods of microbiological analysis. Other biospecific compounds and alternative immobilization systems need to be investigated, particularly in combination with novel physical methods for microbial separation.

Acknowledgements

The authors gratefully acknowledge the financial support of the Ministry of Agriculture, Fisheries and Foods.

References

BARONDES, S.H. 1988 Bifunctional properties of lectins: lectins redefined. *Trends in Biochemical Sciences* **13**, 480–482.

CUDJOE, K.S., THORSEN, L.I., SORENSEN, T., RESELAND, J., OLSVIK, O. & GRANUM, P.E. 1991 Detection of *Clostridium perfringens* type A enterotoxin in faecal and food samples using immunomagnetic separation (IMS)-ELISA. *International Journal of Food Microbiology* **12**, 313–322.

DOYLE, M.P. & SCHOENI, J.L. 1987 Isolation of *Escherichia coli* 0157:H7 from fresh meats and poultry. *Applied and Environmental Microbiology* **53**, 2394–2396.

ENTIS, P. 1990 Improved hydrophobic grid membrane filter method, using EF-18 agar, for the detection of salmonella in foods: collaborative study. *Journal of the Association of Official Analytical Chemists* **73**, 734–742.

ETZLER, M.E. 1985 Plant lectins: molecular and biological aspects. *Annual Reviews of Plant Physiology* **36**, 209–234.

GRANT, K.A., DICKINSON, J.H., PAYNE, M.J., CAMPBELL, S., COLLINS, M.D. & KROLL, R.G. 1993 Use of the polymerase chain reaction and 16S rRNA sequences for the rapid detection of *Brochothrix* spp. in foods. *Journal of Applied Bacteriology* **74**, 260–267.

LIS, H. & SHARON, N. 1986 Lectins as molecules and as tools. *Annual Review of Biochemistry* **55**, 35–67.

PATCHETT, R.A., KELLY, A.F. & KROLL, R.G. 1991 The adsorption of bacteria to immobilised lectins. *Journal of Applied Bacteriology* **71**, 277–284.

PAYNE, M.J., CAMPBELL, S. & KROLL, R.G. 1992 The use of immobilised lectins in the separation of *Staphylococcus aureus*, *Escherichia coli*, *Listeria* and *Salmonella* spp. from pure cultures and foods. *Journal of Applied Bacteriology* **73**, 41–52.

PAYNE, M.J., CAMPBELL, S. & KROLL, R.G. 1993a Lectin magnetic separation can enhance conventional methods for the detection of *Staphylococcus aureus*, *Salmonella enteritidis* and *Listeria monocytogenes*. *Food Microbiology* **10**, 75–83.

PAYNE, M.J., CAMPBELL, S. & KROLL, R.G. 1993b Separation of bacteria using agglutinins isolated from invertebrates. *Journal of Applied Bacteriology* **74**, 276–283.

PAYNE, M.J. & KROLL, R.G. 1991 Methods for the separation and concentration of bacteria from foods. *Trends in Food Science and Technology* **2**, 315–319.

SKJERVE, E. & OLSVIK, O. 1991 Immunomagnetic separation of salmonella from foods. *International Journal of Food Microbiology* **14**, 11–18.

SKJERVE, E., RORVIK, L.M. & OLSVIK, O. 1990 Detection of *Listeria monocytogenes* in foods by immunomagnetic separation. *Applied and Environmental Microbiology* **56**, 3478–3481.

WIBAWAN, I.W.T., LAMMLER, C. & PASARIBU, F.H. 1992 Role of hydrophobic surface proteins in mediating adherence of group B streptococci to epithelial cells. *Journal of General Microbiology* **138**, 1237–1242.

Wood, J.M. & Gibbs, P.A. 1982 New developments in the rapid estimation of microbial populations in foods. *Developments in Food Microbiology*, ed. Daniels, R., pp. 183–214. London: Applied Science.

2: Immunomagnetic Separation Techniques for the Detection of Pathogenic Bacteria in Foods

K.S.Cudjoe,[1] P.D.Patel,[2] E.Olsen,[3] E.Skjerve[1] and Ø.Olsvik[4]
[1]*Norwegian College of Veterinary Medicine, Department of Food Hygiene, PO Box 8146 Dep, N-0033 Oslo, Norway;*
[2]*Leatherhead Food Research Association, Randalls Road, Leatherhead, Surrey KT22 7RY, UK;*
[3]*Dynal A/S, PO Box 158, Skøyen, N-0212 Oslo, Norway;*
and [4]*Enteric Diseases Branch, MSCO3, Center for Disease Control, Atlanta, GA 30333, USA*

Microbiological analysis of foods is carried out to assure consumers of the safety and quality of foods. This is achieved by examining food materials throughout the production process and monitoring the finished products before distribution. Two main groups of microorganisms, namely spoilage and pathogenic bacteria, are of concern. Pathogenic bacteria like *Salmonella* spp., *Listeria monocytogenes*, *Campylobacter jejuni*, *Yersinia enterocolitica* and enteropathogenic *Escherichia coli* are of concern to producers, manufacturers, consumer interest groups and public health authorities. Methods for the isolation, detection and identification of pathogenic bacteria are based either on conventional cultural techniques or on newer methods, such as immunoassay, nucleic acid hybridization and electrical conductance or impedance measurements. None of these combines simplicity with speed, specificity and sensitivity.

Immunomagnetic separation (IMS), which employs magnetic particles coated with specific antibodies selectively to enrich target bacterial cells, promises to be a useful tool (see also Chapter 1). It is rapid, inherently specific and requires simple and relatively inexpensive equipment. The technique is well documented in clinical medicine for the selection of lymphocytes and other human cell types (Kemshead & Ugelstad 1987; Mattson *et al.* 1991), and the selective recovery and concentration of bacteria (Johne & Jarp 1988; Lund *et al.* 1988; Skjerve *et al.* 1990; Blackburn *et al.* 1991; Skjerve & Olsvik 1991) and bacterial toxin (Cudjoe *et al.* 1991) from food and other

New Techniques in
Food and Beverage Microbiology

0-632-03755-5

TABLE 2.1. *List of some immunomagnetic particles commercially available*

Particles	Size (μm)	Manufacturer
BioMag 4300: superparamagnetic polydisperse particles	0.5–1.5	Metachem Diagnostics Ltd, Northampton, UK
Polystyrene paramagnetic microparticles	1–2	Polysciences Ltd, Northampton, UK
Magnetic polyacrolein/iron oxide particles	1–10	Scipac, Sittingbourne, UK
Polystyrene/divinyl-benzene	0.7	Sepadyn, Indianapolis, USA
Dynabeads ® M-280/M-450	2.8/4.5	Dynal A/S, Oslo, Norway

heterogeneous systems. The technique represents a time-saving replacement and/or supplement for current lengthy selective enrichment protocols. This chapter describes the essential details for establishing immunomagnetic separation techniques for the analysis of foodborne bacteria and protein toxins.

Development of Immunomagnetic Particles

A variety of magnetic particles are available (Table 2.1). Most of these are non-uniform in shape, size and content of magnetic oxides but are collected readily on permanent magnets. Moreover, they tend to autoagglutinate, which makes handling difficult (Blackburn *et al.* 1991). The development of uniformly sized, monodisperse, polymer particles (1–100 μm) by a method of activated swelling (Ugelstad *et al.* 1980) forms the basis for the development of superparamagnetic particles of uniform size and magnetic oxide content (Ugelstad *et al.* 1985).

These superparamagnetic polymer particles become magnets only when exposed to a magnetic field and lose their magnetism when the field is removed. As a result, the particles do not aggregate. Mention of magnetic particles in this chapter refers to those marketed under the trade name Dynabeads® M-280 (Dynal A/S, Oslo, Norway). They are 2.8 μm in diameter, with an outer surface area of $3–8\,m^2/g$, and can be chemically modified to bind different groups, including amino and carboxyl groups. Covalent binding of proteins, such as specific antibodies against surface antigens of eukaryotic or prokaryotic cells to the particles, allows specific isolation of these analytes from human tissues and heterogeneous food suspensions.

Principles of Immunomagnetic Separation

IMS is based on the principle that bacteria or cells bound to magnetic particles by specific antibodies can be isolated from heterogeneous suspen-

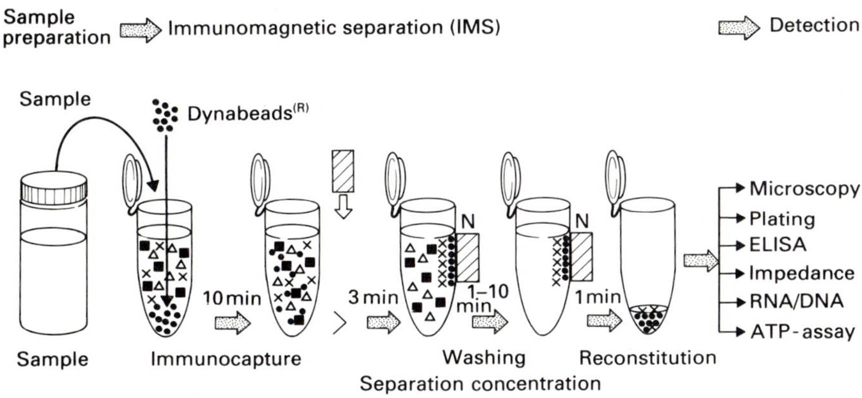

FIG. 2.1. Diagrammatic scheme to show the immunomagnetic separation process.

sions by the application of a magnetic field. Bacteria immunologically bound to beads can subsequently multiply when nutritional requirements are provided. The procedure involves mixing the antibody-coated particles with the prepared sample and, after incubation, the bead–bacteria complex is extracted with a magnet. Particles coated with specific polyclonal or monoclonal antibodies have been used. The isolated bead–bacteria complex must be washed before transfer to suitable growth media or used with other detection systems (Fig. 2.1). The captured antibodies can be linked to the beads covalently or indirectly to beads precoated with antiantibodies.

IMS has several advantages over other separation and concentration methods. The target bacteria are separated from the source and from other bacteria not bound to the particles and may be concentrated by resuspension in a small volume suitable for subsequent cultivation or detection by other means. Furthermore, possible growth-inhibitory substances in the sample are removed, so facilitating cultivation. IMS is not a detection system, but may provide a significant time-saving for existing methods based on selective enrichment. It also allows the flexibility of applying different end-point detection techniques. Depending on requirements, detection techniques including microscopy, plating, enzyme-linked immunosorbent assay (ELISA), polymerase chain reaction (PCR) and impedance methods can be used (see below). The technique is, however, limited by the requirement of antibodies directed against surface antigens of the organism. The presence of high concentrations of dead or lysed cells in the sample, which would include free surface antigens, adversely affects retrieval of viable bacteria.

TABLE 2.2. *Magnetic particle-based products from Dynal A/S, Oslo, Norway*

Product name	Abbreviated name used in text
Dynabeads® M-280 Sheep anti-Mouse IgG	Sheep antimouse IgG beads
Dynabeads® M-280 Sheep anti-Rabbit IgG	Sheep antirabbit IgG beads
Dynabeads® M-280 Tosylactivated	Tosyl-activated beads
Dynabeads® M-280 Streptavidin	Streptavidin beads
Dynabeads® M-280 Anti-Salmonella	*Salmonella* beads
Magnetic particle concentrators (e.g. MPC®-1, MPC®-E or MPC®-M)	MPC

Equipment and reagents

Equipment and reagents used are as follows:

1 Specific antibodies or proteins to be bound to magnetic particles. Alternatively commercial antitarget magnetic particles (Table 2.2).

2 Magnetic particles and magnetic particle concentrator (e.g. MPC®-1, MPC®-E or MPC®-M, Dynal A/S).

3 Apparatus to allow both gentle tilting and rotation during incubation (e.g. Blood Mixer, Robbins Scientific, Mt. View, CA, USA).

4 Buffers/solutions:
- (a) 0.2 mol/l phosphate buffer, pH 7.4.
- (b) 0.2 mol/l borate buffer, pH 9.5.
- (c) 0.15 mol/l NaCl in 0.01 mol/l phosphate buffer, pH 7.5 (PBS).
- (d) PBS containing 0.1% bovine serum albumin (PBS/BSA).
- (e) PBS containing 0.05% Tween-20, wash buffer (PBS/Tween).
- (f) 1.0 mol/l $NaHCO_3$, pH 8.0.

All reagents should be of analytical grade. For microbiological work, the magnetic particle concentrator for microtubes of the Eppendorf type (MPC®-M) is recommended. It is designed as a rack to hold 10 Eppendorf tubes with a removable sliding plate with embedded cobalt–samarium magnets. The MPC rack without the magnetic plate can be used to incubate samples containing particles. Sedimentation of particles in the MPC during a standard time of 2 min is implied throughout the following protocols.

Binding of Antibodies to Magnetic Particles

Two types of multipurpose magnetic particles with covalently bound sheep antimouse IgG or sheep antirabbit IgG are available. These are intended respectively for use with specific mouse and specific rabbit IgG-class antibodies against defined target surface antigens. Target-specific particles can be made by binding specific antibodies directly on to the particles. Alternatively, other species-specific antiantibodies can be bound to the particles to generate new multipurpose particles. Such particle types can be used in direct or indirect techniques for the specific isolation of target organisms. In the *direct technique*, rabbit antibodies against a defined bacterial target surface antigen are immunoadsorbed to the sheep antirabbit IgG beads, before the suspension is incubated with the target or analyte. In the *indirect technique*, the target is pretreated with specific rabbit antibodies so that all binding sites are blocked. Then the antibody-coated target is collected by centrifugation and washed to remove excess antibody, before the sheep antirabbit IgG beads are added. Bacterial targets will bind to the beads by specific antigen–antibody interactions and can then easily be removed from a heterogeneous suspension. So far, only the direct technique has been used in food microbiology.

Protocol for covalent binding of antibodies to tosyl-activated magnetic particles

Magnetic particles available for microbiological work, tosyl-activated beads, are intended primarily for coating with antibodies directed against another antibody or directly against a target but they may be coated with any protein ligand or receptor as follows:

1 Resuspend tosyl-activated beads (10 mg/ml) by gentle vortexing.

2 Pipette the required volume of beads into a tube and concentrate to 20 mg/ml by sedimentation in an MPC for 2 min, pipette off the supernatant and resuspend in half the original volume of 0.05 mol/l borate buffer, pH 9.5.

3 Dissolve protein or antibody in 0.05 mol/l borate buffer, pH 9.5, to a concentration of 400 μg/ml.

4 Mix equal volumes of concentrated particles and protein to give 20 μg/mg beads protein or antibody.

5 Incubate at 37 °C for 24 h or at 4 °C for 48 h with slow tilting and rotation.

6 Sediment the protein-coated beads. Aspirate and discard the supernatant while the tube is in the MPC.

7 Wash five times with 1 ml of PBS/BSA in Eppendorf tubes: three times at room temperature for 10 min, once at room temperature for 30 min and once at 4 °C for 24 h with gentle tilting and rotation.

8 Sediment the fully coated and washed particles, discard the supernatant and resuspend in PBS/BSA to the desired concentration, for example, 10 mg coated beads/ml. The immunomagnetic particles are now ready for use and can be stored for up to 6 months at 4 °C.

Protocol for immunoadsorption of antibodies on to immunomagnetic particles

This describes the procedure with sheep antirabbit IgG beads, but the instructions are equally valid for sheep antimouse IgG beads.

1 Resuspend sheep antirabbit IgG beads (10 mg/ml) well by gentle vortexing.
2 Add rabbit IgG or purified antibodies to the desired volume of beads. The exact amount of antibody required for optimal binding depends on the affinity of the antibody and must be determined in each case.
3 Incubate the mixture either at room temperature for 2 h or at 4 °C for 24 h, with slow tilting and rotation.
4 Sediment the coated beads in the MPC.
5 Aspirate and discard the supernatant while the tube is in the MPC. Remove the tube from the MPC and resuspend the beads in PBS/BSA.
6 Wash the beads at 2–4 °C for 30 min in PBS/BSA, pH 7.5. Sediment the beads in the MPC, discard supernatant and repeat three times.
7 After the last washing, resuspend the beads in original volume of storage buffer, PBS/BSA.
8 The immunomagnetic particles are now ready for use.

Protocol for binding of biotinylated antibodies to magnetic particles coated with streptavidin

A third way of attaching antibodies to immunomagnetic particles is by using streptavidin-coated beads to bind biotin-labelled antibodies. This principle can also be employed to separate other biotinylated compounds such as proteins, sugars, lectins, DNA or RNA. For the biotinylation protocol see Coffer *et al.* (1990). Streptavidin beads are commercially available and can be used to bind biotinylated antibodies as follows:
1 Calculate the amount of biotinylated antibody needed. Between 5 and 10 μg antibody/mg of particles is sufficient to saturate the streptavidin, assuming a 100% biotinylation of the antibody.
2 Mix the biotinylated antibody with the streptavidin beads and incubate at room temperature for 30 min with gentle rotation of the tube.
3 Sediment the coated streptavidin beads in the MPC.
4 Wash five times in PBS/BSA.
5 Resuspend to the desired concentration.
6 The streptavidin beads coated with biotin-labelled antibodies can now be used to extract the target organisms of interest (see below).

For all the above protocols, a final concentration of 0.02% sodium azide can be added as a preservative for suspensions of coated particles, which should be stored at 4 °C. The azide must be removed by washing before use.

Isolation of Bacteria from Foods by Immunomagnetic Separation

The following protocol describes the use of the immunomagnetic separation technique for bacteria with *Salmonella* as a model. The IMS protocol consists of (i) sample preparation; (ii) immunomagnetic separation (IMS); and (iii) detection.

Sample preparation

This is critical for the successful use of IMS, and reference must be made to the specific International Standard appropriate to the product concerned.

1 For dry powdered samples, weigh 25 g aseptically into a 300 ml screw-cap bottle and, while mixing thoroughly, add 255 ml of buffered peptone water (1:10, w/v).

2 For moist lumpy samples, weigh 25 g aseptically into a special stomacher bag with a filter and add 255 ml of buffered peptone water (1:10, w/v). Macerate by stomaching for 1 min and transfer the filtered suspension aseptically into a 300 ml screw-cap bottle.

3 Some food types require special enrichment broths, because of the presence of substances that may be toxic for the target bacteria or for other reasons connected with the composition of the food (Anon. 1991).

4 Incubate statically at 37 °C for 18–22 h with the cap slightly loose.

5 Remove enriched sample from the incubator and adjust to refrigeration temperature for at least 1 h. This sediments particulate matter while fat forms a surface layer.

6 Carefully remove 5–10 ml of clear sample supernatant, free from particulate matter. Duplicate 1-ml subsamples should be taken for IMS.

7 For samples that are known to have a high background flora and/or much particulate matter or that form colloidal suspensions, it is recommended that 1-ml subsamples are taken for IMS analysis from a 10-fold dilution of the clear sample supernatant. Otherwise, 5 ml of the enriched sample should be centrifuged for 5 min at 140 *g*, before taking the subsamples.

8 Refrigerated sample aliquots must be adjusted to room temperature before IMS analysis.

Immunomagnetic separation procedure

Washing immunomagnetic particles before immunomagnetic separation

1 Suspend the Dynabeads® by gentle vortexing and remove the volume required for the whole analysis into a sterile Eppendorf tube.

2 Place the Eppendorf tube into the MPC®-M for 2 min.

3 Aspirate and discard the supernatant and then remove the magnetic plate.

4 Add 1 ml of wash buffer (PBS/Tween) and resuspend the particles by inverting several times and replace the magnetic plate for another 2 min.

5 Aspirate and discard the wash buffer supernatant and remove the tube from the MPC®-M rack.

6 Resuspend the particles in the original volume of wash buffer.

Immunomagnetic separation

1 Dispense 20-μl volumes of *Salmonella* beads into Eppendorf tubes.
2 Add 1 ml of subsample equilibrated to room temperature and seal the tubes.
3 Invert twice to mix the contents and incubate for 10 min at room temperature on a tilting rotating device to prevent settling. Otherwise invert the tubes intermittently throughout the incubation period.
4 Sediment particles with the MPC. Tilt the MPC with tubes back and forth twice to collect the particles in the cap on to the tube wall nearest the magnet. With a Pasteur pipette, carefully aspirate and discard the sample supernatant from the tube wall opposite the magnet, as well as any remaining liquid in the tube cap, and remove the magnetic plate.
5 Add 1 ml of wash buffer to the tubes, mix and repeat Step 4 twice. For heavily contaminated samples an additional wash may be necessary.
6 Remove magnetic plate and resuspend the particles in 100 μl of wash buffer.
7 The resuspended particle–bacteria complex is then used in a detection system.

Care must be taken to avoid the development of aerosols while aspirating sample supernatants or wash buffer supernatants or when adding wash buffers. When opening Eppendorf tubes, the index finger should be placed over the cap and the thumb used gently to lift it upwards; otherwise an opening device for Eppendorf tubes (Tubease^{TM}, US Biochemicals, Cleveland, Ohio, USA) should be used.

Factors that Affect the Performance of Immunomagnetic Particles

In practice, the performance of immunomagnetic particles directed against bacterial surface antigens depends on a number of factors, including the type of coating antibody, its functional affinity, the concentration of coating antibody, the number of particles, incubation time, mode of incubation, type of sample, temperature and the washing regimes employed. If the coating antibodies are too restricted in specificity, as may be the case for monoclonal antibodies, specificity is restricted to bacteria with common antigenic epitopes. Target bacterial epitopes may only loosely be associated with the bacterial cell and, in old cultures, free antigen from dead or lysed cells can interfere with the binding abilities of the particle-bound antibodies, particularly if the recovery of viable cells is intended. Some capsulate bacterial strains are 'sticky' and attach non-specifically to the particles. This may interfere with the

TABLE 2.3. *The effect of mixing during incubation and sample particulate matter on the recovery of seeded* Salmonella Give *(400 cells/ml) by IMS plating*

	% Recovery		
		Whey milk powder	
Type of mixing	Broth	Whole suspension	Clear supernatant
Continuous mixing	33.0	4.0	12.0
Intermittent mixing	20.0	3.0	5.0
Without mixing (static)	11.0	2.0	2.0

binding of the antibodies to the specific targets and may result in isolation of mixed bacterial populations. High coating-antibody concentrations may lead to aggregation of bound targets. Use of too many particles and too long an incubation time normally increases non-specific binding, while the use of suboptimum particle concentrations reduces sensitivity. During incubation, intermittent mixing enhances binding kinetics, resulting in the isolation and separation of higher levels of target organisms than when static incubation is used. However, continuous mixing during incubation gives even better binding kinetics, resulting in increased recovery of target organisms. The presence of particulate matter and/or fat micelles results in loss of particles during washing with concomitant reduced recovery of target organisms (Table 2.3). Too much washing results in loss of target bacteria, while underwashing

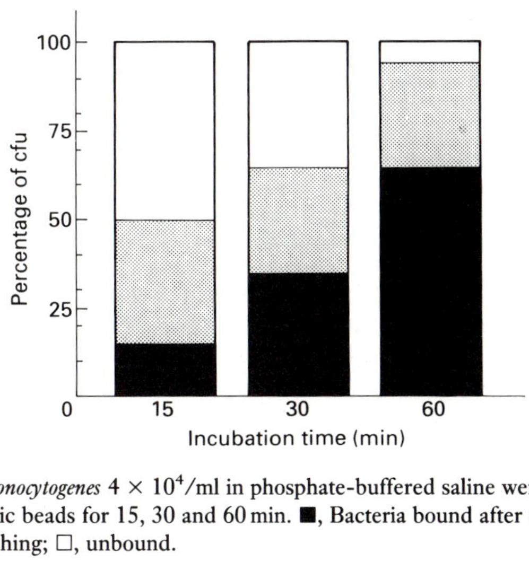

FIG. 2.2. *Listeria monocytogenes* 4×10^4/ml in phosphate-buffered saline were incubated with 5×10^6 immunomagnetic beads for 15, 30 and 60 min. ■, Bacteria bound after six washes; ▨, bound but removed by washing; □, unbound.

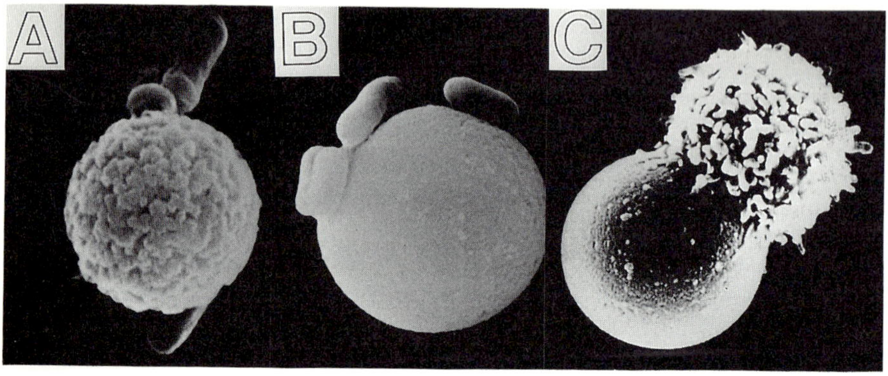

FIG. 2.3. Bacterial target cells bound to immunomagnetic particles. (a) Dynabeads M-280 (diameter 2.8 μm), (b) Dynabeads M-450 (diameter 4.5 μm) and (c) human tissue cell bound to Dynabeads M-450.

may not remove non-target bacteria that are non-specifically bound to the particles and so may hinder detection, particularly if growth is used for detection (Blackburn *et al.* 1991). According to Skjerve *et al.* (1990) an increase in incubation time from 15 to 60 min reduces the number of unbound bacteria from 50% to 5%, and 30% of the initially bound bacteria are washed away, to leave about 65% bound after six washings. For shorter incubation times of 15 and 30 min, 15% and 35% respectively are bound (Fig. 2.2).

The recovery of target organisms varies from 10 to 90% depending on which of the above factors is dominant. However, with eukaryotic cells almost 100% recovery is sometimes achieved. The reason for this difference is not clear. The presence of a rigid cell wall in bacteria and its absence in human tissue cells could be contributory factors (Fig. 2.3). The manner of binding of the larger human tissue cells as compared with bacteria may explain the different level of recovery of the two cell types.

Detection Systems After Immunomagnetic Separation

Bacteria attached to beads can be stained with acridine orange and visualized under a fluorescence microscope (Lund *et al.* 1988). This is a rapid and simple approach for presumptive identification of specific organisms. Others have identified bacteria by electron microscopy (Johne *et al.* 1989; Olsvik *et al.* 1991). To visualize *E. coli* 0157:H7 bound to immunomagnetic particles, Fratamico *et al.* (1992) incubated the particles further with FITC-labelled polyclonal antiserum against the organism before examination by fluorescence microscopy.

Both plate and broth cultivation have been carried out on several bacterial species bound to beads; the attachment apparently has no effect on growth (Johne & Jarp 1988; Blackburn & Patel 1989). However, several cells attached to a cluster of beads can give rise to a single colony on agar media (Skjerve *et al.* 1990), thus making counting semiquantitative at best. For the detection of *Salmonella* a moderate and highly selective agar medium with indicators is recommended, to discriminate target colonies from other contaminating flora as follows:

1 Washed and resuspended bead-bound bacteria (50 μl) are plated on each of two selective media, such as xylose lysine deoxycholate (XLD) agar or brilliant green agar (BGA).

2 Plates are incubated at 37 °C for 20–24 h and confirmed as positive red colonies with black centres on XLD agar or pinkish-red colonies on BGA by slide agglutination, with a *Salmonella* polyvalent serum.

3 Further biochemical and serological confirmation should be carried out, if necessary.

In a study with immunomagnetic particles directed against *Salmonella*, these were effectively isolated from several artificially contaminated and control raw and processed foods, during 24-h enrichment (Patel & Blackburn 1991). Skjerve *et al.* (1990) demonstrated that *Listeria monocytogenes* could be extracted by the IMS technique from pure liquid cultures and heterogeneous food suspensions. Furthermore, Cudjoe & Skjerve (1992) reported increased sensitivity of IMS plating over a commercial ELISA procedure and the traditional method for isolating and detecting *L. monocytogenes* from fresh salmon samples. Both intact bacteria and their soluble antigenic determinants can be detected with a second antibody in a sandwich format after magnetic extraction from the test sample (IMS-ELISA). Cudjoe *et al.* (1991) described such a procedure for the detection of *Clostridium perfringens* type A enterotoxin in food and human faecal samples. IMS followed by PCR has been successfully used to detect *Y. enterocolitica* in foods (Vardlund *et al.* 1992). This combination may reduce assay times to hours, while increasing both specificity and sensitivity. Detection of viable *Salmonella* spp. in 20 min, with a combination of IMS and bioluminescence measurement of intracellular microbial ATP, shows the effectiveness of immunomagnetic particles for rapid separation (Blackburn *et al.* 1991). Patel (1991) used the bead–bacteria complex after IMS for a conductimetric analysis of *Salmonella*, and showed a reduction in the *Salmonella* detection time by about 4 h as compared with the untreated culture. At present, the use of electrical instruments for reliable detection of *Salmonella* is restricted to a few foods, such as confectionery products. Methods for other pathogens are largely at a research stage. Since commercially available automated systems have several other limitations, including a relatively high level of false positives and negatives in certain food categories, such as raw meats, their combination with IMS is an interesting proposition.

Conclusion

IMS appears to have potential use in food microbiology, because of the increased sensitivity it offers for recovering very low levels of target organisms from foods with a high background flora. It also provides a significant time-saving over existing methods based on selective enrichment and allows flexibility in the application of different end-point detection techniques. For naturally contaminated foods, a short incubation time in a non-selective or low-selective enrichment medium, followed by IMS, is necessary to obtain at least a $2\log_{10}$ unit increase in target organisms to allow specific detection. IMS has been shown to work in most food matrices, but non-specific binding of competing flora to the particles may affect specific detection in heavily contaminated samples. Proper sample preparation, by careful removal of solid and particulate matter by sedimentation before IMS, appears to be critical for the successful use of the technique. For routine use with several samples for analysis, the method, despite overall savings in time, is labour-intensive. This is due to the many different manipulations, in the absence of automation, during the washing procedures.

References

ANON. 1991 Salmonella bacteria, detection in foods. *Nordic Committee on Food Analysis*, 4th edn, No. 71. Esbo, Finland.

BLACKBURN, C. DE W. & PATEL, P.D. 1989 Interaction of salmonella with immunomagnetic particles. *Proceedings of Society for Applied Bacteriology, Conference Booklet*, p. 50.

BLACKBURN, C. DE W., PATEL, P.D. & GIBBS, P.A. 1991 Separation and detection of salmonellae using immunomagnetic separation. *Biofouling* **5**, 143–156.

COFFER, A., SINNET-SMITH, J. & ROZENGURT, E. 1990 Bombesin receptor from Swiss 3T3 cells. Affinity chromatography and reconstitution into phospholipid vesicles. *Federation of European Biochemical Societies* **275**, 159–164.

CUDJOE, K.S. & SKJERVE, E. 1992 Immunomagnetic separation and detection of pathogenic bacteria from foods 1992. *Proceedings, 3rd World Congress, Foodborne Infections and Intoxications*, Vol. 1, pp. 540–542. Berlin: Robert von Ostertag Institut.

CUDJOE, K.S., THORSEN, L.I, SØRENSEN, T., RESELAND, J., OLSVIK, Ø. & GRANUM, P.E. 1991 Detection of *Clostridium perfringens* type A enterotoxin in faecal and food samples using immunomagnetic separation (IMS)-ELISA. *International Journal of Food Microbiology* **12**, 313–322.

FRATAMICO, P.M., SCHULTZ, F.J. & BUCHANAN, R.L. 1992 Rapid isolation of *Escherichia coli* 0157:H7 from enrichment cultures of foods using immunomagnetic separation method. *Food Microbiology* **9**, 105–113.

JOHNE, B. & JARP, J. 1988 A rapid assay for protein-A in *Staphylococcus aureus* strains using immunomagnetic monosized polymer particles. *Acta Pathologica, Microbiologica, et Immunologica Scandinavica* **96**, 43–49.

JOHNE, B., JARP, J. & HAAHEIM, L.R. 1989 *Staphylococcus aureus* exopolysaccharide in vivo demonstrated by immunomagnetic separation and electron microscopy. *Journal of Clinical Microbiology* **27**, 1631–1635.

KEMSHEAD, J. & UGELSTAD, J. 1987 Magnetic separation techniques: their application to medicine. *Molecular and Cellular Biochemistry* **67**, 11–18.

LUND, A., HELLEMANN, A.L. & VARTDAL, F. 1988 Rapid isolation of K88+ *Escherichia coli* by using immunomagnetic particles. *Journal of Clinical Microbiology* **26**, 2572–2575.

MATTSON, L., BONDJERS, G. & WIKLUND, O. 1991 Isolation of cell populations from arterial tissue, using monoclonal antibodies and magnetic microspheres. *Atherosclerosis* **89**, 25–34.

OLSVIK, Ø., WASTESON, Y., LUND, A. & HORNES, E. 1991 Pathogenic *Escherichia coli* found in food. *International Journal of Food Microbiology* **12**, 103–114.

PATEL, P.D. 1991 Advances in methods for rapid microbiological food analysis. *Food Technology Europe* 235–239.

PATEL, P.D. & BLACKBURN, C. DE W. 1991 Detection of food-poisoning agents using immunomagnetic particles. *Magnetic Separation Techniques Applied to Cellular and Molecular Biology*, ed. Kemshead, J.T., pp. 93–107. Bristol: Wordsmith Conference Publications.

SKJERVE, E. & OLSVIK, Ø. 1991 Immunomagnetic separation of *Salmonella* from foods. *International Journal of Food Microbiology* **14**, 11–18.

SKJERVE, E., RØRVIK, L.M. & OLSVIK, Ø. 1990 Detection of *Listeria monocytogenes* in foods using immunomagnetic separation. *Applied and Environmental Microbiology* **56**, 3478–3481.

UGELSTAD, J., MØRK, P.C., KAGGERUD, K.H., ELLINGSEN, T. & BERGE, A. 1980 Swelling of oligomer-polymer particles. New methods of preparation of emulsions and polymer dispersions. *Advances in Colloid and Interface Scicence* **13**, 101–140.

UGELSTAD, J., MFUTAKAMBA, H., MØRK, P.C., ELLINGSEN, T., BERGE, A., SCHMID, R., HOLM, L., JØRGEDAL, A., HANSEN, F.K. & NUSTAD, K. 1985 Preparation and application of monodisperse polymer particles. *Journal of Polymer Science* **72**, 225–240.

VARDLUND, T., KAPPERUD, G., SKJERVE, E., MICHAELSEN, T. & HORNES, E. 1992 Detection of *Yersinia enterocolitica* in food by immunomagnetic separation and polymerase chain reaction. *Proceedings, 3rd World Congress, Foodborne Infections and Intoxications*, Vol. 1, pp. 543–545. Berlin: Robert von Ostertag Institut.

3: Rapid Separation and Detection of Foodborne Yeasts and Moulds by Means of Lectins

P.D.Patel,[1] D.W.Williams[1] and S.D.Haines[2]
[1]*Leatherhead Food Research Association, Randalls Road, Leatherhead, Surrey KT22 7RY, UK;*
and [2]*bioMérieux UK Ltd, Grafton House, Grafton Way, Basingstoke RG22 6HY, UK*

Yeast and moulds are widely distributed environmental contaminants of air, water, soil and dust. Although only a small proportion are responsible for food spoilage, they are a significant cause of economic loss to food manufacturers.

The Howard Mould Count (Howard 1911) is the officially recognized method for the examination of tomato and other products for the presence of moulds. The method relies on a correlation between the proportion of decayed fruit in the raw material and the occurrence of mould fragments, as measured by microscopy. The limitations of the technique are that it is labour-intensive and time-consuming and that inaccuracies may arise from operator fatigue and difficulty in distinguishing hyphae from plant fragments.

Developments in other areas of microbiology have stimulated interest in the possible application of indirect methods for the analysis of moulds. Assays for components of mould mycelium, chitin (Ride & Drysdale 1972) and ergosterol (Seitz *et al.* 1979) have been proposed. Immunological methods (Notermans & Heuvelman 1985) and electrical monitoring of growth (Williams 1989) have also been applied to mycological studies.

A further technique is immunofluorescence microscopy (Jarvis *et al.* 1983). This is, however, limited by the non-availability of specific antibodies from commercial sources. A possible alternative to antibodies is lectins, which are defined as proteins or glycoproteins of non-immune origin that bind to specific sugar residues. Previous studies have shown the ability of lectins to detect food-grade galactomannans (Patel & Hawes 1988) and food-spoilage or pathogenic bacteria (Williams & Patel 1990).

Blackburn *et al.* (1991) demonstrated the use of antibody-coated magnetic

New Techniques in
Food and Beverage Microbiology

0-632-03755-5

particles for the specific separation and detection of *Salmonella* from mixed cultures and food enrichment broths. Lectins may also be used as alternatives to antibodies for the rapid physicochemical enrichment of yeasts and moulds. This chapter describes and discusses methods for the rapid separation and detection of a range of food-spoilage moulds and yeasts with labelled lectins.

Reagents

Preparation of lectin reagents

Fluorescein isothiocyanate (FITC)-labelled *Triticum vulgaris*, *Phytolacca americana* and *Lycopersicon esculentum* lectins, the biotinylated lectins *T. vulgaris* and Concanavalin A (Con A, 1 mg, Sigma) were reconstituted in 10 ml of phosphate-buffered saline pH 7.2 (PBS, Oxoid).

Separation Methods

Preparation of yeast and mould spore cultures

Mould spores from *Mucor* spp., *Aspergillus flavus* and *Penicillium aurantiogriseum* were produced by culture at 25 °C for 5 days on rose Bengal chloramphenicol agar (RBCA, Oxoid). The resulting spore crop was resuspended in maximum recovery diluent (MRD, Oxoid) and the spores were counted by preparing dilutions in MRD, which were plated on RBCA and incubated at 25 °C for 24–72 h.

Yeast cultures (*Metchnikowia pulcherrima*, *Kluyveromyces lactis*, *K. marxianus*, *Candida pseudotropicalis*, *C. holmii* and *Saccharomyces bisporus*) were grown in malt extract broth (MEB, Oxoid) at 30 °C for 24–48 h. The yeasts were counted by preparing dilutions in MRD and plating on malt extract agar (MEA; Lab M). The plates were then incubated at 30 °C for 24–48 h.

Magnetic separation from broth cultures

Phosphate-buffered saline (1 ml), containing 5% bovine serum albumin (BSA, Sigma), was added to sterile glass tubes and 300 µl to the wells of a Microstrip (Labsystems). After approximately 30 min incubation at ambient temperature, the PBS/BSA from the tubes and Microstrip was discarded. A diluted culture (200 µl) of yeast or mould spores was added to a glass tube and allowed to react with 50 µl of biotinylated lectin for 15 min. After incubation, 100 µl of a diluted (100 µl in 900 µl of PBS containing 0.5% gelatin) suspension of magnetic streptavidin particles were added and allowed to react for a further 15 min. A portion (250 µl) of this mixture was transferred to a well of the

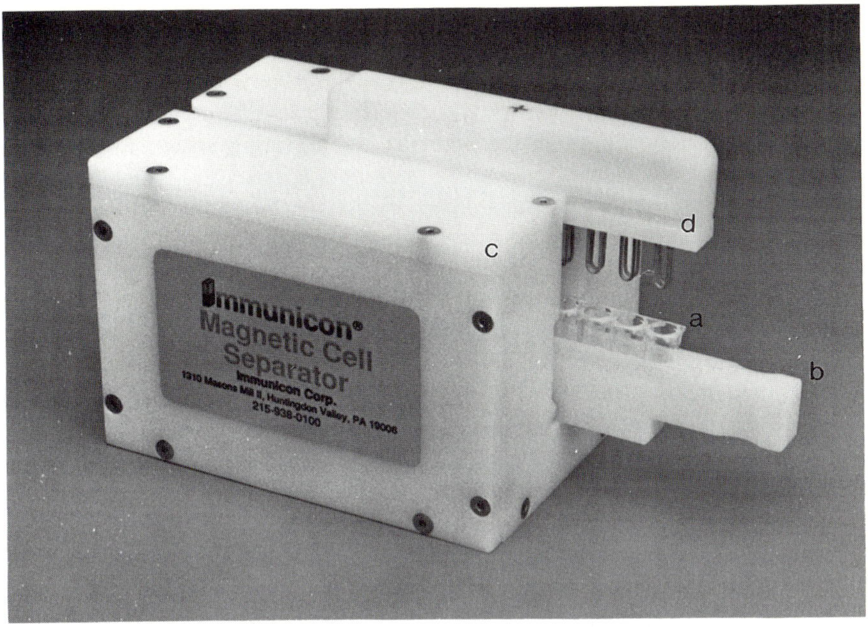

FIG. 3.1. The 'Immunicon' magnetic particle separation system, consisting of microstrip (a), microtitre well-holder (b), magnetic separator (c) and cell collector (d).

Microstrip. This was then placed in a well-holder and inserted into a magnetic separator. A cell collector was then placed in the magnetic separator so that its metal pins were immersed in the wells. The complete system (Immunicon) is shown in Fig. 3.1.

After a 5-min incubation, the collector was removed and the concentration of yeast cells/mould spores remaining was determined. The particle-bound cells were eluted by agitation of the pins in wells containing 300 µl PBS/BSA. The yeast cells and mould spores were then counted as described above.

Magnetic separation of yeast from fruit juice

Yeast cultures were diluted 100-fold in either apple or orange juice purchased locally. In the case of orange juice, a further 10-fold dilution in PBS/gelatin was made. The cell suspensions were then treated with biotinylated Con A and magnetic streptavidin particles and the separation efficiency was assessed as described above.

Detection Methods

Preparation of fungal cultures

Penicillium chrysogenum, *P. aurantiogriseum*, *A. flavus*, *A. ripens*, *Cladosporium cladosporoides*, *C. herbarum*, *Alternaria alternata*, *Geotrichum candidum*, *Mucor piriformis* and *Rhizopus* spp. were grown on MEA slopes at 25 °C for 5 days. The mould growth was washed from each slope with 5 ml of PBS containing 0.05% v/v Tween 20 (PBS/Tween). The suspensions were centrifuged at 2000 *g* (MSE Centaur 2) for 10 min. The supernatant was discarded and the pellet washed twice and resuspended in 5 ml of PBS.

Candida pseudotropicalis was grown at 30 °C for 24 h in MEB, of which 5 ml was centrifuged and washed as described above.

Preparation of food samples

Samples of tomato purée, matured Cheddar cheese, strawberry jam (low solids), cheese and mushroom pizza, beef pie and ground black pepper were purchased from a local supermarket. They were stored at 25 °C for 5 days before sampling.

A 20-g sample of each food was homogenized with 30 ml of PBS/Tween (Colworth 400 stomacher, A.J.Seward, Blackfriars, London) for 2 min. An 8-ml portion was centrifuged at 2000 *g* for 10 min. The supernatant was discarded and the pellet washed twice in 8 ml of PBS/Tween and resuspended in 8 ml PBS/Tween.

Reaction of FITC-labelled lectins with fungi

A previously washed mould culture or food sample (1 ml) was mixed with 1 ml of FITC-labelled lectin. The suspension was mixed gently on a shaker (Multimix Major, Luckham Ltd) for 30 min at room temperature and the mixture centrifuged at 2000 *g* for 10 min. The supernatant was discarded and the pellet washed twice in PBS/Tween. The washed pellet was resuspended in 1 ml of PBS/Tween and a drop (*c.* 50 µl) placed on to a microscope slide and covered with a coverslip. The slides were then viewed with an Olympus BH2 epifluorescent microscope with UV illumination. Photomicrographs were taken with an Olympus OM10 camera and 400 ASA Kodak Ektachrome film. Controls containing FITC not coupled with lectin (1 mg/ml; Sigma) and *N*,*N*′,*N*″-triacetyl chitotriose (1 mg/ml; Sigma; essentially a trimer of *N*-acetyl glucosamine) as a specific inhibitor of the lectin mould interaction were also included.

Candida pseudotropicalis was also examined. In this case, 50 µl of the

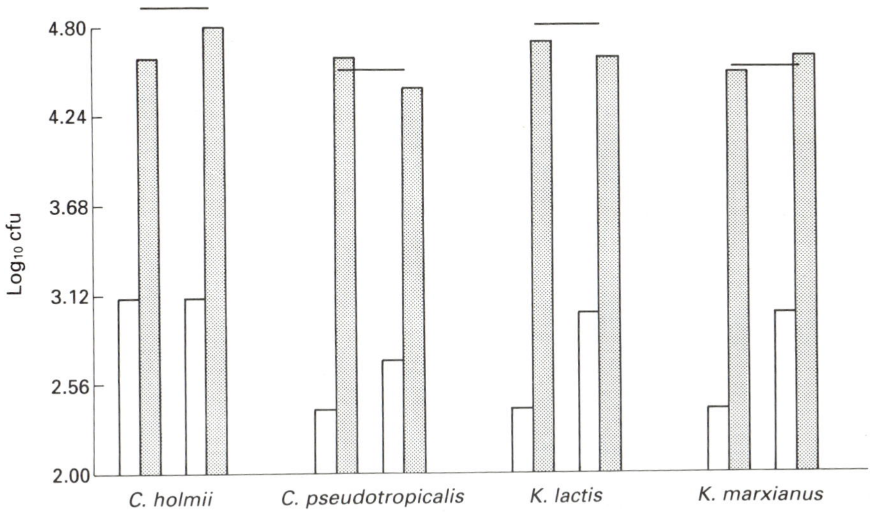

FIG. 3.2. Magnetic separation of *Candida holmii*, *C. pseudotropicalis*, *Kluyveromyces lactis* and *K. marxianus* from broth cultures with biotinylated Con A and magnetic streptavidin particles. The recovery was >95% in each case. □, Cells remaining; ▨, cells eluted; —, Inoculum level.

prepared yeast culture was reacted with 450 µl of FITC-labelled Con A lectin. The yeasts were then detected as described above.

Magnetic separation of yeasts

The results for the separation of yeasts with biotinylated Con A and magnetic streptavidin particles are summarized in Fig. 3.2. In the case of yeasts in broth cultures (approximately 4.5 $\log_{10}$ cfu/well), more than 95% of the cells were recovered in each case. Similar results were obtained for apple juice contaminated with a range of food-spoilage yeasts (Fig. 3.3). In this case, approximately 75–98% of the cells were recovered (from starting levels of 10^3–10^4 cfu/well) depending on the yeast species.

In the case of orange juice contaminated with yeasts, the recovery was approximately 58% for *M. pulcherrima* and 5% for *C. pseudotropicalis* from the undiluted juice (Fig. 3.4). From a 10-fold diluted orange juice, the recovery increased to 90% and 41% respectively.

Magnetic separation of mould spores

The results obtained with *T. vulgaris* lectin for the separation of mould spores from aqueous suspensions are summarized in Fig. 3.5. In these experiments,

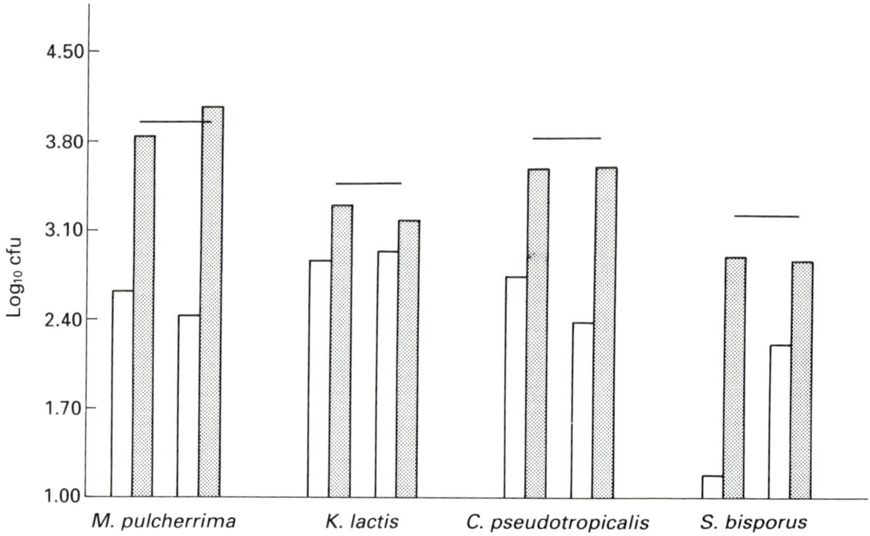

FIG. 3.3. Magnetic separation of *Metschnikowia pulcherrima*, *Kluyveromyces lactis*, *Candida pseudotropicalis* and *Saccharomyces bisporus* from apple juice with biotinylated Con A and magnetic streptavidin particles. The recovery was 75–98%. □, Cells remaining; ▨, cells eluted; —, Inoculum level.

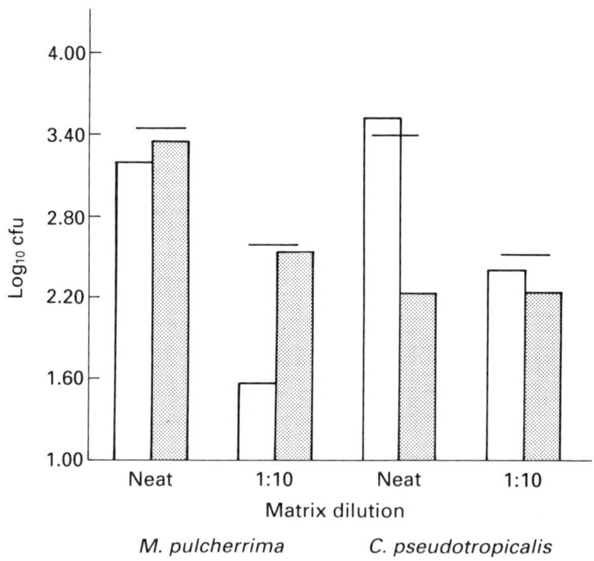

FIG. 3.4. Magnetic separation of *Metschnikowia pulcherrima* and *Candida pseudotropicalis* from undiluted orange juice with biotinylated Con A and magnetic streptavidin particles. The recovery was 58% for the former but only 5% for the latter. □, Cells remaining; ▨, cells eluted; —, Inoculum level.

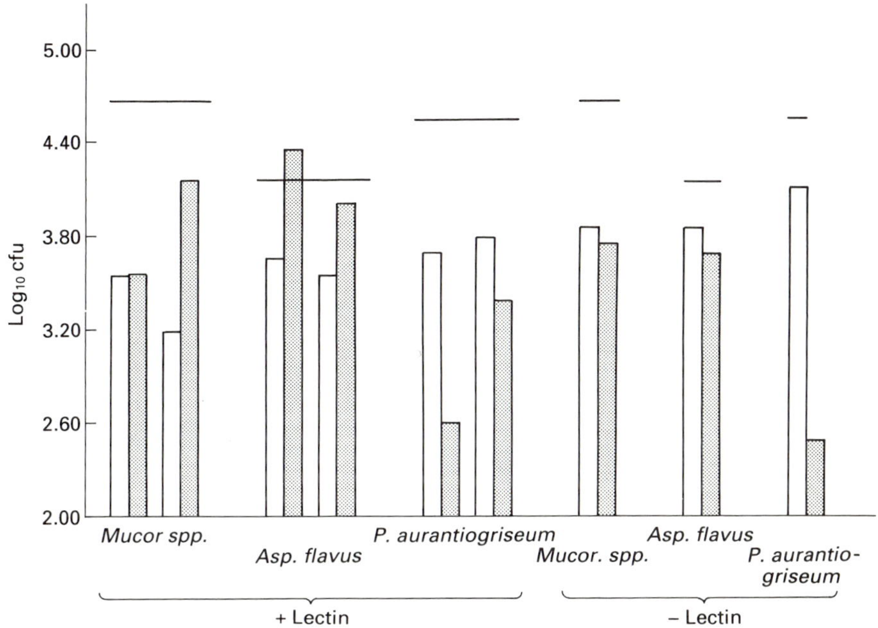

FIG. 3.5. Separation of mould spores from aqueous suspensions with *Triticum vulgaris* lectin. The recovery was 17.5–76.8%, depending on the spore type; recovery in control experiments with samples not treated with lectin was 2.3–44%. □, Cells remaining; ▨, cells eluted; —, Inoculum level.

spores were recovered at levels between 17.5% and 76.8%, depending on the spore type. Control samples not treated with lectin gave recoveries ranging from 2.3% to 44%.

Detection of yeasts and moulds from pure cultures

Of the *N*-acetyl glucosamine-specific lectins tested, the strength of binding to the mould cultures in decreasing order was *T. vulgaris*, *P. americana* and *L. esculentum*. These lectins stained all the mould cultures tested but the *T. vulgaris* FITC-lectin showed the greatest intensity of fluorescence. Figure 3.6 shows a photomicrograph of a typical reaction of the lectin with a mould culture. This lectin also reacted with mycelial septa and mould spores. Similarly, *C. pseudotropicalis* reacted strongly with the FITC-labelled lectin Con A (Fig. 3.7). Mould preparations treated with free FITC and *N,N′,N″*-triacetyl chitotriose showed no reaction with the mould cultures.

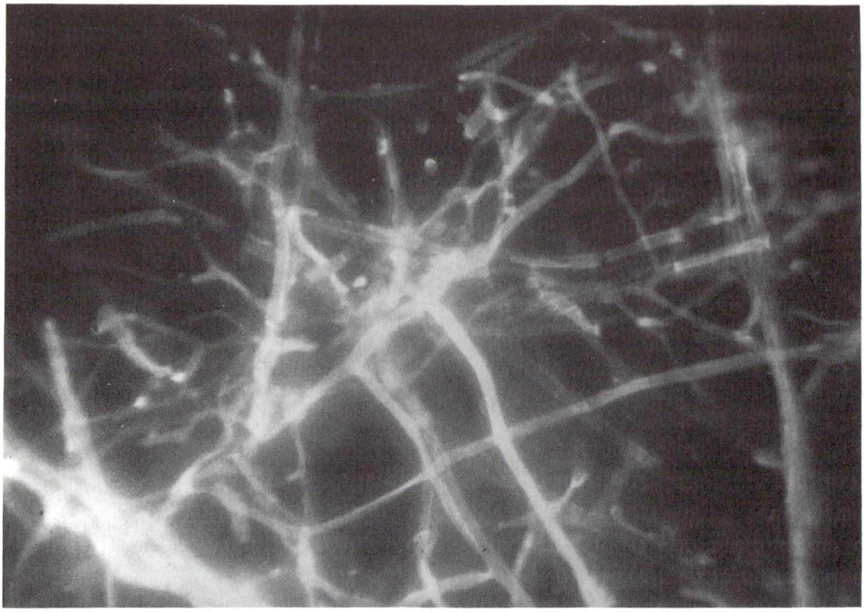

FIG. 3.6. Reaction of FITC-labelled *Triticum vulgaris* lectin with a mould (magnification ×800).

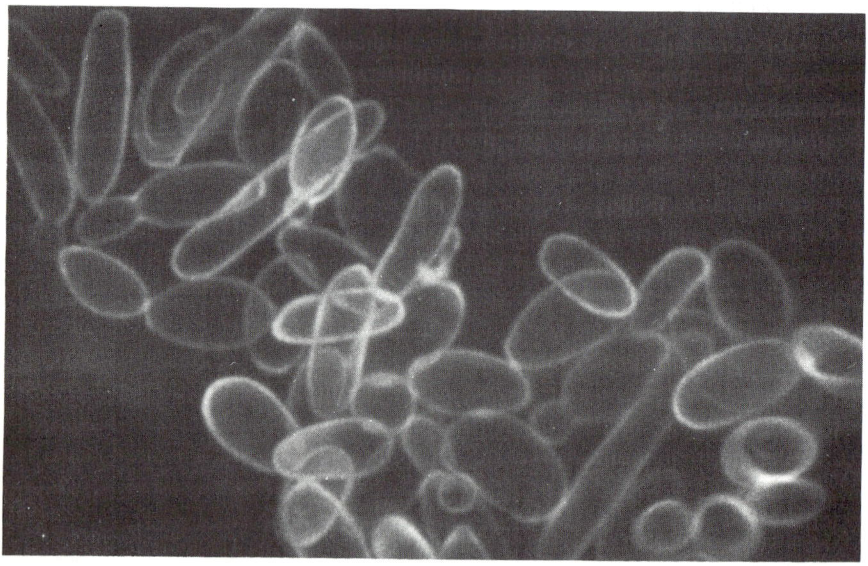

FIG. 3.7. Reaction of FITC-labelled Con A lectin with *Candida pseudotropicalis* (magnification ×2310).

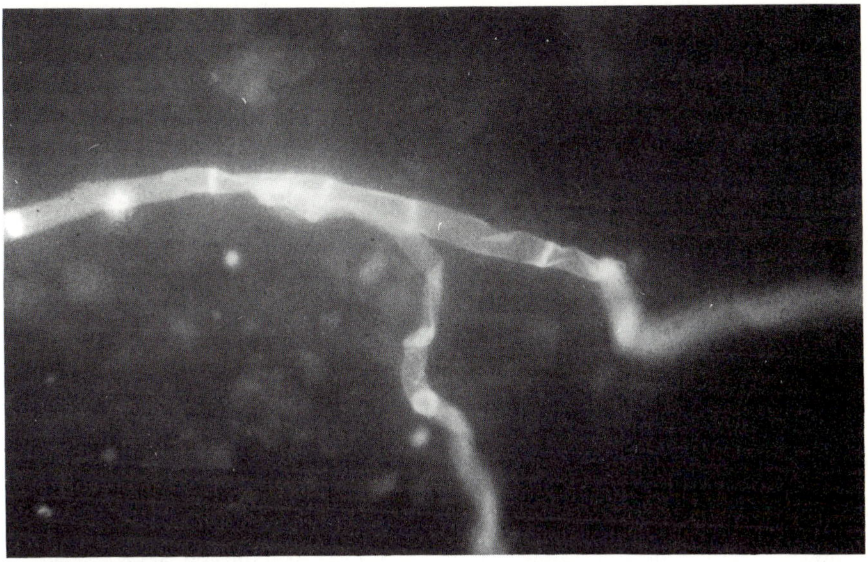

FIG. 3.8. Reaction of FITC-labelled Con A lectin with a mould fragment in tomato purée (magnification ×800).

Reaction of FITC-labelled lectins with moulds in foods

An example of a mould detected in the tomato purée sample is shown in Fig. 3.8. Mould fragments were detected in all the foods tested except strawberry jam. Mould was, however, detected in strawberry jam after artificial contamination with *P. chrysogenum*. In the case of black pepper, a higher level of background fluorescence was apparent, making distinction of mould fragments more difficult.

Conclusion

Lectins have found increasing use in microbiological studies, which have exploited their stereospecific properties to aid in the identification of microorganisms. In the case of Gram-positive bacteria, the lectin from *Helix pomatia* has been used to identify group C streptococci (Kohler *et al.* 1973). Concanavalin A lectin is also able to interact with teichoic acids derived from *Staph. aureus* and *Staph. epidermidis* (Reeder & Ekstedt 1971). In the case of Gram-negative organisms, the lectins from *Perseau americana* and *Mangifera indica* have been used in a slide agglutination procedure to identify strains of *Campylobacter jejuni* and *C. coli* (Corbel & Gill 1987). In addition, lectins have been used in epidemiological investigations, such as in the characterization of

Neisseria gonorrhoeae (Schalla *et al.* 1985). We have examined lectins for their potential in both detecting and separating yeast and moulds.

Conventional mycological examination of foods is time-consuming, cumbersome and highly subjective. A number of rapid methods have been developed but these also have significant limitations, including time-consuming culture for detection at low levels by immunological techniques, high capital costs of automated electrical techniques and lack of viability data for chemical and immunological techniques. Several rapid methods are available commercially for the analysis of yeasts and moulds, including bioluminescence (e.g. Sonco Ltd), automated electrical systems (e.g. bioMerieux), flow cytometry for mycological examination of fermented milk products (Chemunex) and latex agglutination to detect antigens of heat-treated *Penicillium* and *Aspergillus* spp. derived from foods (Holland Biotechnology).

The speed, efficiency and reliability of modern rapid methods may be enhanced significantly by inclusion of a physicochemical separation step to 'enrich' the target cells or antigens from the 'background' interference due to non-target organisms and the food matrix. The 'enriched' target can then readily be detected and estimated. The use of magnetic particles coated with specific binding proteins has many advantages over other separation techniques, such as gel filtration and density gradient centrifugation. It is rapid, requires simple and relatively inexpensive equipment and lends itself to automation. Studies at the Leatherhead Food Research Association have shown the value of antibody-coated magnetic particles for the separation of *Salmonella* (Patel & Blackburn 1991).

In our investigations, lectins in combination with magnetic streptavidin particles have demonstrated their potential for separating yeast in broth cultures and from apple and orange juices. In the case of orange juice, sample pretreatment may be required since the undiluted juice appeared to inhibit the separation process. This may have been due to interference of juice components with the lectin binding, or the low pH of the juice may have had an inhibitory effect on the Con A–yeast interaction. *Triticum vulgaris* lectin demonstrated potential for the separation of mould spores in pure culture, particularly those from *Mucor* spp. and *A. flavus*. However, further work is required to develop and optimize these separation techniques.

The results show that fluorescent-labelled lectins can be used to detect moulds in tomato purée, pizza, beef pie, cheese and jam. The technique takes longer to complete than the Howard Mould Count (*c.* 2 h compared with 1 h) but the distinction between mould fragments and food particles is easily made. Thus, a major drawback of the Howard Mould Count is obviated by means of the fluorescent lectin assay. Indeed, the distinction between mould fragments and food particles in this assay is such that an automated count of mould by image analysis may be possible.

Overall, the value of labelled lectins for the rapid separation and detection of yeasts and moulds has been demonstrated. With further optimization, a combination of magnetic separation and the use of appropriate lectins, it should be possible to develop 'real-time' techniques with desired end-points, such as bioluminescence, chemiluminescence and electrical signals, for mycological examination of foods.

Acknowledgements

The authors are grateful to the UK Ministry of Agriculture, Fisheries and Food for financial support and to Jon Mellor for technical assistance.

References

BLACKBURN, C. DE W., PATEL, P.D. & GIBBS, P.A. 1991 Separation and detection of salmonellae using immunomagnetic particles. *Biofouling* **5**, 143–156.

CORBEL, M.J. & GILL, K.P.W. 1987 Lectin agglutination of thermophilic *Campylobacter* species. *Veterinary Microbiology* **15**, 163–173.

HOWARD, B.J. 1911 Tomato ketchup under the microscope with practical suggestions to insure a cleanly product. *US Department of Agriculture Bulletin* No. 581.

JARVIS, B.J., SEILER, D.A.L., OULD, A.J.L. & WILLIAMS, A.P. 1983 Observations on the enumeration of moulds in food and feedingstuffs. *Journal of Applied Bacteriology* **55**, 325–336.

KOHLER, W., PROKOP, O. & KUHNEMUND, O. 1973 Routine identification of group-C streptococci by means of an agglutinal (protectin) from the albumin gland of the edible snail, *Helix pomatia*. *Journal of Medical Microbiology* **6**, 127–130.

NOTERMANS, S. & HEUVELMAN, C.J. 1985 Immunological detection of moulds in foods by using the enzyme-linked immunosorbent assay (ELISA): preparation of antigens. *International Journal of Food Microbiology* **2**, 247–258.

PATEL, P.D. & BLACKBURN, C. DE W. 1991 Detection of food-poisoning agents using immunomagnetic particles. *Magnetic Separation Techniques Applied to Cellular and Molecular Biology*, ed. Kemshead, J.T., pp. 93–105. Badgworth: Wordsmiths' Conference Publications.

PATEL, P.D. & HAWES, G.B. 1988 The potential of lectins for analysis of food-grade galactomannans by double-diffusion lectin assay. *Food Hydrocolloids* **2**, 179–185.

REEDER, N.J. & EKSTEDT, R.D. 1971 Study of the interaction of concanavalin A with staphylococcal teichoic acids. *Journal of Immunology* **106**, 334–340.

RIDE, J.P. & DRYSDALE, R.B. 1972 A rapid method for the chemical estimation of filamentous fungi in plant tissue. *Physiology of Plant Pathology* **2**, 7–15.

SCHALLA, W.D., RICE, R., BIDDLE, J.W., JEAN LOUIS, Y., LARSEN, S.A. & WHITTINGTON, W.L. 1985 Lectin characterisation of gonococci from an outbreak caused by penicillin-resistant *Neisseria gonorrhoeae*. *Journal of Clinical Microbiology* **22**, 481–483.

SEITZ, L.M., SAUER, D.B., BURROUGHS, R., MOHR, H.E. & HUBBARD, J.P. 1979 Ergosterol as a measure of fungal growth. *Phytopathology* **69**, 1202–1203.

WILLIAMS, A.P. 1989 Fungi in foods – rapid detection methods. *Progress in Industrial Microbiology. 26. Rapid Methods in Food Microbiology*, eds Adams, M.R. & Hope, C.F.A., pp. 255–272. Oxford: Elsevier Applied Science.

WILLIAMS, D. & PATEL, P.D. 1990 Detection of food pathogens by novel lectin binding assays. *Leatherhead Food Research Association Research Report* No. 674.

4: Separation of *Salmonella* from Foods with Magnetic Antibody-based Biosorbents

J.A.Bobbitt, J.P.Steer, J.L.Jones and R.P.Betts
Campden Food and Drink Research Association,
Chipping Campden, Gloucestershire GL55 6LD, UK

Conventional methods for the detection of *Salmonella* in food samples involve lengthy and manually intensive procedures. These usually include an initial non-selective preenrichment stage (Anon. 1990) to allow injured *Salmonella* to recover, but this also allows the background competitor organisms to do likewise. Samples of preenrichment broth are transferred to selective enrichment where *Salmonella* can grow but the competitors are subjected either to a bacteriostatic or a bactericidal effect from the medium components. *Salmonella* are then isolated from the selective enrichment broths on selective and/or differential agars and confirmed with biochemical and serological tests.

The development of rapid detection techniques, such as immunoassays, nucleic acid hybridization probes and electrical methods, has led to some reduction in the analysis time. All of these methods, however, have a relatively low sensitivity (e.g. $10^5 - 10^6$ cells/ml) and so still require an enrichment step to allow the target organism to reach a detectable level. These methods can also be affected by interference from food debris present in the sample, which even further reduces their sensitivity.

Techniques for the rapid separation/concentration of microorganisms from food could obviate the need for lengthy culture procedures and allow maximum benefit from rapid detection. A number of approaches can be taken to separate and concentrate microorganisms. Filtration is potentially the simplest of these and can separate or concentrate specific components on the basis of physical size or morphology. There have been several investigations of the use of membranes to remove or trap microorganisms from a variety of sources including foods (Tanny *et al.* 1980; Pettipher 1983; Shields *et al.* 1985; Sharpe & Peterkin 1988; Bobbitt & Betts 1992).

New Techniques in
Food and Beverage Microbiology

0-632-03755-5

Filtration with large pore-size filters can separate microorganisms from food debris by trapping food particles in the membrane and allowing the bacteria to pass through. By combining different pore-size membranes, a sample can be 'cleaned up', followed by a concentration step before microbiological analysis, as in the ATP analysis of fresh meat products (Littel *et al.* 1986). Similarly, serial filtration can be used in the direct epifluorescent filter technique (DEFT; Pettipher 1983). Here, when analysing herbs and spices (Manninen & Sjoberg 1991), there is prefiltration with a Whatman No. 4 filter, followed by a 5–10-μm Gelman filter, to remove dried food debris and finer dust particles. Finally, the cells are captured on a 0.6-μm polycarbonate membrane before staining.

Electropositive filters have been used to isolate bacteria from pure cultures, vegetables and milk (Kroll 1985). These filters exploit the net negative charge of bacteria. Bacterial adsorption to these filters is very efficient, but release of the cells is poor because the cells are physically trapped within the membrane.

Density gradient centrifugation has been used to separate cells from milk (Foss Electric, Denmark), before detection by fluorescence microscopy (Kaereby & Asmussen 1989). However, many foods contain their own density gradients, which vary between samples, and, with fatty foods, some cells become trapped in the upper fatty layer and at interfaces during centrifugation.

Combinations of centrifugation and filtration have also successfully been applied as a pretreatment for foods before flow-cytometric analysis for microorganisms by Chemunex (France) (see Chapter 7). A small filter is placed in a holder in a centrifuge tube and the sample is placed on the filter. The complete tube/filter system is centrifuged to force the sample through the membrane, thus removing large food debris. Continued centrifugation then pellets the microbial cells and concentrates them (Bankes *et al.* 1991).

Two-phase aqueous systems, usually of dextran and polyethylene glycol, have been used to separate a mixture of three strains of *Escherichia coli* and one strain of *Chlorella pyrenoidea* (Albertsson & Baird 1962). In this way Stendahl *et al.* (1973a, b) were able to separate R and S mutants of *Salmonella typhimurium*.

Other workers (Tuley 1992) have investigated cell separation and concentration by dielectrophoresis, which exploits differences in the dielectric properties of cells. It has been shown that microorganisms concentrate at the internodes of the electrode within 2–5 min of applying a high-frequency dielectric field.

In contrast, antibodies and lectins can be used to separate components of a mixture on the basis of their chemical properties. Thus antibodies have a high affinity for the antigens that elicited their production, while lectins have a high affinity for specific saccharides. Biosorbents are composed of such affinity agents immobilized on a solid phase such as a bead or a column. The affinity agent attaches specifically to its target, which can then be separated from the mixture by collection of the solid phase.

Introduction of biosorbent methods into the conventional detection methods for pathogens has the potential to reduce the analysis time by several days, depending on the stage of the procedure used. This selective separation of pathogenic microorganisms from heterogeneous suspensions with biosorbents has previously been described for the removal of *Staphylococcus aureus* from milk (Johne & Jarp 1988), the detection of *Listeria monocytogenes* (Skjerve *et al.* 1990), the isolation of *Salmonella* from foods (Skjerve & Olsvik 1991; Vermunt *et al.* 1992) and of *E. coli* 0157:H7 from foods (Fratamico *et al.* 1992).

This chapter describes the separation of *Salmonella* from food samples containing normal background microflora with reduced enrichment times and by means of biosorbents. The biosorbent consists of *Salmonella*-specific antibodies, immobilized on magnetic beads. Beads were used as the solid phase, because it is easy to attach antibodies to them, they provide a large surface area for antibody–cell interactions and they can easily be collected from suspensions by applying a magnetic field.

Construction and Development of Biosorbents

Antibodies (Dynatech, Billingshurst, UK) were immobilized on magnetic beads by an avidin–biotin bridge. Goat anti-*Salmonella* antibodies were labelled with biotin by means of *N*-hydroxysuccinimide–long-chain (NHS-LC) biotin and biotinylation reagents according to the supplier's instructions (Pierce, Luton, UK). Antibody (10 mg) was reconstituted in 1 ml of 0.1 mol/l phosphate buffer (pH 7.2) and 100 μl was added to 7 μl of NHS-LC biotin (10 mg/ml). After incubation for 30 min at 22 °C, with gentle agitation on a rotary shaker, the mixture was diluted with 900 μl phosphate-buffered saline (PBS) (g/l: NaCl 8, KH_2PO_4 0.2, $Na_2HPO_4 \cdot 12H_2O$ 2.9, KCl 0.2) and transferred to pre-wetted dialysis tubing (molecular weight cut-off 12 000–14 000 Da) and dialysed twice (24 h each) against 2 l fresh PBS at 4 °C. The dialysed antibody was then transferred to an Eppendorf tube, cleared by centrifugation (3000 *g* for 4 min) and the supernatant stored at 4 °C. The degree of biotinylation of the antibodies was determined with the 2-[4′-hydroxyazobenzene] benzoic acid (HABA) assay according to the instructions of the supplier (Pierce).

Biotinylated antibody was immobilized on commercial streptavidin-coated magnetic beads (BioMag, Metachem Diagnostics, Northampton, UK) by mixing 200 μl beads (5×10^8 particles) with an equal volume of biotinylated anti-*Salmonella* antibody (25 μg/ml in PBS). After incubation for 3 h at room temperature, the beads were collected with a magnetic separation unit

(Metachem Diagnostics Ltd, UK) (5 min) and washed twice (2 × 2 ml) with PBS containing 1% (w/v) bovine serum albumin (PBS/BSA).

Determination of cell-trapping efficiency

Antibody-coated BioMag beads were resuspended and mixed with 1-ml suspensions of *Salmonella* cells (10^8 cells/ml). After incubation for 30 min at room temperature, the beads were collected (magnetic separation 5 min) and washed twice in PBS/BSA, and the degree of adsorption was determined by one of the following two methods:

1 Detection of adsorbed cells with antibody was achieved by resuspending the beads in peroxidase-labelled anti-*Salmonella* antibody (Dynatech, Billingshurst, UK) diluted 1:500 (v/v) in PBS/BSA (1 ml). This was incubated for 30 min at room temperature, washed and resuspended in 0.5 ml peroxidase substrate solution (2,2′-azino-*bis*-(3-ethylbenzthiazoline-6-sulphonic acid) (10 mg), citrate phosphate buffer, pH 4.1 (20 ml) and hydrogen peroxide (4 μl). After 15 min incubation at 37 °C, the reaction was stopped by adding 0.5 ml sodium fluoride (1 mg/ml) and the optical density of the supernatant at 405 nm was determined. Beads not coated with antibody were used as negative controls.

2 Beads were resuspended in maximum recovery diluent (MRD, Oxoid CM733) (1 ml) and plate counts were performed by the spread-plate technique of both the beads and the combined wash/supernatant fractions. Nutrient agar (NA, Oxoid CM3) was used for the total bacterial count and xylose lysine desoxycholate agar (XLD, Oxoid CM469) for a *Salmonella* count. All plates were incubated at 37 °C for 24 h.

Beads coated with antibody and exposed to *Salmonella* had strong enzyme activity, indicating that *Salmonella* had been trapped and the detection antibody attached to them. Beads not exposed to cells had only low levels of enzyme activity. This was due to a low level of non-specific binding of the detection antibody to the beads.

The extent to which this adsorption was serotype-dependent was then determined to ensure that subsequent experiments were based on the best combination of antibody and serotype, in order to explore the full potential of the biosorbent. Adsorption experiments were performed with cells of several *Salmonella* serotypes at a range of cell concentrations between 10^0 and 10^8 cells/ml with adsorption of cells again determined by enzyme-labelled antibody detection. The results indicated that, of the serotypes tested, *S. enteritidis* and *S. hadar* interacted to the greatest extent and *S. typhimurium* and *S. kedougou* to the least extent (Fig. 4.1). On this basis *S. enteritidis* was used in subsequent experiments.

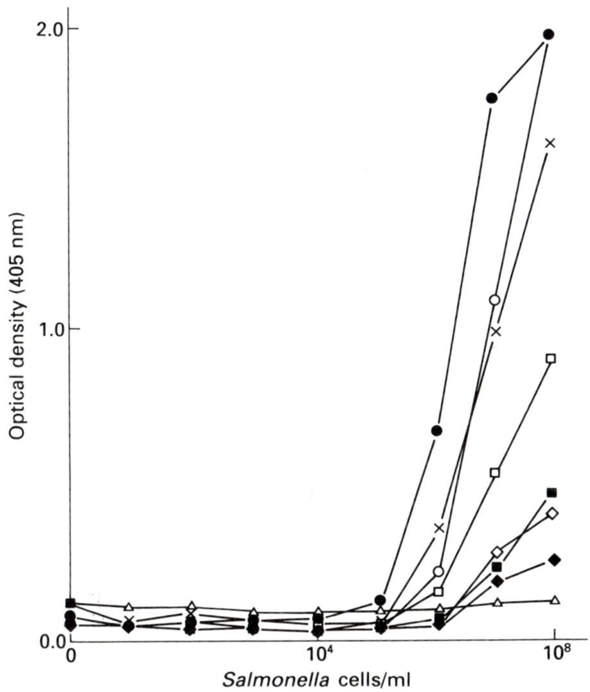

FIG. 4.1.Capture of various *Salmonella* spp. by immunomagnetic separation from nutrient broth. Detection was by means of an enzyme-labelled anti-*Salmonella* antibody. ●, *S. enteritidis*; ○, *S. hadar*; ×, *S. newport*; □, *S. infantis*; ■, *S. heidelberg*; ◇, *S. virchow*; ◆, *S. typhimurium*; △, *S. kedougou*.

Adsorption of cells from milk

An experiment was carried out to determine whether the immunomagnetic biosorbents could adsorb bacterial cells in skimmed milk and nutrient broth (NB, Oxoid CM67). The anti-*Salmonella* biosorbent adsorbed cells from skimmed milk to the same extent as from nutrient broth. This shows that skimmed milk did not interfere with the binding of cells to the beads.

Integration of Rapid and Conventional Detection of Pathogenic Microorganisms in Foods

For the biosorbent to work effectively when introduced to preenrichment broth, a threshold number of cells must be present such that they will bind to the antibody and subsequently be detected. Very few data are available on the growth kinetics of *Salmonella* in pre-enrichment media. This was investigated in a preliminary way by monitoring the growth kinetics with pure cultures and a range of foods that contain competitor organisms. Since the effect of different food matrices on the action of the beads was unknown, the immunomagnetic separation was introduced after 20 h pre-enrichment, when kinetic data indicated that sufficient target cells were available to ensure adsorption.

Samples (25 g) of skimmed milk powder, raw chicken, raw minced beef, chicken tikka ready-meal and raw steak were added aseptically to 225 ml buffered peptone water (BPW, Oxoid CM509). The samples were inoculated with 1–10 *S. enteritidis* cells/25 g, as estimated by DEFT and confirmed by plate counts on NA incubated at 37 °C for 24 h. Immediately after inoculation, the samples were plated on NA for total counts and on XLD for the *Salmonella* count.

The samples were incubated at 37 °C for 20 h. At the end of the pre-enrichment period, samples were diluted 1:100 in MRD to minimize interference from large food debris. They were then subjected to immunomagnetic separation.

Separation protocol

Enriched food samples (1 ml) are exposed to *Salmonella* antibody-coated beads (200 μl; 5×10^8 particles) for 30 min at room temperature. The beads are separated from the suspension for 3 min in the magnetic separation unit and the supernatant is retained. The beads are washed with PBS/BSA (1% w/v) and separated again. The supernatant is removed and combined with the previous supernatant. The beads are resuspended and serially diluted in MRD, and bacteria are enumerated on NA and XLD spread plates. Bacteria in the supernatant are similarly enumerated. The initial sample was also enumerated for both total bacterial count and for *Salmonella*. For a control, this was also done with beads without attached antibody. Plates were incubated at 37 °C for 24 h.

This procedure was carried out with a number of foods with the following results. Skimmed milk powder enriched in BPW for 20 h had a microbial population of 10^8–10^9 cells/ml, of which 10^7–10^8 were *Salmonella*, making up approximately 10% of the total population. Over 90% of the total population bound to the beads was *Salmonella* (Table 4.1), whereas <5% of the population in the supernatant, after separation of the beads, was *Salmonella*. This is an effective concentration of *Salmonella*, because <10% of the cells captured by the beads were non-*Salmonella* and the degree of non-specific binding was very low.

Raw chicken gave very variable results. *Salmonella* formed only 1% of the population in the pre-enrichment broth at 20 h incubation and approximately 10% of the cells captured by the beads consisted of *Salmonella* (Table 4.1).

TABLE 4.1. *Capture of* S. enteritidis *by immunomagnetic separation from foods, pre-enriched in buffered peptone water for 20 h*

Food	*S. enteritidis* as % of total population in pre-enrichment broth	*S. enteritidis* as % of total population captured on beads	*S. enteritidis* as % of total population in supernatant
Skimmed milk powder	11.80	91.40	1.42
Raw chicken	0.58	8.47	1.97
Chicken tikka	52.53	100.00	8.38
Raw minced beef	0.02	0.29	0.08
Raw steak	5.36	41.08	0.26

Again, the *Salmonella* in the pre-enrichment broth underwent a 10-fold concentration after exposure to the antibody-coated beads, but in this case non-specific binding was extremely high.

Excellent trapping by the beads was found with chicken tikka. The *Salmonella* made up approximately 50% of the population after pre-enrichment for 20 h. All of the cells that bound to the antibody-coated beads were *Salmonella* and less than 10% of the cells in the supernatant were *Salmonella* (Table 4.1). Of the *Salmonella* population of 10^7-10^8 cells/ml, only 10^5-10^6 (1%) bound non-specifically to the uncoated beads. This degree of non-specific binding is negligible.

With minced beef, poor capture of *Salmonella* was observed and there was a great deal of non-specific binding. Although there were only 10^5 *Salmonella* in the total population of 10^8 cells, the total number of cells binding to the beads was actually 10^7 cells/ml, which suggests that a great many non-*Salmonella* were being captured (Table 4.1). *Salmonella* made up less than 0.1% of the total population in the pre-enrichment broth. There was some degree of concentration after treatment with the beads, so that *Salmonella* made up around 0.5% of the population captured. However, this means that the remaining 95.5% of the captured cells were non-*Salmonella* binding non-specifically to the biosorbent. This was probably due to the high fat content of this food matrix. The fat may interact non-specifically with the beads, to trap high concentrations of non-*Salmonella*, thus giving a degree of non-specific binding. The variation of the results with raw chicken could also be attributed to high fat content in the product. Vermunt *et al.* (1992) demonstrated the isolation of *Salmonella* from minced meat pre-enrichment broths, incubated for 6 h. They reported that *Aeromonas hydrophila* and *Enterobacter agglomerans* showed considerable cross-reaction with the immunomagnetic particles, possibly because of an interaction with fat. This effect was considerably reduced by the addition of Tween 20 to the sample before immunomagnetic separation. Incorporation of chemical or physical pretreatments to solubilize fat in the product could therefore increase the range of food types suitable for immunomagnetic separation by reducing non-specific binding.

Raw steak, which contained less fat than minced beef, gave excellent results (Table 4.1). *Salmonella* made up less than 10% of the microbial population of the sample after preenrichment. Of the 3.7×10^7 *Salmonella* present in the pre-enrichment broth, 2.97×10^7 were captured on the beads, leaving only 1.75×10^6 in the supernatant. Again there was good concentration and the *Salmonella* were captured specifically, such that they comprised approximately 40% of the population on the beads and less than 0.3% of the population in the supernatant. A fairly high percentage of non-*Salmonella* bound to the beads but, nevertheless, an efficient concentration of the target cells was achieved. Only about 1% of *Salmonella* bound non-specifically to the uncoated beads. The results obtained with steak indicate that, in a low-fat product, the beads can bind efficiently to their target organisms and separate these from suspension. Foods with a high fat content can reduce the efficiency of binding, because the fat coats the beads and this results in the capture of high numbers of non-*Salmonella*.

Having established that the antibody-coated beads function well in skimmed milk powder, chicken tikka and raw steak after a 20-h pre-enrichment, it was necessary to establish whether the beads were as efficient after a shorter period of pre-enrichment. Initially *S. enteritidis* was studied in BPW without food material. After only 4 h incubation, there was a population of 9 *Salmonella*/ml. Of these approximately 7 bound to the beads, leaving approximately 4 cells/ml in the supernatant (Table 4.2). The sum of the cells on the beads and in the supernatant would be expected to equal the total count. The anomaly seen is due to experimental variation, exaggerated by the very low counts. After 5 h incubation, the total count increased to approximately 100 cells/ml. Over 80 cells bound to the antibody-coated beads, while only 1 cell bound to the uncoated beads. This demonstrates the low degree of non-specific binding in pure culture. After 6 h incubation, cell numbers were 7.93×10^2 cells/ml and 5.78×10^2 bound to the beads, leaving 1.66×10^2 in the supernatant. Again the non-specific binding was negligible.

TABLE 4.2. *Capture of* S. enteritidis *by immunomagnetic separation from buffered peptone water*

Pre-enrichment period (h)	Cells captured on beads (%)	Cells in supernatant (%)
4	62.90	37.10
5	69.57	30.43
6	77.69	22.31

Clearly, it is possible, by bead separation, to detect the cells after only 4 h incubation. However, when the growth kinetics of *Salmonella* in inoculated food samples is considered, the introduction of immunomagnetic separation after 4 h incubation would not be practicable because of the low cell numbers. Beads were, therefore, used after 7 h pre-enrichment, by which time a sufficient number of *Salmonella* would be present to ensure detection.

Dilutions of 1:100 were made with 20-h broths to reduce the amount of particulate food material and to lessen the interference with the beads. After 7 h incubation, however, there were insufficient cells in the sample to allow a dilution of 1:100 to be made and the broths had to be tested undiluted. The presence of food debris did not interfere with cell binding so that a 'clean-up' step, such as a coarse filtration to remove the large food particles, was not required.

The chicken tikka had a total microbial population of 1.08×10^5 cfu/ml after 7 h pre-enrichment, of which 1.10×10^2 were *Salmonella*. The number of *Salmonella* that bound to the beads was 1×10^2 cells/ml, leaving less than 10 *Salmonella* cells in the supernatant (Table 4.3). The total population of the 7-h pre-enrichment broth contained 0.1% *Salmonella*. The beads concentrated the *Salmonella*, so that they made up 2.1% of the cells captured. Thus 98% of the remaining cells were non-*Salmonella*, giving a high level of non-specific binding. However, the 20-fold concentration of *Salmonella* that was achieved cannot be ignored, as it clearly demonstrates an active separation/concentration system with the potential to increase efficiency.

Some results showed an anomaly between the total number of *Salmonella* in the enrichment broth and the number that appeared on the beads. For example, with raw steak (Table 4.3), there were 1.27×10^2 cfu/ml *Salmonella* bound to the beads but apparently only 15 *Salmonella* present in the original broth before separation. This was probably due to the suppression of *Salmonella* growth on XLD plates by competitor microorganisms in fractions not subjected to affinity separation, and this led to artificially low plate counts in these cases. Moreover, excellent concentration of *Salmonella* from raw steak was

TABLE 4.3. *Capture of* S. enteritidis *by immunomagnetic separation from foods pre-enriched in buffered peptone water for 7 h*

Food	*S. enteritidis* as % of total population in pre-enrichment broth	*S. enteritidis* as % of total population captured on beads	*S. enteritidis* as % of total population in supernatant
Chicken tikka	0.10	2.10	0.002
Raw steak	0.0001	0.05	0.009
Skimmed milk powder	0.003	0.08	0.002

demonstrated. Since only 1–10 *Salmonella* cells were inoculated into the food sample at the start of pre-enrichment, after 7 h incubation they were only 0.0001% of the total population. After treatment with the beads, however, they made up 0.05% of the total population. Though this was efficient concentration, there was a great deal of non-specific binding because of the high proportion of non-*Salmonella* in the sample.

Similar results were seen with skimmed milk powder (Table 4.3). After 7 h incubation the pre-enrichment broth contained less than 0.01% *Salmonella*. Of the cells captured about 0.08% were *Salmonella*, giving an eightfold increase in numbers. Again, a high proportion of cells bound to the beads non-specifically but concentration of *Salmonella* was still achieved.

Conclusions

The results indicate that antibody-coated beads offer considerable promise for the separation of *Salmonella* from inoculated food samples as an alternative to selective enrichment. Conventional methods for *Salmonella* detection and confirmation can take up to 7 days to complete. By including this biosorbent method, the detection time could be reduced by up to 72 h, depending on where in the enrichment the separation step is incorporated and on the end-point detection method employed. Plate counts were used in these experiments but more rapid methods such as nucleic acid probes, ELISA, impedance methods or the polymerase chain reaction (PCR) could be used. By using a detection method that is rapid, yet sensitive, such as PCR, it is possible that in the future low levels of *Salmonella* could be detected in food samples within the working day.

One limitation of this technique is that it cannot be used efficiently with foods of high fat content, such as raw minced beef and chicken. It is possible, however, that a chemical treatment to solubilize the fat could be incorporated to reduce non-specific binding. Vermunt *et al.* (1992) added Tween 20 to minimize non-specific binding, Morgan *et al.* (1991) prevented the non-specific attachment of *Pseudomonas putida* to antibody-coated magnetic beads with detergents and Fratamico *et al.* (1992) reported that the presence of surface-active agents in media, sodium lauryl sulphate in lauryl tryptose broth and bile salts in EC broth with novobiocin, reduced the level of non-specific binding.

Overall, these results demonstrate the potential of this simple biosorbent system and provide information about the application of immunomagnetic methods to cell separation and concentration. The technique needs further development so that it can be applied to a wide range of organisms to provide a rapid detection method that will be of great benefit to the food industry.

Acknowledgements

The authors acknowledge the support of the Ministry of Agriculture, Fisheries and Food in providing financial support for this project.

Information emanating from this Research Association is given after the exercise of all reasonable care and skill in its compilation, preparation and issue, but is provided without liability in its application and use.

References

ALBERTSSON, P.-Å. & BAIRD, G.D. 1962 Counter-current distribution of cells. *Experimental Cell Research* **28**, 296–232.

ANON. 1990 *Methods for Microbiological Examination of Food and Animal Feeding Stuffs. Detection of* Salmonella. BS5763: Part 4. London: British Standards Institution.

BANKES, P., ROWE, D. & BETTS, R.P. 1991 The rapid detection of yeast spoilage using the Chemflow system. *CFDRA Technical Memorandum* No. 621. Chipping Campden: Campden Food and Drink Research Association.

BOBBITT, J.A. & BETTS, R.P. 1992 The removal of bacteria from solutions by membrane filtration. *Journal of Microbiological Methods* **16**, 215–220.

FRATAMICO, P.M., SCHULTZ, F.J. & BUCHANAN, R.L. 1992 Rapid isolation of *Escherichia coli* 0157:H7 from enrichment cultures of foods using an immunomagnetic separation method. *Food Microbiology* **9**, 105–113.

JOHNE, B. & JARP, J. 1988 A rapid assay for protein-A in *Staphylococcus aureus* strains using immunomagnetic monosized polymer particles. *Acta Pathologica, Microbiologica et Immunologica Scandinavica* **96**, 43–49.

KAEREBY, F. & ASMUSSEN, B. 1989 Bactoscan. Five years of experiences. *Rapid Methods and Automation in Microbiology and Immunology*, eds Ballows, A., Tilton, R.C. & Turano, A., pp. 739–745. Brescia: Brixia Academic Press.

KROLL, R.G. 1985 Electropositively charged filters for the concentration of bacteria from foods. *Food Microbiology* **2**, 183–186.

LITTEL, K.J., PIKELIUS, S. & SPURGASH, H. 1986 Bioluminescent ATP assay for rapid estimation of microbial numbers in fresh meat. *Journal of Food Protection* **49**, 18–22.

MANNINEN, M. & SJOBERG, A.M. 1991 Evaluation of a microbiological method for detection of irradiation of spices. *Zeitschrift für Lebensmittel Untersuchung und Forschung* **192**, 226–229.

MORGAN, J.A.W., WINSTANLEY, C., PICKUP, R.W. & SAUNDERS, J.R. 1991 Rapid immunocapture of *Pseudomonas putida* cells from lake water by using bacterial flagella. *Applied and Environmental Microbiology* **57**, 503–509.

PETTIPHER, G.L. 1983 *The Direct Epifluorescent Filter Technique for the Rapid Enumeration of Microorganisms*. Letchworth: Research Studies Press.

SHARPE, A.N. & PETERKIN, P.I. 1988 *Membrane Filter Food Microbiology*. Letchworth: Research Studies Press.

SHIELDS, P.A., BERENFELD, S.A. & FARRAH, S.R. 1985 Modified membrane filter procedure for concentration of enteroviruses from tap water. *Applied and Environmental Microbiology* **49**, 453–455.

SKJERVE, E. & OLSVIK, O. 1991 Immunomagnetic separation of *Salmonella* from foods. *International Journal of Food Microbiology* **14**, 11–18.

Skjerve, E., Rorvik, L.M. & Olsvik, O. 1990 Detection of *Listeria monocytogenes* in foods by immunomagnetic separation. *Applied and Environmental Microbiology* **56**, 3478–3481.

Stendahl, O., Tagesson, C. & Edebo, M. 1973a Partition of *Salmonella typhimurium* in a two-polymer aqueous phase system in relation to liability to phagocytosis. *Infection and Immunity* **8**, 36–41.

Stendahl, O., Magnusson, K.E., Tagesson, C., Cunningham, R. & Edebo, L. 1973b Characterisation of mutants of *Salmonella typhimurium* by countercurrent distribution in aqueous two-polymer phase system. *Infection and Immunity* **7**, 573–577.

Tanny, G.B., Mirelman, D. & Pistole, T. 1980 Improved filtration technique for concentrating and harvesting bacteria. *Applied and Environmental Microbiology* **40**, 269–273.

Tuley, L. 1992 Saving time with rapid methods. *Laboratory Practice* **41**, 11–13.

Vermunt, A.E.M., Franken, A.A.J.M. & Beumer, R.R. 1992 Isolation of salmonellas by immunomagnetic separation. *Journal of Applied Bacteriology* **72**, 112–118.

5: Ultrasonic Standing Waves: Microbiological Applications

A.F.Zamani, R.W.Owen and D.J.Clarke
Sensor Development Group, Centre for Applied Microbiology and Research, Public Health Laboratory Service, Porton Down, Salisbury, Wiltshire SP4 0JG, UK

In a liquid that supports a standing-wave ultrasonic field, the pressure has a non-zero time-averaged component. Therefore, pressure differentials exist within the liquid and any particles suspended in it will move to positions where the pressure gradient is zero.

In a cylindrical container, where planar sound waves travel along the axis of symmetry, the loci of the zero pressure gradients are planes perpendicular to the axis of symmetry and are spaced half a wavelength of the sound apart. When the particles trapped in these planes are viewed from the side of the container, the profile will appear as bands. This phenomenon is termed *ultrasonic banding.*

Applications of Particle Banding

The ability of ultrasound to cause the migration and stasis of particles to preferential regions of a sound field can be useful in applications where the cells must be stationary for observation of a biological effect. Alternatively, the cells can temporarily be removed from suspension at points where measurements can be made of the cleared media, perhaps by optical or electrical means. In a flow cell, it is also possible to pass various reagents, such as stains or other markers, between the stationary cells. The method can also be used to concentrate cells.

Geometry of the Transducer

Ultrasonic waves are produced by applying alternating voltages of appropriate frequency to electrodes coated on either side of discs of piezoelectric crystals.

New Techniques in
Food and Beverage Microbiology

0-632-03755-5

Discs 10–15 mm in diameter and 1–4 mm thick were used for the studies to be described. The resonant frequencies of the crystals are inversely proportional to their thickness but they can be driven at integer multiples of the lowest resonant frequency. The polycrystalline ceramic material most often used is lead zirconium titanate (PZT), which is available with various characteristics denoted by a number (e.g. PZT4). One end of a coaxial cable is soldered to the electrodes of the crystal and is connected to a radio frequency power amplifier at the other end. The power rating varies between 12 and 150 W, with 25 W as a convenient power rating. The power amplifier must be driven by a source signal generator, which must provide control over the frequency and amplitude at fine precision, to allow the necessary adjustments. The resonant frequency of PZT crystals varies with temperature and the nature of the liquid to which it is coupled. A means is necessary to indicate the resonant condition, so that the resonance and the intensity of the sound can be maintained. Then, if a cathode ray oscilloscope is connected across the electrodes, when the resonance conditions are reached the waveform on the screen will show a dip in the amplitude. The addition of a phase and frequency modulation facility to the signal generator will expand the range of the manipulation. The crystal can be glued to a circular opening in a metal box for convenience and integrity (see Fig. 5.1). The container of cell suspension can either be directly mounted on the crystal by means of silicon

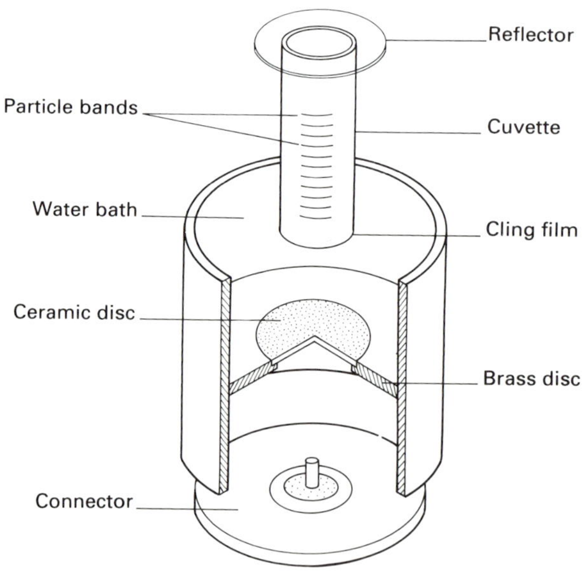

FIG. 5.1. The banding cell (for details see text).

rubber, or it can be coupled by way of a liquid bath. In the first case, the crystal will form the bottom of the cylinder, while in the second a layer of an acoustically transparent material such as 'cling film' can be used to cover the same (Fig. 5.1). A reflector or another crystal is necessary at the opposite end of the cylinder. Care must be taken to ensure that the source and the reflector are as parallel as possible. The water bath, necessary to couple the sound waves into the suspension, can either be a separate vessel stuck to the transducer or a drop of liquid between the cling film at the bottom of the cylinder and the face of the transducer. If capillary microslides are to be used, it is sufficient to put a drop of the suspension on the crystal surface and allow capillary action to draw some of this into the slide.

The flow cell, designed as a general-purpose ultrasonic standing-wave vessel, is shown in Fig. 5.2. The piezoelectric ceramic disc (Ferroperm, Denmark) is 25 mm in diameter and 1 mm thick. The lowest nominal resonant frequency of the ceramic disc is 2 MHz. The disc is mounted flush in a circular plastic disc 75 mm in diameter by means of a silicone adhesive. The electrical connection to the silver electrode is by means of solder at the rear and by silver ink (RS Components, UK) painted on the plastic disc at the front. The plastic disc is provided with an opening for the attachment of chambers of various volumes. The cylindrical chamber has an internal diameter of 20 mm and is terminated at one end by a matching plastic disc

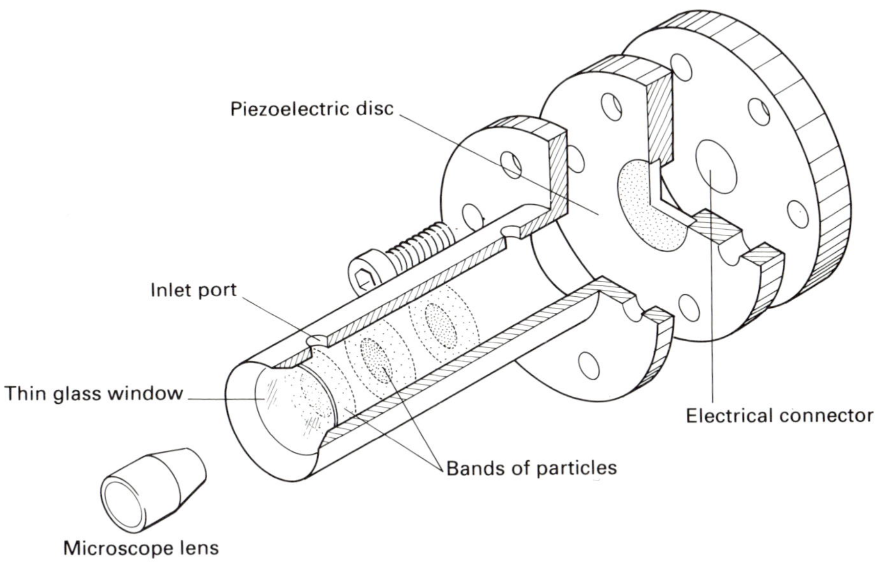

FIG. 5.2. The ultrasonic flow cell (for details see text).

fastened to the disc mounting and at the other by a microscope slide cover 22 mm in diameter. The chamber is provided with inlet and outlet ports at both ends. The connections to the piezoelectric disc are brought out via a standard BNC connector on the back plate.

The flow cell was tested by measuring the percentage efficiency for retaining yeast particles. A suspension of *Rhodotorula glutinans* (ATCC 10788) was prepared in the concentration range 1×10^3 to 1×10^7 organisms/ml in PAAL medium No. 62. The suspensions were passed through the flow cell at a rate of 1 volume change/min. After the flow had been established, time was allowed for the particles in the flow to be trapped in the ultrasonic nodes. Samples of the liquids flowing into and out of the flow cell were collected and were plated in the standard way. The retention rate ranged from 49–65% for fluid containing 1×10^7 particles/ml to 86–96% for fluid containing 1×10^3 particles/ml. The lower efficiency of the flow cell in holding particles at the higher concentration is predictable. On visual inspection of the shape of the bands formed by the trapped particles, the available volume seems to be a small fraction of the total and, therefore, at the higher concentrations there is physically no further volume available for trapping the later-arriving particles, i.e. the bands will fill up rapidly. At lower cell concentrations, it can be argued that the bands will be filled first at positions where the retention is greatest. In other words, the particles will be moved to the points near the centre of the bands, near the axis of symmetry of the flow cell. Particles arriving later will be held by progressively weaker forces.

After sampling, the particles were brought to the thin glass window by the frequency modulation method (Whitworth 1989). The cells were concentrated at the centre of window and no clumped cells moving together could be observed.

The increase in the efficiency of particle retention is encouraging the use of this method for the isolation of small numbers of organisms from large sample volumes. It may be possible to filter organisms selectively on the basis of specific criteria such as density, size or compressibility, if the physical parameter of the flow cell or the field can be tailored to produce substantially different forces for the organisms involved. It is, therefore, necessary to consider the forces that act on cells in an ultrasonic field and the variables available to the experimenter.

Forces Acting on the Particles

The forces responsible for the formation and disruption of ultrasonic bands fall into three categories. First are the forces that move and confine particles in the pressure nodes. Second are the forces that move particles and the suspending liquid without a confinement effect, such as gravity, Brownian motion, viscous drag and cavitation. Third are those that keep the particles together after the other forces have formed the bands. These forces apply only to the particles that are close together.

King (1933) showed that, in a standing-wave field, particles smaller than the wavelength of the sound experience a larger radiation force under standing-wave conditions than in a travelling-wave field. Weiser & Apfel (1982) introduced consideration of the effect of gravity on the levitation of the μm-sized particles. Gould & Coakley (1974) combined the effects of Archimedean buoyancy, streaming and radiation forces. Higashitani *et al.* (1981) considered diffusion of the particles and its effect on the banding. Coakley *et al.* (1989) analysed the forces on the particles, in order to predict the effect of different variables on the banding of particles at the half-wavelength periods.

Theoretical Treatment

Background

The ultrasonic radiation force

Nyborg (1967) developed the theoretical basis that allows calculation of the radiation forces on small, rigid, spherical particles. It contains two components. The first results from the pressure of the radiation on the inclusion of the sound field and the second results from the finite compressibility of the object. The combined terms give:

$$F = VP_o^2\, k \sin(2kz)\{[(1/s\sigma^2) - (5s - 2)]/(2s + 1)\}/4dC^2 \qquad (1)$$

where σ is the ratio of the sound velocity in the sphere and in the fluid medium, s is the ratio of the density of the particles and of the suspending liquid, P_o is the standing-wave pressure amplitude, F is the radiation force acting on the particles, d is the density, C is the velocity of the sound in the particle, V is the volume of the particle and $k = 2\pi/\lambda$, where λ is the wavelength of the sound.

Acoustic streaming

The non-linear effects in a liquid irradiated with ultrasound can produce a net flow of liquid called *acoustic streaming*. The origin of these non-linearities is that, for most liquids, the pressure dependence of density is not a linear relationship. Expansion in a Taylor's series gives:

$$P = P_o + (sP/sd)(d - d_o) + (s^2P/sd^2)(d - d_o)^2 + \ldots \qquad (2)$$

Excluding higher terms this can be written as:

$$P - P_o = A(d - d_o)/d_o + B[(d - d_o)/d_o]^2/2 \quad (3)$$

where

$$A = d_o C_o^2 \text{ and } B = 2Ad_o C_o(\delta C/\delta P)_s \text{ at } P = P_o.$$

The equation of the motion of the liquid can be written as:

$$U(z, t) = U_o \sin[\omega t - (\omega z/C_o)(1 + BU/2AC_o)^{-2(A/B)-1}] \quad (4)$$

where d is the density and C is the sound velocity in the liquid and the subscript o denotes the values for unperturbed liquid. This shows that the propagation velocity will increase as the liquid velocity and displacement increase. The ratio A/B is a function of the nature of the propagation medium and plays an important role in the non-linear effects. The non-zero time average of the contribution of higher harmonics will create a net flow of the liquid (Nyborg 1967). Nyborg (1955) had shown that, for liquids in a channel, streaming currents will flow between the axis and the walls in eddies at every quarter-wavelength between the velocity nodes and antinodes. Gould & Coakley (1974) showed that the ratio of the radial to axial streaming velocity is:

$$U_r/U_z = kr\{1 - (r/R)^2\}[1 - 2(r/R)^2] \quad (5)$$

Therefore, at higher frequencies the radial streaming is the more dominant component.

The total force on the particles

The condition for banding of the particles, zero axial terminal velocity, was shown by Gould & Coakley (1974) to be:

$$P_o^2 \sin(2kz) = G/[RF + MSF] \quad (6)$$

where the term RF is related to maximum radiation pressure and MSF is related to streaming forces. The term G relates to Archimedean buoyancy. The lowest values of P_o for which equation (5) holds is at points where $\sin(2kz)$ has the maximum value of 1.

The effects of variables

The effect of the frequency

Coakley *et al.* (1989) calculated the effect of sound frequency, particle radius, density and sound velocity on the threshold banding sound pressure. Over a range of 1–9 MHz and particle radius of 0.2–100 μm the threshold banding pressure was plotted. For particles of radius >10 μm, the minimum banding sound pressure is proportional to the square root of the ultrasonic wave frequency. The effect of frequency is reduced as the particle size is reduced. For larger particles the dominant term is the radiation pressure, which is frequency-dependent, but for the smaller particles the dominant term is the microstreaming, which is independent of the frequency. In the case where the banded particles are aggregated, the effective particle size will increase accordingly.

The effect of density and sound velocity

The radiation force on the particles depends on their density and that of the suspending liquid. As the density of the particles approaches that of the liquid, the minimum banding sound pressure falls, but at lower ultrasonic pressures the time necessary for the banding will increase. The magnitude of the sound radiation force on a particle is also dependent on the velocity of the sound in that particle. This effect is more marked for larger particles with lower density (Coakley *et al.* 1989).

Bernoulli effect

When a liquid flows from a larger to a smaller cross-section area, it undergoes a fall in pressure. This is the Bernoulli principle. When a liquid flows between two particles that are close together, the decrease in pressure will be equal to a net attractive force. Dyson *et al.* (1971) suggested that the sharply defined edge of the bands observed may not be due solely to the standing-wave formation and that Bernoulli attraction may be involved.

Brownian motion

According to Batchelor (1976), the random movement of particles out of concentration zones, due to the Brownian motion, is the equivalent of diffusion by a force:

$$F_b = kT \,\mathrm{grad}(\ln p(z)) \tag{7}$$

where $p(z)$ is the probability density of the particle position, k is the Boltzmann constant and T is the temperature. The ratio of the number of particles at a node to that of particles at the antinode can be calculated from equation (7) (Higishitani *et al.* 1981). This includes terms that contain the particle radius and its compressibility. The radius at which diffusion will be the limiting factor in banding is 0.2 μm.

Experimental Studies and Applications

Early experiments

The earliest observations of the banding of particles in a standing ultrasonic wave were reported by Kundt & Lehmann (1874). Sollner & Bondy (1936) showed that quartz particles, with a density of 2.66, band at the nodes and toluene, with a density of 0.87, bands at the antinodes.

Later developments

The invention of ultrasonic scanners initiated a new interest in the study of the safety of ultrasonic waves on human embryos. Dyson *et al.* (1971) reported the banding of red blood cells (RBC) in a 3.5-day-old chick embryo placed in a saline bath and exposed to a 3-MHz standing-wave ultrasonic field. They observed that the RBCs aggregated in bands at half-wavelength spacing in both veins and arteries. In arteries parallel to the direction of the sound stasis is readily established, while in those at right angles to the sound the RBCs flow easily. The RBCs disbanded and flowed again when the sound field was removed. Standing waves at 1 MHz, in a cylindrical cavity, led to a similar banding of RBCs and, when a blood vessel was moved, the position of the bands remained fixed and did not follow that of the vein (Baker 1972). Gould & Coakley (1974) demonstrated agreement between the measured banding times and the calculated values for a 1% RBC suspension in saline.

Biological Applications

The ability to bring biological particles into close proximity and restrict them to a fixed position has been exploited in a variety of ways. It is possible to fuse biological cells by application of a μs-pulse of electrical field (Zimmerman 1982). For fusion to take place, it is necessary to bring cells into membrane contact, which can be achieved by various means. Applications of a standing-wave ultrasonic field to bring the fusing cells together has the advantage that it is independent of the electrical properties of the suspending liquid. Vieneken *et al.* (1985) exposed myeloma cells in a low-ionic-strength medium suspension between two concentric stainless-steel tubes that acted as electrodes. Banding under ultrasonic influence brings the cells into close contact before a pulse is administered. The fusing cells are held in the banding position for some time after the pulse, in order to inhibit the disturbing effects of the movement of the settling cells. Bardsley *et al.* (1990) reported fusion of murine hybridomas by electrofusion of cells in sugar solutions and in media of high ionic strength (115 mmol/l NaCl).

By levitating cells in suspension with an ultrasonic standing wave, it is possible to observe cell-to-cell interaction for a long period of time, without the cells moving away from the focus of the microscope objective. Tilley *et al.* (1987) observed the behaviour of RBCs in bands in standing-wave fields in microcapillaries. The cell suspension was drawn into a capillary from a cuvette coupled to a water-cooled piezoelectric crystal at a resonance frequency of 1 MHz. The capillary was observed with a horizontal microscope, which allowed observation of the polymer-induced adhesion of cells in real time. Cells can be held in the field of view and within the depth of focus for up to 4–5 min while cell-to-cell adhesion is recorded.

Ability to bring suspended cells into close contact can also be used to enhance sedimentation and separation. Thus, Kilburn *et al.* (1989) reported enhancement of the sedimentation of human histiocytic lymphoma cells from culture media. Acoustic radiation force can also accelerate agglutination of cells by antiserum. Jepras *et al.* (1989) observed that the agglutination of *Legionella pneumophilia* by diluted whole rabbit antiserum can be accelerated by at least two orders of magnitude at antiserum concentrations 512-fold more dilute than in a standard microwell assay. Similarly, Grundy *et al.* (1989) reported a sixfold acceleration of detection time when a commercial haemagglutination test for hepatitis B virus was subjected to ultrasonic banding.

Transport and harvesting of cells

Kilburn *et al.* (1989) suggested that, to reduce the risk of aerosol formation and contamination, cells could be harvested in an enclosed environment. Human cell lines were first concentrated in bands and then allowed to fall in clumps by removing the sound field. The clumps were disrupted by hydrodynamic drag during sedimentation and only 75% of the cells were removed. In a horizontal cylinder, the clumps may be sedimented through a shorter distance and the efficiency can be as much as 95%.

It is possible to move the bands of cells by moving the position of the standing-wave nodes. A variety of means to achieve the manipulation of the position of nodes have been proposed. Thus, Whitworth (1989) attempted to move the bands by changing the driving frequency. It is necessary to form a single axial column of particle bands, if their movement is to be smooth and without disruption. He demonstrated that rotation of the ultrasonic waveguide results in the formation of a single central column. This is, however, not the only method and Whitworth & Coakley (1992) showed that, by adapting a method first proposed by Du & Breazeale (1985), special patterns of electrodes on the piezoelectric crystal can be created. These result in a Bessel function radial distribution of the sound field, which allows near zero field at the container wall and a single axial column of particles. Whitworth *et al.* (1991)

presented a theoretical basis and a practical demonstration of cell harvesting with the Bessel sound field transducers. They introduced a small frequency difference between two transducers at the top and bottom of a thin-walled tube. The standing-wave nodes then drifted slowly towards one and away from the other transducer and the bands of cells followed. If, in addition, the two frequencies are modulated in 1.5 cycles, the bands can be moved to any position along the central axis of the cylindrical cell. Hager *et al.* (1991) proposed a method similar to the frequency modulation method, which they called the drifting resonance method. Here, frequency modulation is achieved in discrete steps to allow the standing-wave pressure nodes to drift towards the transducers. The geometry of the vessel is different; the transducers are in the middle of the vessel, mounted back to back, and the liquid flows parallel to the surface of the transducers. An alternative to moving the standing-wave pressure nodes is to move the liquid through the stationary nodes. The observation by Baker (1972) has already been noted above: that, if the RBCs in an artery are banded, these will remain in a fixed position relative to the vessel even when the artery itself is moved. Peterson *et al.* (1986) applied this method to the separation of red cells from whole blood that was passed through a tube submerged in a bath in an ultrasonic standing wave. Schram & Rendell (1989) proposed the removal of cells by moving the vessel holding the cells and the suspending liquid, through an ultrasonic bath that consists of a transducer and a reflector.

Biological Effects

It is possible that cells exposed to ultrasonic fields may be damaged. Kilburn *et al.* (1989) studied the relationship between the frequency and intensity of the sound and the viability and cell removal efficiency. They reported an increased cell removal rate with increase in the sound intensity, up to a level where a sudden decrease in viability and a constant cell removal rate are observed. Above this intensity threshold, the viability of the removed cells will rapidly approach zero. The value of the sound intensity at which damage to cells begins increases with the sound frequency. The cause of the damage is cavitation, which is the production and subsequent violent collapse of bubbles of gas, due to the ultrasonic field. This is the effect normally used to produce cell disruption in commercial sonicators.

Ultrasonic separation of the cells is usually carried out in enclosed systems, where reflection is from the walls of the container. If the reflection is at an air–liquid interface, atomization of the liquid will occur and a fine mist of small droplets will be formed. These droplets may carry bacteria and become a health hazard. Grundy *et al.* (1990) collected samples of mist above liquid during sonication, by covering the top of the vessel with a Petri dish, and found that, at a sound intensity that produces visible droplet ejection from the liquid, the mist above the vessel will contain only the airborne organisms near the vessel. At 1.25 times this sound intensity, a general increase in the number of organisms was observed. When the sound level was increased, so that both large droplets and mist were visible above the liquid, a marked increase in organisms was observed. The number of isolates decreased with an increase in the frequency of the sound from 1 to 7 MHz. It is clear that a solid–liquid reflector such as a microscope slide cover is preferable to avoid production of particle-bearing droplets.

Safety Considerations

Apart from standard microbiological safety procedures, ultrasonics aspects justify some attention. Ultrasound at the frequencies of interest, that is from 1 to 9 MHz, will couple to air extremely poorly. Damage to hearing is, therefore, not to be expected and ear protection is unnecessary. The proximity of the power amplifier to conducting liquids should be kept in mind and all electrical connections must be made according to the relevant safety regulations. If parts of the liquid being irradiated are in contact with air, the possibility must be considered that organisms may be carried in the aerosols that are produced. Sound will easily propagate through the walls and reflectors etc., so that drops of the suspension on these may also generate aerosols and therefore must be removed.

References

BAKER, N.V. 1972 Segregation and sedimentation of red blood cells in ultrasonic standing waves. *Nature* **239**, 393–399.

BARDSLEY, D.W., LIDDEL, J.E., COAKLEY, W.T. & CLARKE, D.J. 1990 Electroacoustic production of murine hybridomas. *Journal of Microbiological Methods* **129**, 41–47.

BATCHELOR, G.K. 1976 Brownian diffusion of particles with hydrodynamics interaction. *Journal of Fluid Mechanics* **74**, 1–29.

COAKLEY, W.T., BARDSLEY, D.W., GRUNDY, M.A., ZAMANI, F. & CLARKE, D.J. 1989 Cell manipulation in ultrasonic standing wave fields. *Journal of Chemical Technology and Biotechnology* **44**, 43–62.

DU, G. & BREAZEALE, M.A. 1985 The ultrasonic field of a Gaussian transducer. *Journal of the Acoustical Society of America* **78**, 2083–2086.

DYSON, M., WOODWARD, B. & POND, J.B. 1971 Flow of red blood cells stopped by ultrasound. *Nature* **232**, 572–573.

GOULD, R.K. & COAKLEY, W.T. 1974 The effects of acoustic forces on small particles in suspension. *Proceedings of the Symposium on Finite Amplitude Wave Effects in Fluids*, ed Bjørnø, L., pp. 252–257. Guildford: Pergamon.

GRUNDY, M.A., COAKLEY, W.T. & CLARKE, D.J. 1989 Rapid detection of hepatitis B virus using a haemagglutination assay in an ultrasonic standing field. *Journal of Clinical and Laboratory Immunology* **30**, 93–96.

Grundy, M.A., Coakley, W.T. & Clarke, D.J. 1990 Microbial content of aerosols produced from suspension exposed to megahertz frequency ultrasound. *Ultrasonics* **28**, 415–421.

Hager, F., Benes, E., Bolek, W. & Groschl, M. 1991 Investigation of a new ultrasonic drifting resonance field cell for the refinement of hydrosols. *Proceedings of the 9th FASE Symposium and 10th Hungarian Conference on Acoustics*, pp. 261–269.

Higishitani, K., Fukushima, M. & Matsuno, Y. 1981 Migration of suspended particles in plane stationary ultrasonic fields. *Chemical Engineering Science* **36**, 1187–1192.

Jepras, R.I., Clarke, D.J. & Coakley, W.T. 1989 Agglutination of *Legionella pneumophila* by antiserum is accelerated in an ultrasonic standing wave. *Journal of Immunological Methods* **120**, 201–205.

Kilburn, D.G., Clarke, D.J., Coakley, W.T. & Bardsley, D.W. 1989 Enhanced sedimentation of mammalian cells following acoustic aggregation. *Biotechnology and Bioengineering* **34**, 559–562.

King, L.V. 1933 On the acoustic radiation on spheres. *Proceedings of the Royal Society, Section A* **147**, 212–240.

Kundt, A. & Kehmann, O. 1874 Longitudinal vibrations and acoustic figures in cylindrical columns of liquid. *Annalen der Physikalischen Chemie* (Poggendorff) **153**, 1–5.

Nyborg, W.L. 1955 Acoustic streaming. *Physical Acoustics* Vol. 2B, ed. Mason, W.P., pp. 265–331. New York: Academic Press.

Nyborg, W.L. 1967 Mechanisms for nonthermal effects of sound. *Journal of the Acoustical Society of America* **44**, 1302–1309.

Peterson, S., Perkins, G. & Baker, C. 1986 Development of an ultrasonic blood cell separator. *IEEE/Eighth Annual Conference of Engineering in Medicine and Biology Society (Papers) Fort Worth, USA*. IEEE Catalog No. 86CH2368-9, pp. 154–156.

Schram, C.J. & Rendell, M. 1989 Manipulation of particles in megahertz standing waves. *Ultrasonics International Conference Proceedings*, ed Novak, Z., pp. 262–267. Guildford: IPC Science and Technology Press.

Sollner, K. & Bondy, C. 1936 The mechanism of coagulation by ultrasonic waves. *Transactions of the Faraday Society* **32**, 616–623.

Tilley, D., Coakley, W.T., Gould, R.K., Payne, S.E. & Hewison, L.A. 1987 Real time observation of polylysine, dextran and polyethylene glycol induced mutual adhesion of erythrocytes held in suspension in an ultrasonic standing wave field. *European Biophysical Journal* **14**, 499–507.

Vieneken, J., Zimmerman, U., Zenner, H.P., Coakley, W.T. & Gould, R.K. 1985 Electro-acoustic fusion of erythrocytes and of myeloma cells. *Biochemica et Biophysica Acta* **820**, 259–264.

Weiser, M.A.H. & Apfel, R.W. 1982 Extension of acoustic levitation to include the study of micron sized particles in a more compressible host liquid. *Journal of the Acoustical Society of America* **71**, 1261–1268.

Whitworth, G. 1989 Particle movements in a rotating ultrasonic waveguide. PhD Thesis, University of Virginia, USA.

Whitworth, G. & Coakley, W.T. 1992 Particle column formation in a stationary ultrasonic field. *Journal of the Acoustical Society of America* **91**, 79–85.

Whitworth, G., Grundy, M.A. & Coakley, W.T. 1991 Transport and harvesting of suspended particles using modulated ultrasound. *Ultrasonics* **29**, 439–444.

Zimmerman, U. 1982 Electric field mediated fusion and its related electrical phenomena. *Biochemica et Biophysica Acta* **694**, 227–277.

6: Detection and Enumeration of Viable Bacteria by Flow Cytometry

A.C.PINDER,[1] C.EDWARDS,[2] ROSEMARY G.CLARKE,[1]
J.P.DIAPER[2] AND SANDY A.G.POULTER[1]
[1]*AFRC Institute of Food Research, Norwich Research Park, Colney, Norwich, Norfolk NR4 7UA, UK;*
and [2]*Department of Genetics and Microbiology, Life Sciences Building, University of Liverpool, Liverpool L69 3BX, UK*

The design and applications of flow cytometry have developed greatly. Early work was confined largely to studies of the properties of eukaryotic cells, and in particular mammalian cells. Recently, however, the technique has increasingly been directed to the study of microorganisms (e.g. Pinder *et al.* 1990; Steen 1990). The principles, scope and operation of flow cytometers will not be discussed here because excellent reviews are available (see Shapiro 1988, 1990; Boye & Lobner-Oleson 1990; Edwards *et al.* 1992). The main features of a typical flow cytometer, as relevant to discussions in this chapter, are shown in Fig. 6.1, which also shows the additional feature of cell sorting available with more sophisticated instruments. Two general types of measurement are possible. First, the intrinsic properties of untreated cells can be studied, principally by light scatter, which gives information about the structure and size of cells. Other intrinsic properties include autofluorescence, which has been proposed as an indicator of the energy status of the cell pigmentation (Kell *et al.* 1991), which is useful in the study of photosynthetic organisms, and the presence of naturally fluorescent compounds such as deazaflavin F420 found in methanogens. The second most flexible and informative approach involves the treatment of cells with fluorescence-conjugated molecules or fluorescent dyes specific for target cellular functions or structures (reviewed by Kell *et al.* 1991). DNA-specific dyes, such as Hoechst 33342, which binds to A–T-rich regions, and mithramycin can be used to determine chromosomal number in bacteria (van Dilla *et al.* 1983; Steen *et al.* 1990). Major macromolecular pools, such as protein and RNA, may be monitored

New Techniques in
Food and Beverage Microbiology

0-632-03755-5

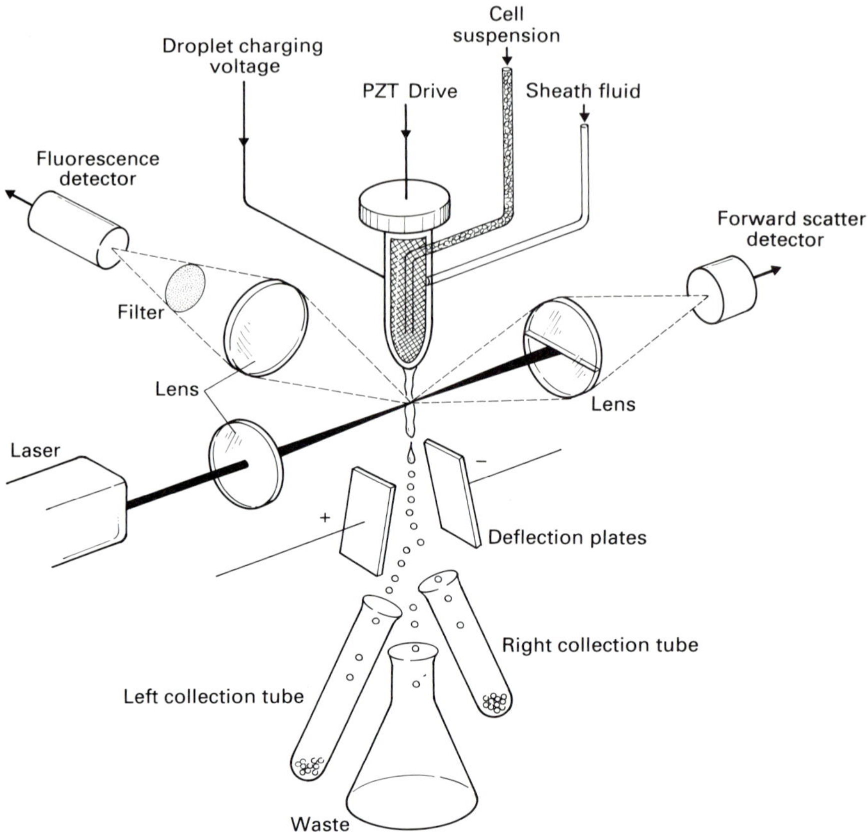

FIG. 6.1.Schematic diagram showing the main features of a typical flow cytometer and cell sorter. PZT, piezoelectric transducer.

qualitatively with fluorescein isothiocyanate and pyronin Y, respectively, and enzyme activities can be measured with fluorescence-labelled substrates.

More recently, fluorescent molecular probes have been used to detect and identify cells in heterogeneous environments. This involves the use of fluorescence-labelled antibodies or oligonucleotides. The latter probes have been designed, principally for 16S rRNA sequences, for the identification of microbial groups, genera, species or even strains. These probes can suffer from low signal intensity when used to identify bacteria from natural environments, and this has hampered their application in flow cytometry (Amann *et al.* 1990a, b, 1992; Zarda *et al.* 1991).

Another major problem, particularly in the application of molecular probes to the identification of microorganisms from natural environments, is that of non-culturability. Some bacteria are thought to adopt a viable but non-culturable state, particularly under environmental stresses, such as those resulting from nutrient limitation (Roszak & Colwell 1987). Therefore, although it is possible to identify or detect the presence of a particular organism with fluorescence-labelled antibodies or nucleic acid probes, it may not be possible with any degree of certainty to decide whether it is viable. Such decisions are crucial in public health, water quality assessment or food technology.

In this chapter new applications of flow cytometry are presented, which attempt to overcome some of the above problems. We describe how fluorescence-labelled antibodies may be used for the rapid identification of target species and how flow cytometry may be used to assess and measure viability, even though a proportion of the population is non-culturable.

Immunofluorescence Flow Cytometry

With flow cytometry it is possible, to a limited extent, to differentiate between organism types on the basis of their relative size and DNA content (Pinder *et al.* 1990). To detect specific groups or species of bacteria, however, a more sophisticated technique is required. Various immunological methods have been developed that use antibodies to detect antigens on the surface of an organism. Although polyclonal antibodies offer some degree of specificity, the development of monoclonal antibodies (MAbs) gives the potential to recognize a single antigen type. The antigen-binding site may be a conformation of five or six amino acids of a protein or five or six sugar residues of a polysaccharide (Kohler & Milstein 1975). To detect the antibodies, and therefore the cells they recognize, they must be labelled with a fluorescent marker.

In order to make the assay quantitative, the sample is injected into the flow cytometer at a precisely controlled flow rate (5–50 μl/min). Electronic discriminator gates are set so that the computer counts only cells that fall within determined size and fluorescence criteria. In this way, the count rate is converted into the concentration of a particular cell type in the sample (Pinder *et al.* 1990). Immunofluorescence flow cytometry, therefore, represents an assay that is both highly specific and relatively rapid and robust.

Fluorescent labels for antibodies

The most widely used fluorescent labels for antibodies and recommendations for excitation source, dichroic beam-splitter and emission band-pass filter are listed in Table 6.1. In choosing a label for a particular application, it is important to consider the wavelength(s) available from the excitation source, the strength of the signal expected and whether two or more different labels

TABLE 6.1. *Fluorescent labels for antibodies*

Fluorescent label	Excitation filter (nm)	Dichroic beam-splitter (nm)	Emission filter (nm)
Fluorescein isothiocyanate (FITC)	480 ± 10 (lamp) 488 (Ar laser)	505	530 ± 20
Tetramethylrhodamine isothiocyanate (TRITC)	535 ± 20 (lamp) 530 (Kr or dye laser)	570	590 ± 20
Rhodamine isothiocyanate (XRITC)	550 ± 15 (lamp) 530 (Kr or dye laser)	575	615 ± 20
Texas red*	580 ± 15 (lamp) 570 (Kr or dye laser)	600	630 ± 15
Phycoerythrin (PE)	535 ± 5 (lamp) 488 (Ar laser)	575	590 ± 20
Phycocyanin (PC)	580 ± 15 (lamp) 633 (HeNe laser)	600	630 ± 15
Allophycocyanin (APC)	590 ± 20 (lamp) 633 (HeNe laser)	620	660 ± 15

* Texas red is a trademark of Molecular Probes, Inc.

are to be used in combination. The light source in a flow cytometer is usually either a mercury lamp or an argon ion laser. Although the laser is more costly, its higher illumination intensity is more effective for the detection of weak immunofluorescence, provided the intensity is not so high as to induce photobleaching of the fluorophores. Of the low-molecular-weight fluorophores, fluorescein is the most popular, with the highest fluorescence/mol in aqueous solution. The quantum yield is quite high, in spite of significant quenching on conjugation with proteins. Fluorescein has an absorption peak conveniently close to the blue (488 nm) wavelength from an argon ion laser.

An important feature of flow cytometry is its ability to differentiate between several types of antigens at the same time, by means of different antibodies, each with its own unique fluorescent label. Multiple fluorescence detectors and appropriate combinations of dichroic beam-splitters and band-pass optical filters are used, so that it is possible, for example, to distinguish between the green, orange and red fluorescent emissions from three specific stains.

A suitable choice of labels for the simultaneous detection of two different antibodies would be fluorescein and a rhodamine derivative, such as rhodamine X isothiocyanate (XRITC), or Texas red (Dean & Pinkel 1978). Rhodamines are generally less efficient than fluorescein but they are also less sensitive to extremes of pH and to photobleaching. Compared with fluorescein, these labels have longer wavelength emission, so that, for multiparameter detection, there is no 'crosstalk' with the fluorescein channel. Unfortunately, the absorption peak is also at a higher wavelength, and this is not available from an argon ion laser. The use of a second krypton ion or dye laser may be considered prohibitive on the grounds of cost and complexity.

Fortunately, dual-colour detection is easier with a class of dyes, the phycobiliproteins, developed from fluorescent algae (Oi *et al.* 1982). These include phycoerythrin, phycocyanin and allophycocyanin. Phycobiliproteins show high absorbance and quantum yield and low susceptibility to fluorescence quenching by internal or external agents. R- and B-phycoerythrin have broad absorbance peaks, suited to either argon ion laser or mercury lamp excitation. They exhibit a large Stokes shift, so that the wavelength of fluorescence emission is well separated from that of fluorescein. For three-colour detection, the phycobiliprotein labels C-phycocyanin or allophycocyanin are of interest. These are best excited by a low-cost helium neon laser. Recently, a new conjugate has been reported that permits three-colour detection with a single argon ion laser source (Festin *et al.* 1990; Russ 1990).

Antibodies can be fluorescence-labelled for flow cytometry by three essentially different methods, indirect, avidin–biotin and direct.

Indirect method for labelling antibodies

Two antibodies are used in the indirect method. The first is raised against the specific pathogen of interest, for example in a mouse. The second antibody is chosen to recognize the Fc fragment of the first antibody (e.g. if the primary is a mouse IgG, an antimouse IgG is chosen). The latter antibody carries the fluorescent label, which therefore becomes indirectly attached to the first antibody. An advantage of this technique is simplicity. While the primary antibody may well be customized and highly specific, it need not be directly labelled. Suitable secondary antibodies, with a more general specificity, are readily available commercially and may be purchased prelabelled.

The procedure involves addition of the first antibody to the sample, incubation, addition of the second antibody, and then a further incubation, before the sample is analysed by flow cytometry. It is necessary to proceed in this order, and to experiment with the ratio of secondary to primary antibody, in order to obtain the optimum fluorescence signal. With this method, the efficiency of labelling is generally high. Although it is not possible to bind more than three molecules of fluorescein, or many other dyes, directly on to an IgG antibody molecule without reduction of its effectiveness (Shapiro 1988), it is possible to attach five or six secondary antibodies to the same primary site. Each antigenic site can, thus, be indirectly labelled with 15–18 dye molecules. Care must be taken to ensure that the secondary antibody is not in excess, because this would result in high background fluorescence.

A serious disadvantage of indirect labelling is that it cannot be used for two-colour detection of two primary antibodies of the same immunoglobulin class. There are problems even with two different antibodies, because IgG antibodies cross-react significantly with IgM.

Avidin–biotin method for labelling antibodies

Another indirect approach makes use of the avidin–biotin system (Bayer & Wilchek 1980). First, the antibody is reacted with *N*-hydroxysuccinimide biotin (NHS-biotin), when free NHS is liberated and the biotin is left bound to the antibody. The biotinylated antibody is then incubated with the sample and, finally, a proprietary avidin–fluorophore conjugate, which forms a strong link between the antibody and the fluorophore, is added.

Since biotin is relatively small, several molecules can be conjugated to an antibody protein without altering its biological activity. Underconjugation results in poor labelling, whereas overconjugation disrupts the antibody and blocks antigen-binding sites (Gretch *et al.* 1987). Avidin is a basic glycoprotein molecule with an extremely high affinity for biotin, the strongest known non-covalent bond; a strong link between the antibody and the fluorophore is therefore rapidly formed. This is more stable than the linkage between two antibodies in the indirect method. The avidin–biotin method produces one fluorophore per biotin. A variant is to use streptavidin, which consists of four subunits, each of which is able to bind one biotin molecule (Chaiet & Wolf 1964). Streptavidin is also available as fluorophore conjugates and may be used as described for avidin. Alternatively, it may be used to form a molecular bridge, in which one streptavidin site is bound to the biotin on the antibody and the other three sites are then free to bind to biotinylated fluorophores. Again, adjustment of the relative concentrations of the reagents is critical.

As with the indirect method, this method is not suitable for multicolour flow cytometry detection, since the biotin–avidin reaction is independent of the antibodies and fluorophores used. It is, however, possible to use mixed antibodies if one is indirectly labelled and the other is biotin–avidin labelled.

Direct method for labelling antibodies

In this case the antibody is conjugated directly to the fluorophore in any of three ways. Although the procedures are comparatively complex and the efficiency of labelling is not as good as with other methods, direct labelling produces covalent and, therefore, stable bonds with the antibody, and multicolour detection of different antibodies poses no major problems, because there are no appreciable cross-reactions between labels.

Conjugation with isothiocyanates

Fluorescent isothiocyanates, such as fluorescein or rhodamine, are reactive reagents for the modification of aromatic amines, such as lysine residues or free terminal amino groups on the antibody protein (Riggs *et al.* 1958). Conditions should be sufficiently alkaline to facilitate deprotonation of the NH_3^+ groups to $-NH_2$, but not sufficiently alkaline to disrupt the antibody, that is a pH between 8.5 and 9.5. The more concentrated the protein, the more protected it is from chemical disruption and, therefore, the antibody should be as concentrated as possible before conjugation. It is necessary to ensure that overconjugation does not occur. Ideally, as many fluorophores as possible should be attached to one antibody to give a large fluorescence signal. If, however, there are too many, the antibody loses specificity. In practice, the optimum number of fluorophores is usually three per antibody molecule. The degree of labelling can be determined by measuring optical absorbance ratios. For example, for fluorescein isothiocyanate the ratio of absorption at 495 nm to that at 280 nm should be in the range 0.5–1.0 (Goding 1986). After conjugation, free fluorophore should be removed by gel filtration. Dialysis is not adequate, as even small amounts will label cell proteins non-specifically.

Conjugation with succinimidyl esters

Another approach to amine modification involves the use of succinimidyl esters, such as those of fluorescein or rhodamine. The reaction is similar to that of biotinylation and liberates free *N*-hydroxysuccinimide. As before, the pH should be sufficiently alkaline to deprotonate the NH_3^+ groups.

Conjugation with phycobiliproteins

The phycobiliproteins represent a different class of fluorescent label, because they are themselves proteins and not small molecules like fluorescein or rhodamine. The process of conjugation is, therefore, not as straightforward. General peptide-linking reagents, such as glutaraldehyde or *N,N'*-dicyclohexylcarbodiimide (DCC), are non-specific and will link any protein to any other, that is, antibody to antibody and label to label, as well as label to antibody. Such linking methods are, therefore, inefficient but they can be improved with thiolation reactions.

The process of disulphide bridge formation between the antibody protein and the phycobiliprotein is illustrated in Fig. 6.2. Rather than using native –SH groups, which would result in a loss of activity, extrinsic –SH groups are introduced. *N*-succinimidyl 3-(2-pyridyldithio)-propionate (SPDP) is used to form 2-pyridyl disulphide groups on both the antibody and the label protein (Carlsson *et al.* 1978). The groups on the label protein are then thiolated with

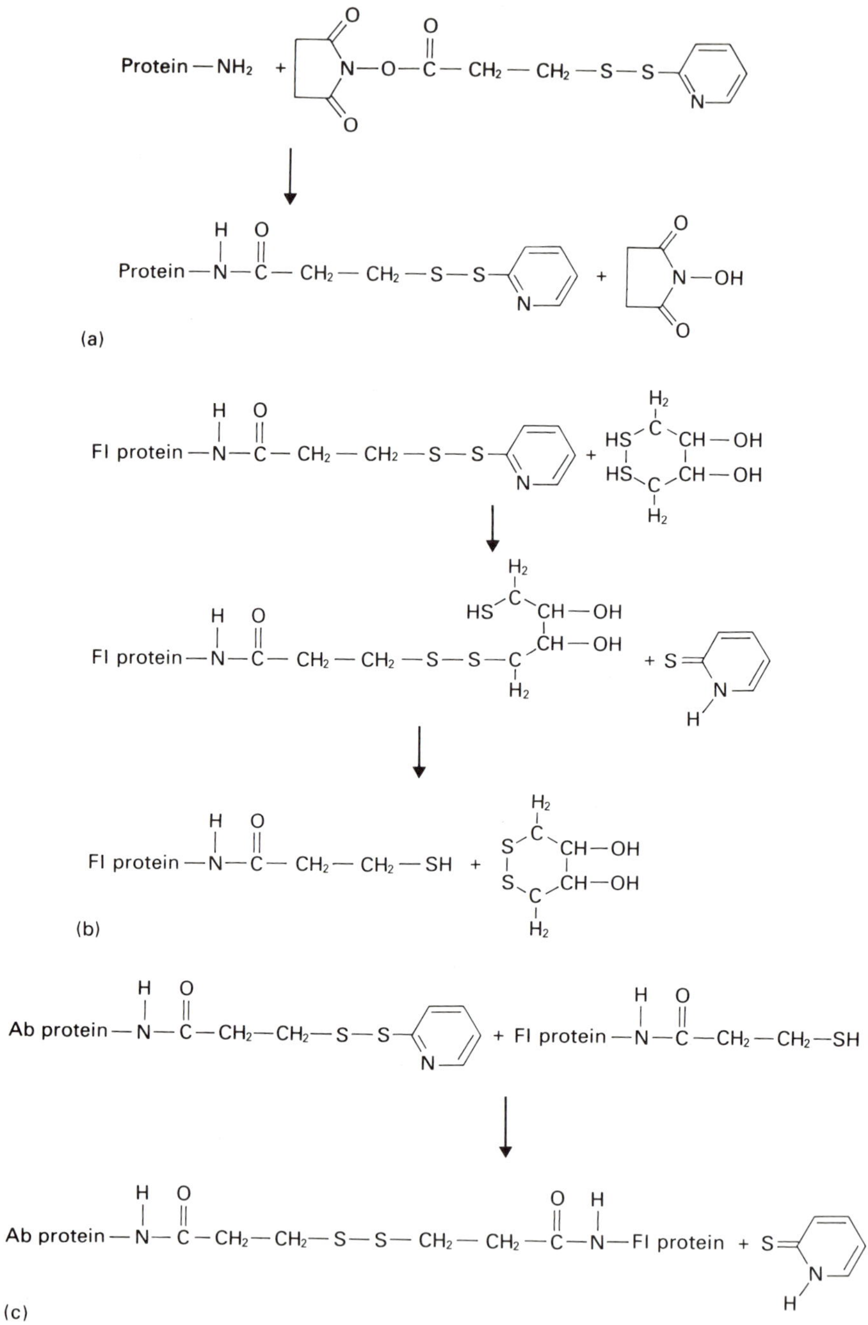

FIG. 6.2. Conjugation of an antibody with a fluorescent phycobiliprotein label. (a) Modification of both antibody and label proteins to form 2-pyridyl disulphide groups. (b) Thiolation of label protein. (c) Formation of disulphide bridge between antibody and label proteins.

dithiothreitol. Finally, the modified label is added to the modified antibody and a disulphide bridge is formed between the two (Kronick & Grossman 1983).

Phycobiliproteins are highly efficient labels and, under comparable conditions, the fluorescence yield may be equivalent to 30 fluorescein or 100 rhodamine molecules.

Application to the detection of food pathogens

Flow-cytometric techniques are being developed to detect bacteria in food. Patchett *et al.* (1991) used a simple nucleic acid staining protocol, which made it possible to enumerate down to at least 10^5 cfu/ml. Below this level problems were encountered from interference by particulate matter in the sample. To

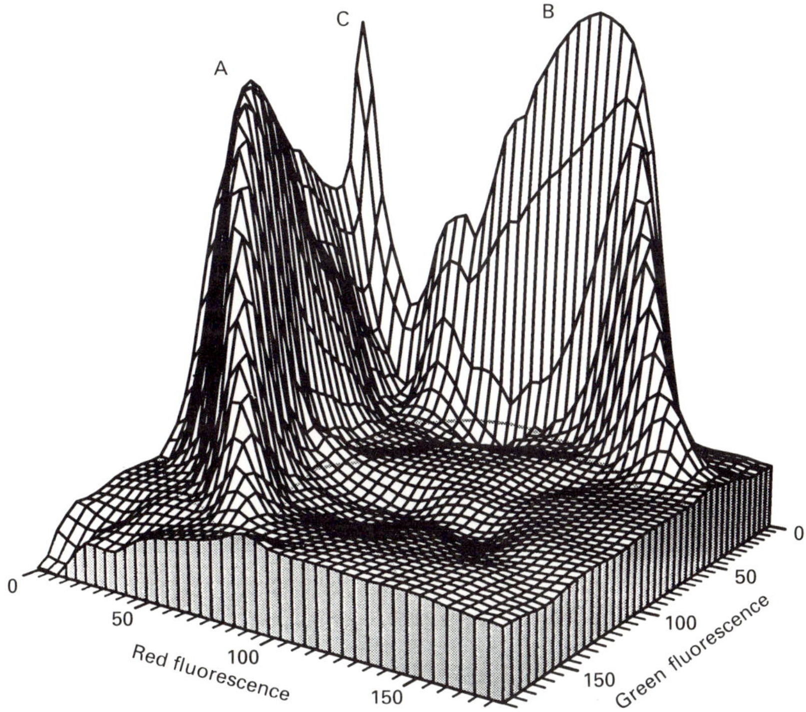

FIG. 6.3. Scattergram showing the analysis of a sample of *Salmonella* bacteria by flow cytometry. The scattergram is plotted as an isometric projection, with the z-axis representing the relative number of cells. Subpopulation 'A' represents *S. enteritidis* cells, which are stained green by FITC-labelled antibody, and 'B' represents *S. typhimurium* cells, stained red by R-PE-labelled antibody. Small amounts of unstained cells and other debris appear in subpopulation 'C'.

circumvent these problems, and at the same time to permit the detection of specific pathogens, polyclonal and monoclonal fluorescence-labelled antibodies are used. Pathogens of interest include *Salmonella* spp., *Listeria monocytogenes*, *Campylobacter jejuni* and *Bacillus cereus* (Tyndall *et al.* 1985; Donnelly & Baigent 1986; Clarke *et al.* 1992). For example, monoclonal antibodies have been developed with relative species specificity for *S. typhimurium* or *S. enteritidis* (Clarke *et al.* 1992).

A mixed *Salmonella* culture was incubated with FITC-labelled antibodies against *S. enteritidis* and R-PE-labelled antibodies against *S. typhimurium*. An isometric projection of a two-colour histogram, with FITC (green) fluorescence plotted along the *x*-axis and R-PE (red) fluorescence along the *y*-axis is shown in Fig. 6.3; the *z*-axis represents the cell abundancy. Three populations, 'A', 'B' and 'C', are visible. 'A' shows strong green fluorescence only and represents the subpopulation of *S. enteritidis* cells. Similarly, 'B' shows only strong red fluorescence and corresponds to the *S. typhimurium* subpopulation. A small region, 'C', which consists of unlabelled cells and possibly cellular debris, is also visible.

Similar analyses were carried out with a range of concentrations of *S. typhimurium* relative to a constant total concentration of mixed *Salmonella* (Fig. 6.4). For comparison, the expected count obtained by the plate method is also

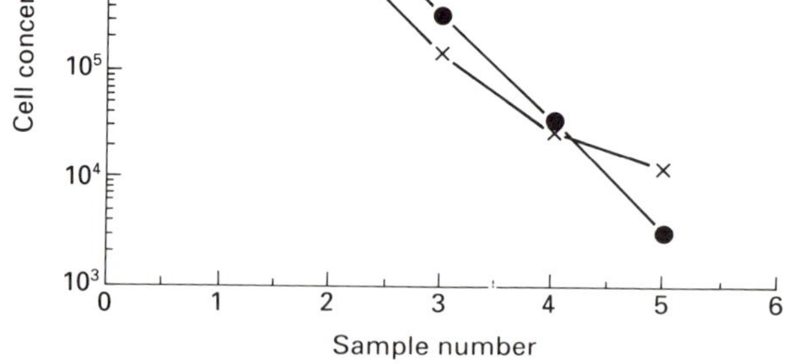

FIG. 6.4.Graph demonstrating the detection of *S. typhimurium* within a mixed *Salmonella* spp. population by flow cytometry. ○, The total cell concentration remains constant while the relative *S. typhimurium* concentration is reduced 10-fold in each sample; ●, shows the expected *S. typhimurium* concentration, as determined by the standard plate count method; ×, shows the counts obtained by flow cytometry.

shown. The immunofluorescence/flow cytometry method gives reliable results for *S. typhimurium* down to below 10^4/ml, even though this represents less than 0.1% of the total bacterial population.

Flow Cytometry for the Assessment of Viability

Problems in assessing viability

A simple experiment (Fig. 6.5) illustrates the problems of determining the numbers of viable cells present in a given sample. The survival of the fish pathogen *Aeromonas salmonicida* in sterile seawater was monitored. Total direct counts determined by flow cytometry remained constant while the plate counts fell rapidly after 8 days. Two explanations may be proposed. Either the counts reflect loss of viability, with the total counts consisting of dead plus live cells, or the counts reflect culturable numbers, with the disparity between the methods due to a subpopulation of viable but non-culturable cells, probably together with an indeterminate number of dead cells.

Viable but non-culturable microorganisms have important implications for

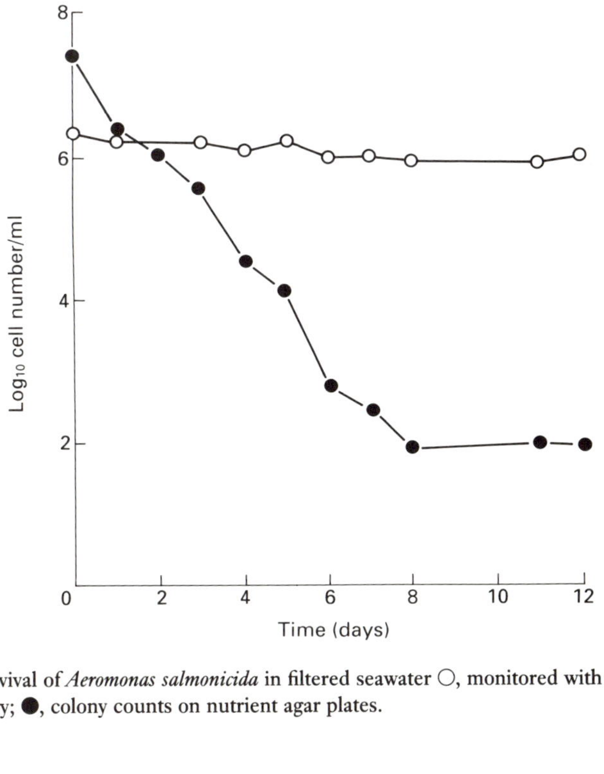

FIG. 6.5.Survival of *Aeromonas salmonicida* in filtered seawater ○, monitored with total counts by flow cytometry; ●, colony counts on nutrient agar plates.

microbiologists in general and particularly for those concerned with public health, water treatment and the food industries. If non-culturability (Roszak & Colwell 1987) is a widespread phenomenon, then, for environmental samples in their widest sense, many methods used to enumerate microorganisms will both fail to identify and underestimate the numbers of viable organisms present. Some species, or a proportion of a population, will remain essentially 'invisible' to traditional methods. Support for this comes from the growing realization that only a small proportion of the microbial biomass (*c.* 1–10%) can be recovered from the natural environment (Pickup 1991). The potential seriousness of this problem has been illustrated in a recent review on pathogenic bacterial species able to adopt a viable but non-culturable state (McKay 1992). Examples include *Legionella pneumophila*, *Campylobacter* spp., *Vibrio* spp., *Escherichia coli*, *Shigella* spp. and *Salmonella enteritidis*. Non-culturability is thought to arise as a result of stress to cells, from nutrient limitation or temperature (McKay 1992). The latter is of particular importance to the food industry. For example, *Vibrio vulnificus*, which is found in shell-fish, becomes non-culturable at low temperatures even under conditions of nutrient excess (Oliver & Wanucha 1989).

Dyes for assessing viability

A number of fluorescent dyes have been described for the flow-cytometric determination of viability in mammalian cells (Shapiro 1988). These may also be suitable for similar measurements in bacteria and they can be divided into two groups (Table 6.2). Some fluorescent dyes are accumulated by cells in response to membrane potential, which is a property of live cells. Examples include rhodamine 123 (Rh123) and cyanine dyes such as 3,3′-dihexyloxacarbocyanine (DiOC6(3)). Another group of dyes are colourless conjugates, usually fluorescein-based, that are cleaved non-specifically within the cell to release the fluorescent portion, which then stains the cell. In dead cells that lack intact membranes, the fluorochrome leaks out of the cell. Such cells may also lack the enzymes necessary for conjugate cleavage (Table 6.2).

Rh123 is concentrated rapidly and its accumulation reflects membrane potential (Kaprelyants & Kell 1992). It appears not to reduce the viability of stained cells, which can be recovered on solid media (Matsuyama 1984). We have found that Rh123 reflects viability for most bacterial species (Diaper *et al.* 1992). DiOC6(3) also stains many bacteria, but its accumulation does not always reflect membrane potential and the dye tends to bind to cell structures, particularly lipids. This non-specific staining makes DiOC6(3) of only limited application although it can be useful to assess the viability of *Pseudomonas putida*. A problem with these two membrane-dependent dyes is that they are not applicable to studies of environmental bacteria because of

TABLE 6.2. *Fluorescent dyes for assessing viability*

Membrane potential-dependent	
Rhodamine 123	Mammalian and plant cells, also works well for many bacterial species
3,3′-dihexyloxacarbocyanine (DiOC6(3))	Many bacterial cells show non-specific staining of cell structures
Intracellular cleavage-dependent	
Fluorescein diacetate (FDA)	Dye conjugate cleaved by most bacterial species; poor retention of fluorescein by most
Carboxyfluorescein diacetate (CFDA)	Similar to FDA but better retention of fluorescein
Calcein acetoxymethyl ester	Slow cleavage – limited application for viability
Chemichrome B	Unknown commercial formulation but works well for the majority of bacterial species
2′,7′-*bis*-Carboxyethyl-5(6)-carboxyfluorescein acetoxymethyl ester (BCECF-AM)	Wider application than other fluorescein conjugates but possibly some non-specific binding

non-specific binding of the dye to non-cellular structures, which results in extensive background fluorescence that masks that of the cells.

The dye conjugates are potentially more useful in this respect, because they are colourless until cleaved within the cell. Fluorescein diacetate (FDA) has been used to assess the viability of marine microalgae (Dorsey *et al.* 1989) and metabolically active soil bacteria (Lundgren 1981). This conjugate is thought to be cleaved by non-specific esterases, which are present in many bacterial species. Unfortunately, the released fluorescein appears not to be retained within the cells, although extensive cleavage of the dye conjugate can be demonstrated (Diaper *et al.* 1992). With *Bacillus subtilis*, however, FDA works exceptionally well. Carboxyfluorescein diacetate, an FDA derivative, shows better retention of released fluorescent dye. A possible explanation is that Gram-negative species tend to cleave FDA in the periplasm so that the fluorescein never enters the cell. Carboxyfluorescein may be less susceptible to enzymic attack and therefore may enter the cell and be subject to intracellular cleavage. Calcein also stains bacteria but its cleavage is much slower than that of the other dye conjugates, and fluorescent cells take some time to appear. Chemichrome B (Chemunex, Cowley Road, Cambridge, UK) is a new commercial reagent for monitoring bacterial viability but its mode of action or active ingredients are not revealed by the manufacturer. We have found it to work well for most bacterial species but not for some *Pseudomonas* spp. It is expensive at present and has a limited shelf life. 2′,7′-*bis*-

Carboxyethyl-5(6)-carboxyfluorescein acetoxymethyl ester (BCECF-AM) also has wide applicability but some species tend to show unacceptable levels of non-specific binding of the released fluorescein.

In summary, many dyes are available that work well for the determination of bacterial viability by flow cytometry but no one dye has universal application. Rh123, carboxyfluorescein, Chemichrome B and BCECF appear to work best. For a given bacterial species, a range of dyes may have to be tested to find the most suitable one.

Experimental details

Membrane potential-dependent staining by Rh123

Rh123 is a lipophilic cationic dye that has an excitation wavelength of approximately 500 nm and light emission between 530 and 550 nm. The procedure for staining bacteria is:

1 Bacteria are washed once with phosphate-buffered saline (PBS), pH 7.3.

2 The cells are resuspended to a cell density of approximately 10^6–10^7 bacteria/ml in either:

(a) for Gram-positive bacteria, PBS containing Rh123, 5 µg/ml; or

(b) for Gram-negative bacteria, Tris-EDTA buffer (10 mmol/l Tris, 1 mmol/l EDTA, pH 7.3) containing Rh123, 5 µg/ml.

The buffers containing rhodamine are prepared on the day of use.

3 Incubate at 37 °C for 30 min.

4 Take 1 ml of the suspension, pellet and wash the cells three times by centrifugation with the appropriate rhodamine-free buffer. This ensures that fluorescence is due to internalized rhodamine and minimizes any surface binding.

5 Analyse the cells either by fluorescence microscopy or by flow cytometry.

6 Settings for the Argus 100 flow cytometer (Skatron Ltd, Newmarket, Suffolk, UK) with the FITC filter block in conjunction with the FITC/phycoerythrin beam-splitter are as follows:

(a) LS1 (low-angle light scatter) PMT, 570 V; gain, 8 to 16.

(b) FL1 (cellular fluorescence) PMT, 650 V; gain, logarithmic.

7 These settings should result in a fluorescence peak channel greater than 50 for stained, that is viable, cells.

8 Appropriate controls should be included, for example heat-killed cell suspensions.

9 The cell counts that comprise the fluorescent populations should match those detected by light scatter to within 5%. If they are significantly lower, this indicates that a proportion of the cells remains unstained.

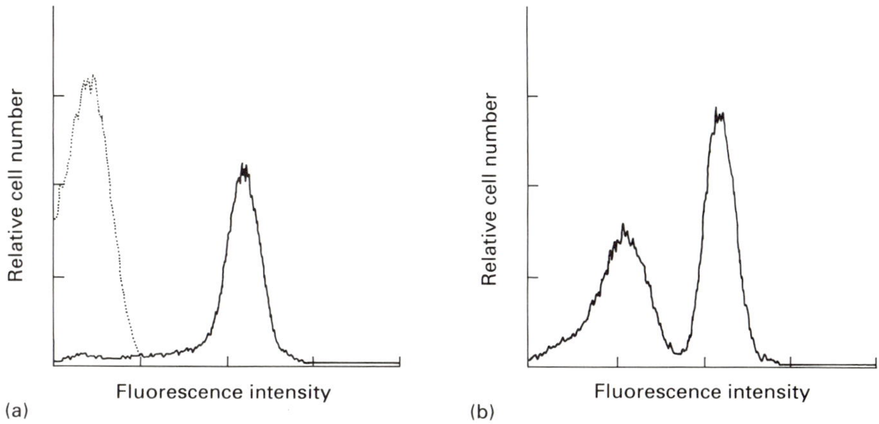

FIG. 6.6.(a) Fluorescence histograms of *Salmonella pullorum stained* with Rh123. Histograms of gramicidin-treated cells are represented by a broken line. (b) Fluorescence histograms of a mixed culture of viable and formaldehyde-fixed cells of *S. pullorum*.

Stained cells are then analysed by flow cytometry in the presence or absence of an ionophore, such as gramicidin. We have found that this is particularly useful in dissipating the membrane potential of bacteria and this treatment confirms whether cell staining has occurred by active membrane potential-dependent uptake or whether cell fluorescence is due to non-specific binding of the dye to cells. Stained *S. pullorum* (Fig. 6.6) can clearly be discriminated (peak channel 110) from those treated with gramicidin (peak channel 25). Further testing of the dye as a viability indicator is possible by mixing suspensions of dead (e.g. washed and formaldehyde-treated cells) and living cells in known proportions. Use has also been made of Rh123 to distinguish between living and dead cells in antibiotic susceptibility studies (Shapiro 1988).

The optimum dye concentration, effects of temperature and time of incubation should also be confirmed. In practice, with Rh123, staining takes place in the first 1–5 min of incubation and is unaffected by temperature. The dye concentration should be sufficient to produce maximum fluorescence, and this depends, to some extent, on the species under investigation. The effect of Rh123 concentration on the fluorescence of cells of *E. coli* and *Staphylococcus aureus* is shown in Fig. 6.7(a). For the former a concentration of up to 5–6 μg/ml was required, while for the latter 2 μg/ml was sufficient to produce maximum fluorescence. This may reflect the difference between Gram-positive and Gram-negative species, in that EDTA is required during the staining period in order to disrupt the outer membrane and allow the rhodamine to enter the cell (Kaprelyants & Kell 1992).

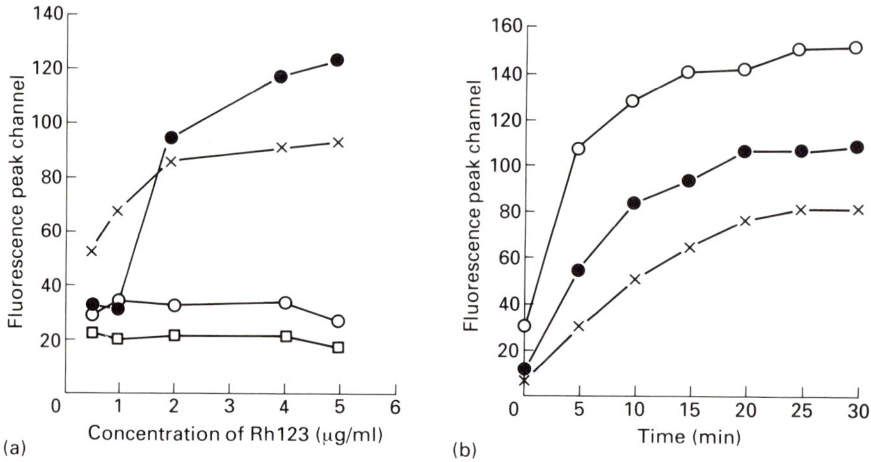

FIG. 6.7.(a) Effects of rhodamine concentration on cellular fluorescence intensity. ×, *Staph. aureus*; □, *Staph. aureus* plus gramicidin; ●, *Escherichia coli*; ○, *E. coli* plus gramicidin. (b) Staining of a viable culture of *Staph. aureus* with fluorescein diacetate derivatives. ●, carboxyfluorescein diacetate; ○, 2′,7′-*bis*-(2-carboxyethyl)-5-(6)-carboxyfluorescein acetoxymethyl ester (BCECF-AM); ×, calcein-acetoxymethyl ester. Increasing cellular fluorescence is represented by higher peak channel numbers. Gramicidin-treated cells had a maximum fluorescent peak channel of less than 50, except BCECF-AM, which had a peak channel of 64.

Staining by colourless dye conjugates

The staining procedure for colourless dye conjugates (Table 6.2) follows that described above for Rh123 except in one major respect: there is no need to wash the cells before passage through the flow cytometer, because these dyes are colourless until cleaved. Again, staining protocols show some species dependence with respect to dye concentration. However, because intracellular fluorescence is a direct consequence of enzymic reaction, there is a greater effect of incubation temperature and time. A typical staining procedure with carboxyfluorescein is described below:

1 Gram-positive bacteria are washed and resuspended to a cell density of 10^6–10^7 ml in PBS, and Gram-negative species are similarly suspended in Tris-EDTA buffer.

2 To 1 ml of cell suspension add 10 μl of a 1-mmol/l stock solution of carboxyfluorescein diacetate in dimethylsulphoxide (DMSO). This stock solution is stable in the dark at room temperature but it should be discarded if any green coloration develops.

3 Incubate for 30 min at 37 °C.

4 Analyse by flow cytometry (or microscopy) as detailed above. *Note*: No washing step is required because the dye is released within the cell.
5 Similar procedures can be used for the other dye conjugates BCECF-AM, calcein-AM and FDA.
The importance of optimizing time of incubation for different dye conjugates tested with *Staph. aureus* is illustrated in Fig. 6.7(b).

Experimental application

The potential for viability assessment by flow cytometry, with the methods described, for monitoring the survival of *Klebsiella pneumoniae* in water was investigated. Cell numbers were determined over a 10-day period by four different methods (Fig. 6.8). Direct counts were determined microscopically from acridine orange-stained cells or direct viable counts after incubation in nalidixic acid and yeast extract. Rh123-stained cells were enumerated by flow cytometry and these three methods were compared with colony counts on

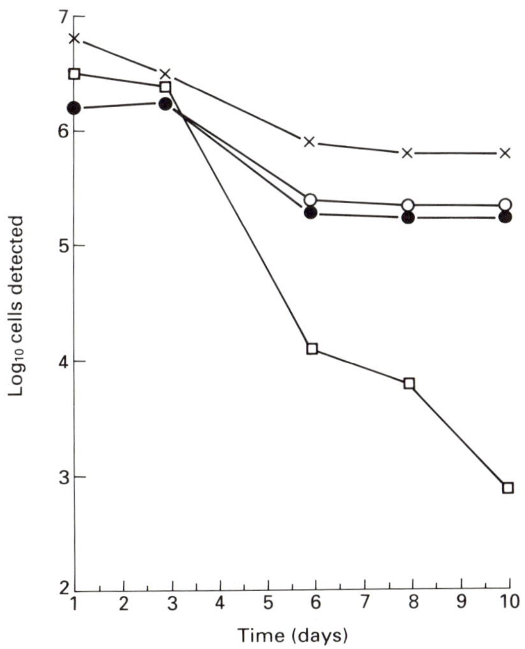

FIG. 6.8. ×, Survival of *Klebsiella pneumoniae* in sterile lake water determined by acridine orange direct counts; □, colony forming units (cfu) on nutrient agar; ○, Rh123-stained viable cells counted with a flow cytometer; ●, direct viable counting by microscopy after nalidixic acid treatment.

nutrient agar plates. All three direct counting methods gave similar results, in that cell numbers decreased by approximately 1 $\log_{10}$ unit, while colony counting gave totally different results, in that numbers fell from *c.* 10^6 cfu/ml to *c.* 10^3 cfu/ml. This experiment illustrates:

1 The weakness of relying on colony counts as a true measure of cell viability.

2 The close correlation between cell viability as determined by flow cytometry and by other methods.

3 The ability of the flow cytometer to give a true representation of viable numbers. Taken together, the results of colony counting and of the other methods imply that *K. pneumoniae* can adopt a viable but not necessarily culturable state.

Conclusion

Flow cytometry can be used as a rapid and sensitive method for the analysis of bacterial populations (Kell *et al.* 1991). Its application for the measurement of bacterial viability is less certain. We have, so far, screened about a dozen different bacterial species with seven different fluorescent stains, and no single stain is universally applicable, although Rh123 works well for most species. Its application for environmental samples is, however, limited because of non-specific binding to particulates, which generates high background fluorescence and masks the cells. Dye conjugates that produce fluorescence after intracellular cleavage have at least the potential to circumvent this problem. Although no single conjugate has universal application, Chemichrome B, which is commercially available, stains the greatest number of different bacterial species. At present, the flow-cytometric assessment of viability should work well in situations where the microbial species are known and for which appropriate, previously tested, dyes can be selected. Another approach may be to formulate a cocktail of different dyes that will stain most viable bacteria present within a sample. However, the methods that have been described provide an excellent way forward for assessing viability of a single species and may, therefore, find application for purposes such as defining the cidal effects of different antimicrobial agents. Finally, another possibility is the use of viability assessment in conjunction with other techniques that can identify target species, such as the use of fluorescent antibodies described in this chapter. The techniques described here are relatively new, and there is no doubt that there will be further developments that will circumvent many of the difficulties that have been described.

Acknowledgements

A.C.P., R.G.C. and S.A.G.P. gratefully acknowledge support from the Ministry of Agriculture, Fisheries and Food, and would like to thank Heather Lee and Matthew Langley for providing the *Salmonella* antibodies. The work on the assessment of viability was funded by grants from the NERC.

References

AMANN, R.I., KRUMHOLZ, L. & STAHL, D.A. 1990a Fluorescent oligonucleotide probing of whole cells for determinative phytogenetic and environmental studies in microbiology. *Journal of Bacteriology* **172**, 762–770.

AMANN, R.I., BINDER, B.J., OLSON, R.J., CHISHOLM, S.W., DEVEREUX, R. & STAHL, D. 1990b Combination of 16S rRNA-targeted oligonucleotide probes with flow cytometry for analysing mixed microbial populations. *Applied and Environmental Microbiology* **56**, 1919–1925.

AMANN, R.I., STROMLEY, J., DEVEREUX, R., KAY, R. & STAHL, D.A. 1992 Molecular and microscopic identification of sulfate-reducing bacteria in multispecies biofilms. *Applied and Environmental Microbiology* **58**, 614–623.

BAYER, E.A. & WILCHEK, M. 1980 The use of the Avidin–Biotin complex as a tool in molecular biology. *Methods in Biochemistry Analysis* **26**, 1–45.

BOYE, E. & LOBNER-OLESON, A. 1990 Flow cytometry: illuminating microbiology. *New Biologist* **2**, 119–125.

CARLSSON, J., DREVIN, H. & AXEN, R. 1978 Protein thiolation and reversible protein–protein conjugation. *Biochemistry Journal* **173**, 723–737.

CHAIET, I. & WOLF, F.J. 1964 The properties of streptavidin, a biotin binding protein produced by streptomycetes. *Archives of Biophysics* **106**, 1–5.

CLARKE, R.G., POULTER, S.A.G., LANGLEY, M., PINDER, A.C. & LEE, H.A. 1992 Detection of salmonellae using fluorescent antibodies and flow cytometry. *Journal of Applied Bacteriology* **73**, xv.

DEAN, P. & PINKEL, P. 1978 High resolution dual laser flow cytometry. *Journal of Histochemistry and Cytochemistry* **26**, 622–627.

DIAPER, J.P., TITHER, K. & EDWARDS, C. 1992 Rapid assessment of bacterial viability by flow cytometry. *Applied Microbiology and Biotechnology* **38**, 268–272.

DONNELLY, C.W. & BAIGENT, G.J. 1986 Method for flow cytometric detection of *Listeria monocytogenes* in milk. *Applied and Environmental Microbiology* **52**, 689–695.

DORSEY, J., YENTSCH, C.M., MAYO, S. & MCKENNA, C. 1989 Rapid analytical technique for the assessment of cell metabolic activity in marine microalgae. *Cytometry* **10**, 622–628.

EDWARDS, C., PORTER, J., SAUNDERS, J.R., DIAPER, J., MORGAN, J.A.W. & PICKUP, R.W. 1992 Flow cytometry in microbiology. *Society of General Microbiology Quarterly* **19**, 105–108.

FESTIN, R., BJORKLAND, A. & TOTTERMAN, T.M. 1990 Single laser flow cytometric detection of lymphocytes binding three antibodies labelled with fluorescein, phycoerythrin and a novel tandem fluorochrome conjugate. *Journal of Immunological Methods* **126**, 69–78.

GODING, J.W. 1986 *Monoclonal Antibodies: Principles and Practice*. 2nd edn. London: Academic Press.

GRETCH, D.R., SUTER, M. & STINSKI, M.F. 1987 The use of biotinylated monoclonal antibodies and streptavidin affinity chromatography to isolate herpesvirus hydrophobic proteins and glycoproteins. *Analytical Biochemistry* **163**, 270–277.

KAPRELYANTS, A.S. & KELL, D.B. 1992 Rapid assessment of bacterial viability and vitality by rhodamine 123 and flow cytometry. *Journal of Applied Bacteriology* **72**, 410–422.

KELL, D.B., RYDER, H.M., KAPRELYANTS, A.S. & WESTERHOFF, H.V. 1991 Quantifying heterogeneity: flow cytometry of bacterial cultures. *Antonie van Leeuwenhoek* **60**, 145–158.

KOHLER, G. & MILSTEIN, C. 1975 Continuous culture of fused cells secreting antibodies of defined specificity. *Nature* **256**, 495–497.

KRONICK, M.N. & GROSSMAN, P.D. 1983 Immunoassay techniques with fluorescent phycobiliprotein conjugates. *Clinical Chemistry* **29**, 1582–1586.

LUNDGREN, B. 1981 Fluorescein diacetate as a stain of metabolically active bacteria in soil. *Oikos* **36**, 17–22.

MCKAY, A.M. 1992 Viable but non-culturable forms of potentially pathogenic bacteria in water. *Letters in Applied Microbiology* **14**, 129–135.

MATSUYAMA, T. 1984 Staining of living bacteria with rhodamine 123. *FEMS Microbiology Letters* **21**, 153–157.

OI, V.T., GLAZER, A.N. & STRYER, L. 1982 Fluorescent phycobiliprotein conjugates for analysis of cells and molecules. *Journal of Cell Biology* **93**, 981–986.

OLIVER, J.D. & WANUCHA, D. 1989 Survival of *Vibrio vulnificus* of reduced temperatures and elevated nutrient. *Journal of Food Safety* **10**, 79–86.

PATCHETT, R.A., BACK, J.P., PINDER, A.C. & KROLL, R.G. 1991 Enumeration of bacteria in pure cultures and in foods using a commercial flow cytometer. *Food Microbiology* **8**, 119–125.

PICKUP, R.W. 1991 Development of methods for the detection of specific bacteria in the environment. *Journal of General Microbiology* **137**, 1009–1019.

PINDER, A.C., PURDY, P.W., POULTER, S.A.G. & CLARK, D.C. 1990 Validation of flow cytometry for rapid enumeration of bacterial concentrations in pure cultures. *Journal of Applied Bacteriology* **69**, 92–100.

RIGGS, J.L., SEIWALD, R.J. & BUCKERHALTER, J.H. 1958 Isothiocyanate compounds as fluorescent labelling agents for immune serum. *American Journal of Pathology* **34**, 1081–1086.

ROSZAK, D.B. & COLWELL, R.R. 1987 Survival strategies of bacteria in the natural environment. *Microbiological Reviews* **51**, 365–379.

RUSS, W.D. 1990 Simultaneous, three colour immunofluorescence analysis by flow cytometry. *Focus* **12**, 96–99.

SHAPIRO, H.M. 1988 *Practical Flow Cytometry*, 2nd edn. New York: Alan R. Liss.

SHAPIRO, H.M. 1990 Flow cytometry in laboratory microbiology: new directions. *ASM News* **56**, 584–588.

STEEN, H.B. 1990 Flow cytometric studies of microorganisms. *Flow Cytometry and Sorting*, eds Melamed, M.R., Lindmo, T. & Mendelsohn, M.L., pp. 605–622. New York: Wiley-Liss.

STEEN, H.B., SKARSTAD, K. & BOYE, E. 1990 DNA measurements of bacteria. *Methods in Cell Biology* **33**, 519–526.

TYNDALL, R.L., HAND, R.E., MANN, R.C., EVANS, C. & JERNIGAN, R. 1985 Application of flow cytometry to detection and characterisation of *Legionella* spp. *Applied and Environmental Microbiology* **49**, 852–859.

VAN DILLA, M.A., LANGLOIS, R.G., PINKEL, D. & HADLEY, W.K. 1983 Bacterial characterisation by flow cytometry. *Science* **220**, 620–622.

ZARDA, B., AMANN, R.I., HALLNER, G. & SCHLEIFER, K.H. 1991 Identification of single bacterial cells using oligoxygenin-labelled rRNA-targeted oligonucleotides. *Journal of General Microbiology* **137**, 2823–2830.

7: Industrial Application of Flow Cytometry for the Rapid Detection of Microorganisms

M.A. Brailsford[1] and S. Gatley[2]
[1]*Chemunex UK, Opas Centre, St John's Innovation Park, Cowley Road, Cambridge CB4 4WS, UK;*
and [2]*Chemunex SA, 41 Rue du 11 Novembre, 94700 Maisons Alfort, Paris, France*

The primary limitation of standard microbiological techniques for the detection of a single bacterial or fungal cell is that the detection signal is represented by the growth of that cell into a colony on a plate. This process can be relatively slow, and may take from 1 to 7 days depending upon the organism and growth conditions. In the food and beverage industries, such a delay in detection means that information on the microbiological status of raw materials, the state of plant hygiene and the microbiological quality of final product is often retrospective to production schedules.

The development of rapid methods for the detection of microorganisms has largely been stimulated by these limitations and has led to the development of methods based on a wide range of principles including microcalorimetry (Boling *et al.* 1973; Lampi *et al.* 1974; Beezer *et al.* 1976), radiometry (Previtte 1972; Stewart *et al.* 1980), impediometry (Richards *et al.* 1978; Evans 1982; Firstenberg-Eden & Eden 1984), bioluminescence (Sharpe *et al.* 1970; Williams 1971) and direct epifluorescence (Pettipher 1982; Rodrigues & Pettipher 1984), amongst others. In the majority of cases these systems depend on indirect detection methods centred around the generation of global signals. These are limited in the amount of information they can provide and, in mixed populations, it may be difficult to differentiate between the signal generated by the target organism and that from other organisms present. This is of particular importance in food and beverage applications. Optical analysis overcomes the problems associated with the use of indirect methods and allows fast direct detection of microorganisms (Gatley 1989; Brailsford & Gatley 1992).

New Techniques in
Food and Beverage Microbiology

0-632-03755-5

Optical Analysis

In its simplest form, optical analysis can be undertaken with a microscope. This, together with a variety of cell labels, such as monoclonal antibodies, nucleic acid stains and cell activity markers, offers the possibility of multiparametric analysis on a cell-by-cell basis. In this way, optical analysis of micro-organisms can provide a means for the highly specific identification of target organisms, together with detailed information on the size, shape and metabolic state of individual cells. Simple optical systems, such as the fluorescence microscope, however, have the disadvantage that they require a skilled operator and are time-consuming. This limits the use of the microscope to the analysis of small numbers of samples. Microscopy is further restricted by the relatively small volume of sample that can be analysed. The use of image analysers, together with the direct epifluorescent filter technique (DEFT) has made possible a reduction in analysis time and led to increased use of the microscope as an industrial process control tool (Pettipher 1982; Rodrigues & Pettipher 1984; Pettipher *et al.* 1989). Imaging systems are now being developed for total flora analysis but they are only sensitive above 10^4 cells/ml.

Flow Cytometry

Flow cytometry has been widely used in research but has only recently been developed into an easy-to-use, reliable and robust format applicable to industrial use (Gatley 1989, 1990; Desnouveaux *et al.* 1990; Jakobsen & Jespersen 1991; Brailsford & Gatley 1992). The technique allows cell-by-cell analysis of the sample under study and, coupled with the use of fluorescent labels, provides a rapid means of obtaining information on cell numbers and metabolic activity, cell type and growth cycle.

Flow cytometry in the format used in the ChemFlow system (Chemunex UK, Cowley Rd, Cambridge, UK) may be regarded as a form of automated fluorescence microscopy, where, instead of a sample fixed to a slide, the sample is injected into a fluid that passes under the objective via a hydrodynamic focusing flow cell. In this system there is no movement between the stage and the objective, since the flow cell remains in a fixed position as the liquid sample passes through it. In the flow cell, the sample is injected into the path of a sheath fluid that flows continually through the flow cell. This has the effect of focusing the sample stream into a narrow linear flow (Fig. 7.1). The sample then passes through a beam of light that causes the labelled cells to emit fluorescent pulses, and each pulse is detected by a photomultiplier tube. Computer analysis then allows the pulses to be registered as separate counts and graded in terms of their relative intensities of fluorescence.

The Chemunex flow cytometer, which was used in the work described

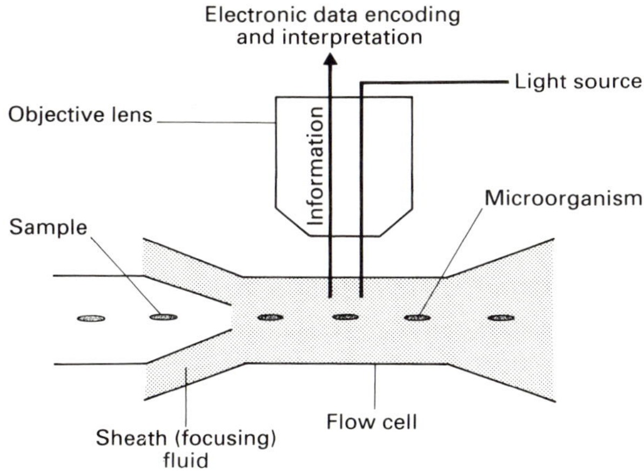

FIG. 7.1.The principle of the ChemFlow analyser. The system uses a single fluorescence detector to count fluorescence-labelled cells by flow cytometry.

below, can detect up to 400 cells/s at sample flow rates of approximately 0.4 ml/min. The process of labelling and counting of cells can be carried out in under 30 min.

Fluorescence Cell-labelling Techniques

A number of fluorescent labels can be used. These include FITC-labelled antibodies, nucleic acid intercalating stains and fluorescent metabolic activity (viability) cell markers. This combination provides a means of obtaining detailed information about cell populations, such as the total viable counts (Dumain *et al.* 1990; Bankes *et al.* 1991, 1993; Pettipher 1992), real-time analysis of growth curves (Jakobsen & Jespersen 1991) and identification and enumeration of specific target organisms in mixed populations (Levasseur *et al.* 1993). These labels in combination with flow cytometry are now providing new methods for industrial quality assurance and research.

Viability labels

Viable counts are performed with substrate molecules, which, after entry into the cell, are cleaved by cytoplasmic enzymes to liberate free fluorochrome into the internal matrix of the cell (Fig. 7.2). The intensity of the resulting intracellular cytoplasmic fluorescence is related to the metabolic activity of the cell and non-viable cells remain non-fluorescent. This fluorescence signal is detected by the analyser, which measures the relative fluorescence intensity of

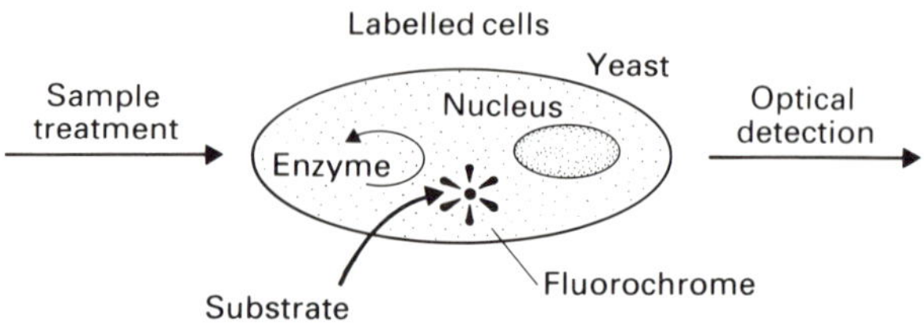

FIG. 7.2. Fluorescence-labelling methods for cell viability.

the individual cells to produce a relative measure of their metabolic activities. It can, therefore, be used to count viable cell numbers, and also to provide information about the activity of the population under study.

Applications in Industrial Microbiology

Analysis of total viable yeast counts in pure culture

Saccharomyces cerevisiae, *Zygosaccharomyces bailii* and a yeast isolated from the factory are individually inoculated at a range of cell concentrations into glucose salts medium (GSM; Oxoid), 7.5% significance broth (Oxoid GSM + 7.5% ethanol) and dry white wine. The inoculated media are incubated at 25 °C for up to 3 days, and subsamples are taken at intervals for analysis by plate count and flow cytometry. Plate counts for yeasts are performed on Wallerstein Laboratories nutrient agar (WLN agar, Difco) at 25 °C. The plates were incubated for 3 days before analysis.

Samples are labelled as follows for flow cytometry. Samples (3 ml) are centrifuged at 1500 *g* for 5 min (Beckmann GP40) to sediment the yeast. The supernatants are removed by pouring and the pellets resuspended in 3 ml of labelling buffer RVY (Chemunex). Yeast viability substrate (Chemunex) (30 μm) is added to each tube and the samples are sonicated with a VibraCell probe sonicator set at 40% of maximum power (BioBlock Scientific, Vanves, Paris, France; 3-mm probe, 25 kHz/50 W). The sonicated samples are incubated at 40 °C for 10 min to ensure maximal fluorescent labelling of viable cells. After incubation and before analysis the samples are placed in a cooled block at 4 °C for 10 min to stabilize the cells.

Labelled samples are analysed with an automated flow cytometer incorporating an excitation filter (450–490 nm) and an emission set at OG 515–FI 515. The automatic sample injection incorporates a filtration step with a nylon membrane filter (Millipore; 15 μm, 25 mm diameter). Samples are run in linear mode at a flow rate of 400 μl/min. The flow cytometer is calibrated against Standard C (2.0 μm fluorescent beads – Chemunex) on channel 170.

Analysis of total viable yeast counts in wine

Wine and process intermediates are analysed by flow cytometry and the standard plate count. Samples (500 ml) are equally divided and filtered through cellulose nitrate membrane filters (Sartorius; 0.45 µm, 47 mm diameter). One filter is placed on WLN agar and incubated at 25 °C for 3–8 days. The other filter is added to a tube containing 20 ml of GSM. The tubes are vortexed for 30 s and incubated at 25 °C. Samples (5 ml) are removed at various intervals and analysed by plate count and flow cytometry as described above.

Analysis of total viable yeast counts in fruit preparations

Fruit samples (10 g) are inoculated with *Z. bailii* and *S. bisporus* at a range of cell concentrations. The inoculated samples are added to Sabouraud broth (Merck) at a ratio of 1 : 9 (sample : Sabouraud) and incubated at 25 °C. Samples are taken at intervals and analysed by plate count and ChemFlow. Plate counts are performed on oxytetracycline glucose agar (Oxoid) incubated at 25 °C for 3 days.

For flow cytometry, 3 ml samples are centrifuged at 1500 *g* for 5 min with a 30-µm centrifilter device (centrifuge tube incorporating a 30-µm mesh filter; Chemunex). After centrifugation the centrifilter is removed, the supernatant discarded by pouring and the pelleted cells labelled as described above. Counter stain CSA (Chemunex) (150 µl) is added to the samples to reduce interference from autofluorescent particles. Analysis parameters are as previously described.

Analysis of total viable bacterial counts in vegetables

Vegetable samples (10 g) are diluted fivefold (w/v) with maximum recovery diluent (Oxoid) in sterile stomacher bags equipped with filters, and homogenized for 1 min in a stomacher (Lab. Blender 80 – Seward Surgical). Stomached samples are analysed by plate count and ChemFlow. Plate counts are performed on plate count agar (Oxoid) incubated at 37 °C for 36 h. Samples for flow cytometry are labelled as follows. Samples (40 ml) are centrifuged at 350 *g* for 8 min to remove large particles. The supernatants are decanted and centrifuged through 30-µm centrifilters at 2450 *g* for 8 min. The supernatants are discarded by pouring and the pelleted cells resuspended in 3 ml of labelling buffer RVB (Chemunex). Bacterial viability substrate (Chemunex; 30 µl) is added to each tube, sonicated and incubated as described above. For cytometry, 150 µl of CSA is added to each sample.

Labelled samples are analysed with an automated flow cytometer with an excitation filter (450–490 nm) and an emission set at OG 515–EM

520. The automation system adds 1 g of ChemRed background reducer (Chemunex) and agitates the sample for 2 min before analysis. Sample injection incorporates a filtration step with nylon membrane filters (Millipore; 15 μm, 25 mm diameter). Samples are run in log 1 mode at a flow rate of 200 μl/min. The cytometer is calibrated against Standard B (0.5 μm fluorescent beads – Chemunex) on channel 100.

Antibiotic sensitivity testing

A culture of *E. coli* (ATCC 25922) is grown overnight in peptone broth incubated at 37 °C. After incubation the culture is subcultured and split into three: one untreated sample is the control and the remaining samples are treated with gentamicin and chloramphenicol, respectively. The samples are placed into a water bath at 37 °C and subsamples are removed at 10 min intervals. These are analysed with a Thomas cell counting chamber and by ChemFlow. Samples for flow cytometry are labelled as described for pure cultures, with the ChemFlow system in log 1 mode and at a flow rate of 200 μl/min.

Biomass monitoring

Samples (1 ml) are removed from a penicillin seed fermenter and added to 2 ml of labelling buffer RVY. These samples are labelled and analysed as described for pure cultures, with and without sonication.

Analysis of filterable products

Flow cytometry for the rapid detection of low-level yeast contamination in wine and cider products was validated in a factory. The initial phase of the validation was to study the growth of several target yeasts in a range of culture media.

Even at the lowest inoculum levels (<2 cfu/ml) both *S. cerevisiae* and a factory yeast isolate were detected after 24 h preincubation in glucose salts medium (Table 7.1). Even with the slow-growing yeast *Z. bailii*, growth in GSM indicated that detection at 48 h was likely even at the lowest inoculum level (<3 cfu/ml).

There was no growth of *S. cerevisiae* in dry white wine at any inoculum level (2–200 cfu/ml). This result corresponded with that in significance broth, which is a salts-based broth containing 7.5% ethanol to mimic the product. It is interesting to note that both the spoilage yeast isolated from the factory and *Z. bailii* were alcohol-tolerant.

With naturally contaminated products (Tables 7.2 and 7.3), positive sam-

TABLE 7.1. *Total viable counts in pure culture by standard plate count and ChemFlow analysis*

Medium/organism	Inoculum (cfu/ml)	Counts after incubation: 6-h incubation ChemFlow (count/ml)	6-h incubation Plate count (cfu/ml)	24-h incubation ChemFlow (count/ml)	24-h incubation Plate count (cfu/ml)	48-h incubation ChemFlow (count/ml)	48-h incubation Plate count (cfu/ml)
GSM/*S. cerevisiae*	2	nd	nd	1.0exp3	434	1.9exp6	8.2exp5
	20	nd	nd	1.9exp4	nd	9.4exp6	1.5exp7
	200	−ve	72	2.1exp5	6.3exp4	nd	nd
Significance broth/*S. cerevisiae*	2	nd	nd	−ve	0	−ve	0
	20	nd	nd	−ve	0	−ve	11
	200	−ve	−ve	−ve	24	−ve	129
GSM/factory yeast	6	nd	nd	3.4exp3	1.0exp3	1.3exp7	3.0exp7
	66	nd	nd	5.0exp4	1.0exp3	nd	nd
	660	−ve	176	6.7exp5	1.0exp3	nd	nd
Significance broth/factory yeast	6	nd	nd	−ve	9	1.7exp3	1.0exp3
	66	nd	nd	670	312	3.4exp3	1.0exp3
	660	580	230	5.9exp3	1.0exp3	6.1exp5	7.1exp5
GSM/*Z. bailii*	3	−ve	2	nd	<200	1.0exp4	1.0exp3
	26	−ve	4	685	152	1.8exp4	1.0exp3
	260	−ve	147	6.9exp3	428	6.5exp4	1.0exp3
Significance broth/*Z. bailii*	3	−ve	1	−ve	6	990	320
	26	−ve	4	685	152	1.8exp4	1.0exp3
	260	−ve	147	6.9exp3	428	6.5exp4	1.0exp3

S. cerevisiae, *Z. bailii* and a factory yeast isolate were inoculated at a range of cell concentrations into glucose salts medium (GSM) and 7.5% significance broth (GSM + 7.5% ethanol). The inoculated media were incubated at 25 °C, with subsamples being taken at various intervals for enumeration on Wallerstein Laboratories nutrient agar, incubated at 25 °C for 3 days, and by ChemFlow analysis.

TABLE 7.2. *Total viable yeast counts in base wine*

Direct plate count (cfu/ml): plate incubation		ChemFlow count after preincubation (count/ml)			Culture confirmation of ChemFlow count (cfu/ml)		
3 days	7–8 days	24 h	48 h	72 h	24 h	48 h	72 h
Conf		5.3exp4	—	—	1.0exp5	—	—
Conf		2.1exp5	—	—	1.3exp3	—	—
Conf		2.6exp5	—	—	2.0exp5	—	—
Conf		2.3exp4	—	—	2.4exp4	—	—
Nil	2m, 1y	−ve	4.8exp3	3.5exp5	Nil	22y	1.0exp3
Conf		3.6exp4	—	—	3.0exp4	—	—
Conf		2.3exp3	1.8exp6	—	624	1.3exp6	—
Conf		5.6exp4	—	—	1.0exp3	—	—
Nil	Nil	−ve	−ve	—	Nil	Nil	—
Conf		1.4exp4	—	—	1.0exp3	—	—
Conf		7.3exp5	—	—	2.4exp6	—	—

Samples of base wine were filtered through two 0.45 μm membrane filters. One filter was placed on Wallerstein Laboratories nutrient agar (WLN) and incubated at 25 °C for 3–8 days. The other was placed in glucose salts medium and incubated at 25 °C for up to 3 days. Samples were removed at various intervals and analysed on WLN agar incubated for 3 days and by ChemFlow analysis. Conf, confluent growth; m, mould; −ve, no count; Nil, no count; y, yeast.

TABLE 7.3. *Total viable counts in filter-sterilized wine*

Direct plate count (cfu/ml)		ChemFlow count after preincubation (count/ml)			Plate count confirmation of ChemFlow count (cfu/ml)		
3–5 days	7–8 days	24 h	48 h	72 h	24 h	48 h	72 h
Nil	Nil	−ve	−ve	−ve	Nil	Nil	Nil
Nil	Nil	−ve	−ve	−ve	Nil	Nil	Nil
Nil	Nil	−ve	−ve	−ve	Nil	Nil	Nil
1y	1y, 1m	−ve	−ve	−ve	Nil	Nil	Nil
Nil	Nil	−ve	−ve	−ve	Nil	Nil	Nil
Nil	Nil	−ve	−ve	−ve	Nil	Nil	Nil
Nil	Nil	−ve	−ve	−ve	Nil	Nil	Nil
Nil	1m	−ve	−ve	3.0exp4	Nil	Nil	1.0exp3
Nil	Nil	−ve	8.3exp3	—	2y	1.0exp3	—
Nil	Nil	−ve	−ve	−ve	Nil	Nil	Nil
Nil	Nil	−ve	−ve	−ve	Nil	Nil	Nil
Nil	Nil	−ve	−ve	−ve	Nil	Nil	Nil

Samples of filter-sterilized wine were filtered through two 0.45-μm membrane filters. One filter was placed on Wallerstein Laboratories nutrient agar (WLN) and incubated at 25 °C for 3–8 days. The other was placed in glucose salts medium and incubated at 25 °C for up to 3 days. Samples were removed at various intervals and analysed on WLN agar incubated for 3 days and by ChemFlow analysis. m, mould; −ve, no count; Nil, no count; y, yeast.

ples were confirmed within 48 h and in the majority of cases after 24 h. Base wine had not been filter-sterilized and thus showed consistent levels of contamination. Where the membrane filtration results did not correlate with the cytometry results, the latter correlated well with the culture confirmation for that sample. This disparity is attributed to the division of the original sample before membrane filtration and labelling.

The results of the validation of process samples are in agreement with those from culture analysis and demonstrate that the ChemFlow system provides a saving of 1–3 days over the standard cultural method.

Analysis of low-level yeast contamination in non-filterable products: fruit preparations

The detection of low-level yeast and mould contamination is a key problem for producers and users of fruit products. In many cases pre-enrichment is used to increase cell numbers in populations which have often been stressed by processing conditions and product formulation.

The time to detection is related to the yeast strain, the initial cell concentration and the product type (Table 7.4). Even with the slow-growing yeast *Z. bailli*, at a concentration of <1 yeast cell/g of product before incubation,

TABLE 7.4. *Total viable yeast counts in fruit products*

Product	Inoculum	Inoculum level (cells/g)	Detection time (h)
Rhubarb	*S. bisporus*	1.7	17
		8.6	16
		17	12
Rhubarb	*Z. bailii*	1.2	33
		6.2	36
		12	27
Raspberry	*Z. bailii*	0.5	40
		2.3	36
		4.7	33
Honey and pecan	*Z. bailii*	0.5	48
		2.7	42
		5.4	37

Fruit samples were inoculated with *Z. bailii* and *S. bisporus* at a range of cell concentrations. The inoculated samples were then added to Sabouraud broth and incubated at 25 °C. Samples were taken at various intervals and analysed on oxytetracycline glucose agar plates incubated at 25 °C and by ChemFlow analysis.

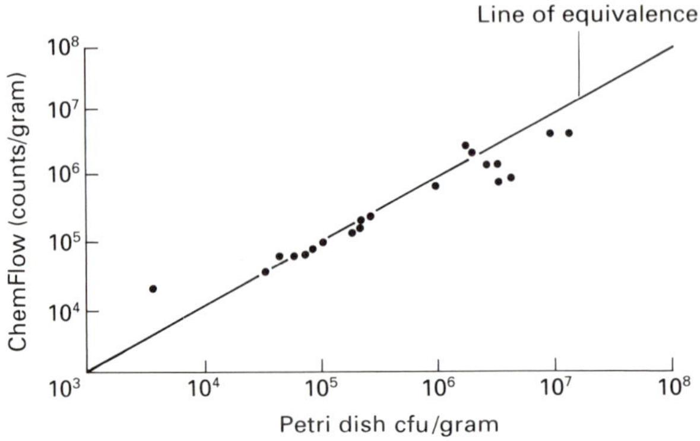

FIG. 7.3. Correlation between standard plate count and ChemFlow count in fruit products. These were inoculated with *Z. bailii* and *S. bisporus* at a range of cell concentrations. Samples were homogenized and enumerated on oxytetracycline glucose agar plates incubated at 25 °C for 3 days and by ChemFlow analysis. $r = 0.93$.

detection could be achieved in less than 48 h. This corresponds to a saving of 2–3 days compared with the standard plate count. With *S. bisporus*, detection was possible at the lowest inoculum after 17 h. A good correlation ($r = 0.99$) was obtained between flow cytometry and the plate count (Fig. 7.3).

While fruit composition has no influence on the yeast detection level, the ability of the yeast to grow can vary according to the type of product tested. The data presented indicate that a 48-h screen will ensure detection of a slow-growing yeast, even at low levels of contamination, regardless of the product type.

Direct analysis of total viable counts: process monitoring

Flow cytometry has been evaluated for the direct analysis of total viable bacterial counts (TVC) during frozen vegetable production for 'real-time' process control.

The results of correlation studies with raw spinach are shown in Fig. 7.4. A good correlation was obtained between the flow cytometry count and the standard plate count throughout the range of samples analysed (2×10^4–10^7 cells/ml). The microbial load was significantly reduced ($3 \log_{10}$ units) after blanching of the fresh spinach, which exhibited levels of total flora consistent with those found in commercial samples obtained from retail outlets. The standard deviation associated with the analysis of the commercial samples was high because of the wide sourcing for these samples.

Comparable data have been produced in a range of products to show the ability of the ChemFlow analyser to detect low-level contamination in under

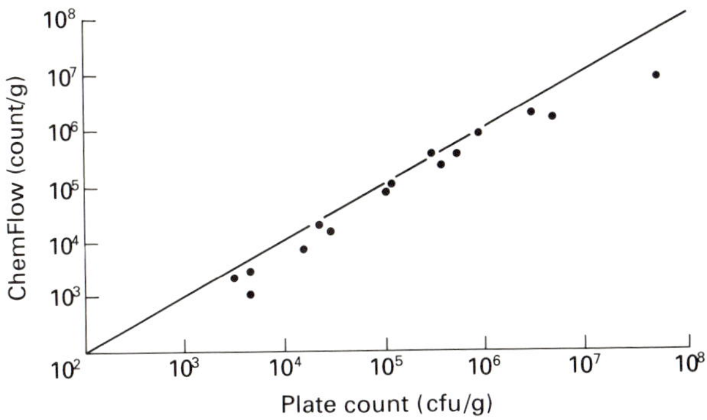

FIG. 7.4. Correlation between standard plate count and ChemFlow count in spinach. Samples of fresh spinach were homogenized with maximum recovery diluent in a stomacher. Samples were then enumerated on plate count agar incubated at 37 °C for 24–28 h and by ChemFlow analysis. $r = 0.99$.

30 min. The analysis of critical control points in industrial processes, from raw materials to final products, is now possible in a system that provides high sample throughput combined with high sensitivity.

Cell growth analysis

The use of optical analysers allows accurate measurement of even a small increase in cell numbers. It can be seen from Fig. 7.5 that in the absence of antibiotics the cell count increases throughout the incubation period. At each data point, taken at 10-min intervals, even small increases in cell numbers could be detected with the ChemFlow analyser. As expected, the cell numbers in the samples treated with gentamicin and chloramphenicol did not increase because they inhibit cell division.

Cell growth has also been followed in a penicillin fermentation. It can be seen from Fig. 7.6 that increases in biomass can be determined even with cell populations that form mycelial growth. As these samples were obtained from a seed fermentation, they did not require sonication before ChemFlow analysis, because the mycelial development is minimal. However, in later stages of fermentation, where significant mycelial masses are present, samples must be sonicated before analysis to avoid blockage of the flow cell. In these cases limited treatment (5 s) does not significantly affect the biomass recovery.

Conclusions

The use of optical analysis combined with fluorescent cell labels now provides the user with the ability to achieve rapid direct cell counts combined with

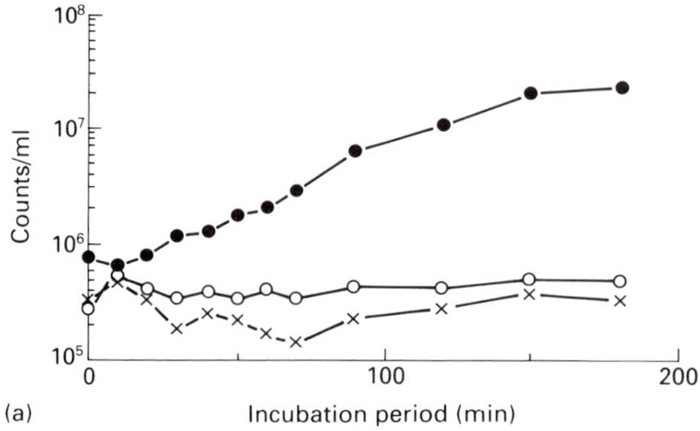

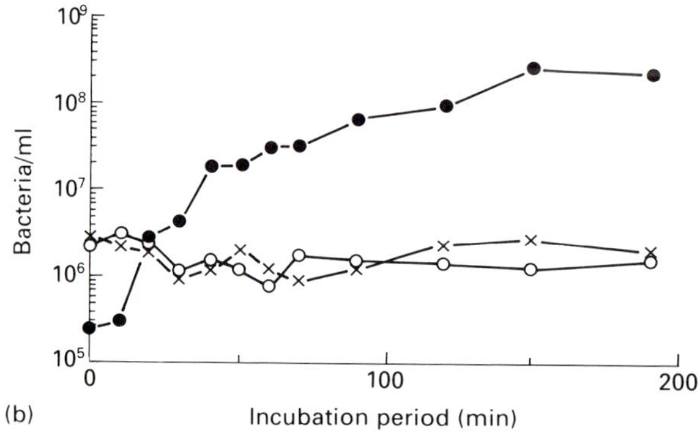

FIG. 7.5.Comparison of antibiotic sensitivity curves generated (a) by ChemFlow analysis and (b) with a counting chamber. Before addition of the antibiotic, a culture of *E. coli* was grown overnight in peptone broth at 37 °C. This was then subcultured and divided to provide: ●, an untreated control and two samples treated with: ○, gentamicin and ×, chloramphenicol, respectively. The samples were incubated in a water bath at 37 °C and samples removed at 10-min intervals for enumeration in a Thomas counting chamber and by ChemFlow analysis.

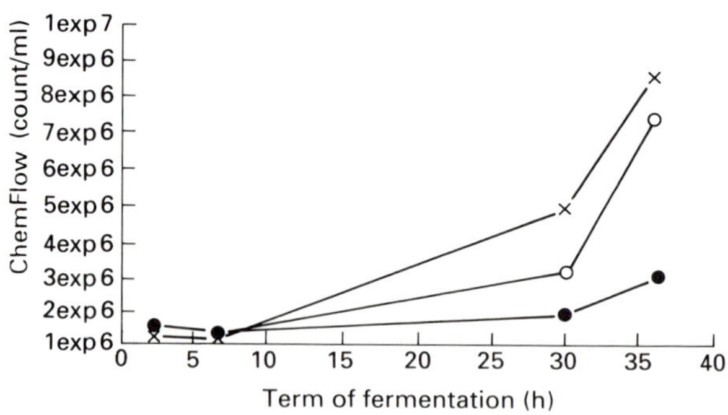

FIG. 7.6.Biomass monitoring in a penicillin seed fermentation. Samples were removed at various times and, before ChemFlow analysis, they were labelled with and without sonication. ●, sonicated for 10 s; ○, sonicated for 5 s; ×, no sonication; e, represents exponential.

information on cell cycle and cell identification. These capabilities combined with the availabilty of a robust industrial system have led to the increasing use of flow cytometry as an industrial process control tool. These systems now provide real-time data throughout industrial processes from raw materials, through process control and also as a final product check. In addition, the ChemFlow system is increasingly used in the research sectors of the pharmaceutical industry for antibiotic sensitivity studies, strain development and fermentation control.

References

BANKES, P., ROWE, D. & BETTS, R.P. 1991 The rapid detection of yeast spoilage using the ChemFlow system. *Campden Food and Drink Research Association – Technical Memorandum* No. 621.

BANKES, P., BRAILSFORD, M.A., GATLEY, S. & BETTS, R.P. 1993 The use of flow cytometry for the rapid detection of microorganisms. *Campden Food and Drink Research Association – Technical Memorandum* In press.

BEEZER, A.E., NEWELL, D. & TYRELL, H.J.V. 1976 Application of microcalorimetry to analytical problems. *Journal of Applied Bacteriology* **41**, 197–207.

BOLING, E.A., BLANCHARD, C.G. & RUSSEL, W.J. 1973 Bacterial identification by microcalorimetry. *Nature* **241**, 472–473.

BRAILSFORD, M.A. & GATLEY, S. 1992 Rapid analysis of microorganisms using sensitive optical analysers. *Biotechnology International* **2**, 186–192.

DESNOUVEAUX, R., LECOMTE, C., DE COLOMBEL, E., PLESSIS, M.C., DUMAIN, P.P., BLOCH, L., FURHMANN, B. & VALERY, S. 1990 Use of flow cytometry for yeast and mould detection in process control of fermented milk products. *Biotech Forum Europe* **3**, 224–229.

DUMAIN, P.P., DESNOUVEAUX, R., BLOCH, L., LECOMTE, C., FURHMANN, B., DE COLOMBEL, E., PLESSIS, M.-C. & VALERY, S. 1990 Use of flow cytometry for yeast and mould detection in process control of fermented milk products: the ChemFlow system – a factory study. *Biotech Forum Europe* **3**, 224–229.

EVANS, H.A.V. 1982 A note on two uses for impediometry in brewing microbiology. *Journal of Applied Bacteriology* **53**, 423–426.

FIRSTENBERG-EDEN, R. & EDEN, G. 1984 *Impedance Microbiology*. Letchworth: Research Studies Press.

GATLEY, S. 1989 Rapid process control with flow cytometry. *Food Technology International Europe* pp. 49–53. London: Sterling Publications.

GATLEY, S. 1990 Digital microbiology – a radical approach to the development of a new rapid microbiology system. *Biotech Forum Europe* **6**, 478–482.

JAKOBSEN, M. & JESPERSEN, L. 1991 Use of flow cytometry for monitoring brewery fermentation and examination of pitching yeast. *Annual Congress of the American Society of Brewing Chemists* 203–207.

LAMPI, R.A., MIKELSON, D.A., ROWLEY, D.B., PREVITTE, J.J. & WELLS, R.E. 1974 Radiometry and microcalorimetry techniques for the rapid detection of food borne organisms. *Food Technology* **28**, 52–55.

LEVASSEUR, S., HUSSON, M.O., LEITZ, R., MERLIN, F., LAURENT, F., PELADIN, F., DROCOURT, J.-L., LECLERC, H. & VAN HOERGARDEN, M. 1993 Rapid detection of Enterobacteriaceae by a monoclonal antibody. *Applied and Environmental Microbiology* In press.

PETTIPHER, G.L. 1982 The direct epifluorescent filter technique. *Kieler Milchswirtschafts Forschung* **34**, 67–73.

PETTIPHER, G.L. 1992 Preliminary evaluation of flow cytometry for the detection of yeasts in soft drinks. *Applied Microbiology* **12**, 109–112.

PETTIPHER, G.L., KROLL, R.G., FARR, L.J. & BETTS, R.P. 1989 DEFT: recent developments for foods and beverages. *Rapid Microbiological Methods for Foods, Beverages and Pharmaceuticals*, SAB Technical Series No. 25, eds Stannard, C.J., Petitt, S.B. & Skinner, F.A., pp. 33–45. Oxford: Blackwell Scientific Publications.

PREVITTE, J.J. 1972 Radiometric detection of food borne bacteria. *Applied Microbiology* **24**, 535–539.

RICHARDS, J.C.S., JASON, A.C., HOBBS, G., GIBSON, D.M. & CHRISTIE, R.H. 1978 Electronic measurement of bacterial growth. *Journal of Physics* **11**, 560–568.

RODRIGUES, U.M. & PETTIPHER, G.L. 1984 Use of direct epifluorescent filter techniques for predicting the keeping quality of pasteurized milk within 24 hours. *Journal of Applied Bacteriology* **57**, 125–130.

SHARPE, A.N., WOODRON, H.N. & KACKSON, A.K. 1970 ATP levels in food contaminated by bacteria. *Journal of Applied Bacteriology* **33**, 758–767.

STEWART, B.J., EYLES, M.J. & MURRELL, W.G. 1980 Rapid radiometric method for the detection of salmonella in foods. *Applied and Environmental Microbiology* **40**, 223–230.

WILLIAMS, M.L.R. 1971 The limitations of the DuPont luminometer in the microbiological analysis of foods. *Canadian Institute of Food Technology* **4**, 187–189.

8: New Techniques for the Detection of Protozoan Parasites in Water

G. Vesey, J.S. Slade, M. Byrne,
K. Shepherd and C.R. Fricker
*Thames Water Utilities, Spencer House Laboratory,
Manor Farm Road, Reading RG2 0JN, UK*

Cryptosporidium and *Giardia* are among the most common protozoan causative agents of diarrhoeal disease in man (Current 1986; Adam 1991) and their importance has recently been emphasized (Badenoch 1990).

The life cycle of both organisms involves a complex process in the host intestine and results in the formation of resistant *Cryptosporidium* oocysts or *Giardia* cysts that are excreted in the faeces. Cysts and oocysts are extremely robust and highly infectious; the ingestion of as few as 10 may lead to the development of disease (Miller *et al.* 1986).

Cryptosporidium oocysts are widespread in untreated sources of drinking water throughout the UK. The main sources appear to be sheep and cattle (Fayer & Ungar 1986), a range of wild animals and also sewage. Infected animals can excrete sufficient oocysts in 1g of faeces to infect a million people; hence water that drains from farm land and activities such as the illegal emptying of slurry tanks into rivers can send enormous numbers of oocysts downstream to water treatment plants.

Contamination of surface water with *Giardia* also takes place in the UK, where the main source of contamination is sewage effluent, but wildlife may also act as reservoirs. The presence of *Cryptosporidium* or *Giardia* in raw water sources poses a major problem for the water industry, mainly because of their resistance to established disinfection procedures. Filtration procedures are the most effective, but these must be optimized to ensure removal, and breakthrough may occur if the filters are challenged with high numbers. Treatment that relies solely on chlorination is of limited use for *Giardia* (Adam 1991) and is totally ineffective for *Cryptosporidium* (Korich *et al.* 1990). The problem

New Techniques in
Food and Beverage Microbiology

0-632-03755-5

is further complicated because the established bacterial indicators of water hygiene do not apply to these organisms. *Giardia* cysts and *Cryptosporidium* oocysts can survive for several months in the aquatic environment and, since most indicator organisms survive for a matter of days or at most for weeks, they are unsuitable as indicators for the presence or absence of *Cryptosporidium* and *Giardia*. Waterborne outbreaks of cryptosporidiosis have been traced to water treatment plants that have met all current statutory microbiological and chemical standards (Badenoch 1990). Since these outbreaks, many water supply companies have considered it necessary to monitor raw water sources for the presence of cysts and oocysts.

The methods used to detect *Cryptosporidium* oocysts and *Giardia* cysts in water depend on concentration of large volumes of water by filtration, staining of the concentrate on a glass slide or a membrane with fluorescence-labelled antibodies and examination of the stained sample by epifluorescence microscopy. These methods are labour-intensive and inefficient. Cysts and oocysts are lost throughout the procedures and, with turbid raw waters, recoveries are often less than 10%. Furthermore, microscopy requires experienced staff and is tedious, and it often takes several hours to analyse one sample.

We have developed alternative methods that overcome many of these problems. A calcium carbonate flocculation method is used to concentrate the sample (Vesey *et al.* 1993a) before analysis by flow cytometry (Vesey *et al.* 1993b). Flocculation is inexpensive, requires little labour, is more efficient and gives recoveries of 69–80%.

Sample concentrates are stained in suspension and analysed with a Coulter Epics Elite flow cytometer, which acts as a highly efficient purification process. The flow cytometer analyses individual particles and compares their fluorescence and light-scattering characteristics with those already defined for cysts and oocysts. Particles with these characteristics are then sorted from the stream of material and collected on a microscope slide for further examination. The sorted particles can easily be scanned because of the absence of contaminating debris. In addition the use of suspensions, rather than dried films, results in improved staining of the cysts and oocysts.

In this chapter we describe these new techniques, as well as the conventional methods. Current developments of new techniques for detecting pathogenic protozoan parasites will also be considered.

Sample Collection and Concentration

Methods for detecting *Cryptosporidium* oocysts and *Giardia* cysts in water must be very sensitive because of the low infective dose. Attempts to achieve such sensitivity have relied on concentration and analysis of very large volumes of water. This was first successfully achieved during an outbreak of waterborne giardiasis in Rome, New York (Shaw *et al.* 1977). During this outbreak the Center for Disease Control used a sand swimming pool filter to concentrate vast volumes of the drinking water supply. Since then, various investigators have developed a variety of filtration methods.

Filtration

Wound cartridge filters

The use of wound cartridge filters is fully described by Rose *et al.* (1986). Sample concentration is carried out on site with a 1-μm nominal pore-size filter (Microwynd, Cuno) connected via a housing to the pressurized water supply. Alternatively, water is pumped through with a portable pump. In the laboratory, the filter is cut lengthwise to its plastic core and the fibres are removed and teased apart. Particles are eluted from the fibres by washing in at least 1 l of Tween 80 (0.1%) with a stomacher. The washwater is squeezed from the fibres and concentrated to a pellet by centrifuging at 3000 *g* for 10 min. The method is labour-intensive and cysts and oocysts can be lost due to passage through the filter and because of poor elution recoveries.

Pleated membrane cartridge filters

Pleated membrane cartridge filters (Vokes Ltd, Guildford, UK) are more efficient than wound cartridge filters. The method remains labour-intensive but loss of cysts and oocysts through the filter is reduced and greater numbers are eluted. Sample collection and filter elution are basically the same as for wound cartridge filters, except that the membrane is cut away from the central core and the different layers are separated before washing.

Membrane filtration

Ongerth & Stibbs (1987) advocated use of 242 mm diameter polycarbonate membrane filters. This involves collection and transportation of 10- or 20-l samples in barrels or tanks to the laboratory. The samples are passed through the membrane filter, particulate matter is eluted by scrubbing the membrane with a rubber blade and rinsing with a stream of water and washings are then pelleted by centrifugation. Recoveries with this technique are good, but problems may be experienced with filter blockage when turbid samples are processed, as protective prefilters cannot be used. This method limits the volume of sample that can be analysed, but large-volume sampling is often unnecessary if good recovery is achieved.

Tangential flow filtration

Tangential flow filtration has been used to concentrate *Giardia* cysts (Isaac-Renton *et al.* 1986) but recoveries are low, especially for *Cryptosporidium* oocysts, and, unless numerous filtration units are available, the number of samples that can be concentrated is restricted.

Flocculation

A calcium carbonate flocculation method has previously been described (Vesey *et al.* 1993a). Briefly, a 10-l sample is collected and transported in barrels or tanks to the laboratory, where it is flocculated by addition of 100 ml of 1 mol/l calcium chloride, 100 ml of 1 mol/l sodium bicarbonate and adjusting the pH to 10 with sodium hydroxide. Calcium carbonate crystals form and enmesh particles in the size range of cysts and oocysts. After 4 h, the calcium carbonate settles, which allows removal of the supernatant by vacuum aspiration, and the precipitate is then dissolved in 10% sulphamic acid. Particulate matter is further concentrated by centrifugation (7000 *g* for 12 min) and is washed sequentially with distilled water, phosphate-buffered saline (PBS) pH 7.2, and 0.01% Tween 20 solution. The method is efficient, is easy to perform, does not require expensive equipment and gives up to 79% recovery.

Other Concentration Methods

Numerous other techniques have been used to concentrate cysts and oocysts, including sand column filtration, continuous-flow centrifugation and vortex filtration. None of these solves the problems associated with the concentration of cysts and oocysts, namely to achieve good recovery without damage and with the minimum of labour and expense.

Clarification Procedures

Density gradient centrifugation is used routinely to purify and concentrate cysts and oocysts from faeces samples before screening. Clarification procedures have been modified and used, with varying success, to purify cysts and oocysts from concentrated water samples. LeChevallier *et al.* (1991) evaluated a range of such procedures and found that the most efficient was Percol–sucrose flotation medium, with 80–87% recovery. A 19-ml sample is underlaid with 30 ml Percol–sucrose of specific gravity 1.10 and centrifuged at 1050 *g* for 10 min. The top layer plus 5 ml of the Percol interface is carefully collected, washed and stained.

Most attempts to develop and evaluate clarification methods have been with seeded samples. In our experience seeded samples are not a good representation of environmental samples. With naturally occurring oocysts and cysts in turbid samples, losses during flotation are very high. This is probably due to the way in which oocysts tend to adhere to other particulates when in the environment. When seeded samples are used, oocysts tend to remain largely free from adherent material, maintain their true density and, therefore, behave as expected in density gradients.

Immunofluorescence Staining

Antibodies

Several antibodies are commercially available for staining concentrated water samples (Table 8.1). We routinely use a combined *Cryptosporidium* and *Giardia* antibody (Cellabs, NSW, Australia), which is marketed in the UK (Bradsure Biologicals, Leicestershire, UK). These antibodies show very little cross-reaction with particulate matter present in water samples, but they do cross-react with *Cryptosporidium* spp. oocysts and *Giardia* spp. cysts that may not be pathogenic to humans.

Staining procedures

Staining on a glass slide

Sample concentrates are distributed on to a glass multiwell slide (Hendley-Essex, Essex, UK) and air-dried at 50 °C. It is important that the sample is

TABLE 8.1. *Commercially available monoclonal antibodies for the detection of* Cryptosporidium *oocysts and* Giardia *cysts*

Antibody	Type	Reactivity	Background staining
Crypto-Giardia Cel	*Giardia* + *Cryptosporidium* FITC-conjugated*	All *Giardia* and *Cryptosporidium*	Low
Hydrofluor	*Giardia* + *Cryptosporidium* Not conjugated†	All *Giardia*; *Cryptosporidium parvum* and *C. meleagridis*	High
I122	*Cryptosporidium* FITC-conjugated‡	*Cryptosporidium parvum*, *C. baileyi* and *C. meleagridis*	Low

* Bradsure Biologicals, Market Harborough.
† Meridian Diagnostics, Ohio, USA.
‡ Northumbria Biologicals.

diluted sufficiently with distilled water to ensure a single, even and thin layer of particulate matter across the slide. Insufficient dilution may result in the sample flaking off during the staining procedure. Antibody is then applied to each well and the slide incubated and washed in accordance with the instructions of the antibody manufacturer. Once the slide has dried, a drop of glycerol mounting fluid containing an antifade agent (Citifluor Ltd, London, UK) is added and a coverslip is gently applied.

Staining on a membrane

LeChevallier *et al.* (1991) recommend that the staining procedure is carried out on a polycarbonate membrane. The sample is passed through a membrane (13 mm diameter, 2 µm pore size) either by applying a vacuum or with a syringe. The bottom of the filter holder is then plugged and 100 µl of antibody is applied to the membrane surface. The sample is incubated and then washed by passing several millilitres of the recommended washing solution through the membrane. The membrane is then dehydrated with an alcohol series (10, 20, 40 and 80% ethanol; 5% glycerol), rinsed twice with 95% ethanol and 5% glycerol and cleared with 0.1% propylgallate-glycerol. Finally, the membrane is carefully removed and placed on a glass slide and mounting medium, and a coverslip are applied.

Staining samples for flow-cytometric analysis

Samples for analysis by flow cytometry are stained in suspension with sufficient monoclonal antibody to resuspend the sample pellet. After incubation at 37 °C for 30 min the sample is immediately analysed; washing is unnecessary.

Flow Cytometry

An introduction to and explanation of the principles of flow cytometry is given by Shapiro (1988).

Instrumentation

Investigations into the use of flow cytometry for the detection of oocysts and cysts were first carried out with a Skatron Argus flow cytometer (Vesey *et al.* 1991). This has a mercury vapour lamp as a light source and the sample is examined as it passes over a curved lens. Although the instrument offers good sensitivity and resolution at a fraction of the cost of laser flow cytometers, it lacks a sorting facility. Without a sorting facility, to collect suspect particles for further examination, the required sensitivity for environmental monitoring could not be achieved.

Analysis

The techniques described in this chapter are designed for use with an Epics Elite flow cytometer (Coulter Electronics Ltd) fitted with a 15-mW argon laser and a 100-μm quartz flow cell, which has had the voltage across the deflection plates increased to 3000 V.

The protocol was developed by analysing a range of environmental samples seeded with oocysts. The parameters and data presentation method that produced the most compact populations of cysts and oocysts with maximum separation from background debris were then chosen and fully evaluated. The chosen parameters were $\log_{10}$ 90° light scatter and $\log_{10}$ fluorescence presented as a scatter plot. An area on this scatter plot graph in which all cysts and oocysts would appear was then defined by analysis of fresh oocysts and cysts and also oocysts and cysts that had been aged by storage in various types of water at a range of temperatures for periods of up to 3 months. This

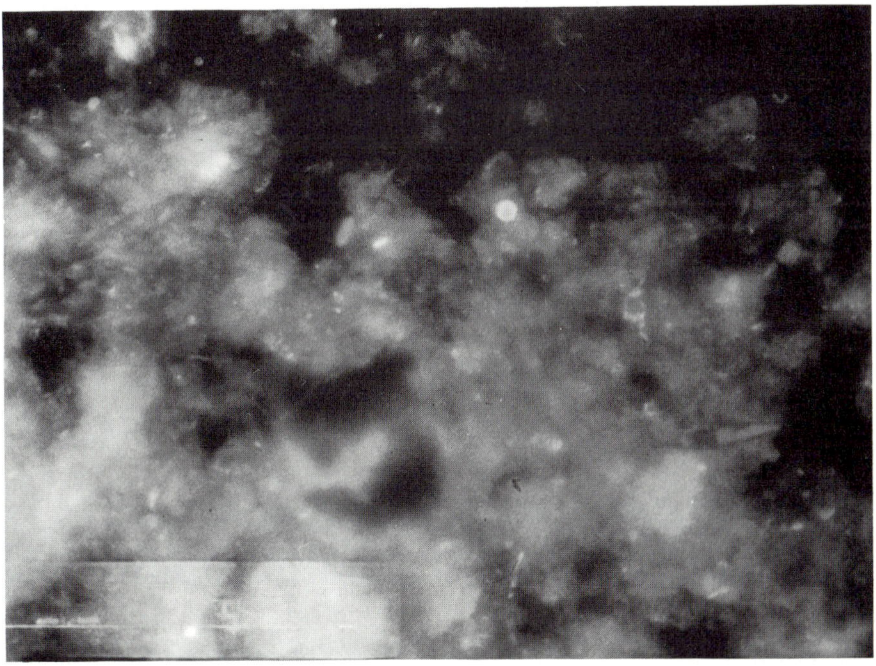

FIG. 8.1. Sample prepared without flow cytometry and viewed by epifluorescence microscopy (magnification ×500).

area was then used to sort cysts and oocysts from environmental samples. It is strongly recommended that these experiments should be repeated unless an identical protocol and sort area are used.

Before sample analysis, the cytometer is calibrated with Coulter Immunocheck beads (Coulter Electronics, Luton, UK), the forward-angle light scatter and the fluorescence signal are examined and the instrument is adjusted until a half-CV (a statistical measurement related to the variation in signal) of less than 2 is obtained for both signals. The fluorescence and 90° light scatter are standardized by analysis of Coulter StandardBrite beads. The high voltage on the fluorescent and the 90° light-scatter photomultiplier tubes is then adjusted until the correct signals are obtained for the beads. The sort delay is calculated by carrying out a sort test matrix with Coulter Immunocheck beads, and sorting is carried out on three droplets to minimize the possibility of particles being missed.

For quality control approximately 30 Immunocheck beads are added to each sample before analysis. If the finished slide does not contain approximately 30 beads, sampling is repeated.

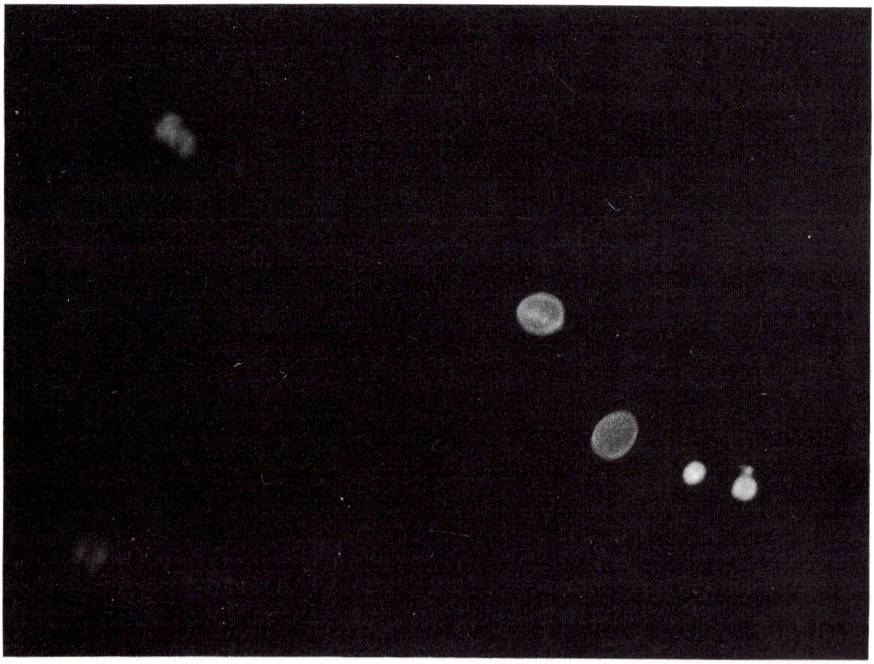

FIG. 8.2.The same sample as in Fig. 8.1 viewed by epifluorescence microscropy after flow cytometry (magnification ×500).

During analysis the camera system is used to view the droplet break-off point and, if any movement occurs, the piezoelectric crystal drive is adjusted to return the break-off point to its original position.

Sorted particles are collected on a glass slide coated with three concentric circles of black Teflon (Hendley-Essex, Essex, UK). Once analysis is complete, the slide is air-dried and a drop of mounting medium and a coverslip are applied and the slide is then examined by epifluorescence microscopy. The cytometer is decontaminated between samples by running through 10% aqueous sodium hypochlorite solution for 1 min followed by Coulter Cleanze for 30 s. Use of flow cytometry at this stage in the procedure removes much of the background debris, as shown in Figs 8.1 and 8.2.

Microscopy

Samples are examined with an epifluorescence microscope fitted with suitable excitation and barrier filters for detecting fluorescence of the appropriate wavelength. Magnifications used by various investigators range from ×200 to ×600. We routinely use ×12 eyepieces and a ×50 water immersion objective for scanning samples, which produces a brighter fluorescent image than dry objectives.

Samples are scanned for objects that fluoresce green and have the morphological characteristics of cysts or oocysts (Table 8.2). The presence of cysts and oocysts is then confirmed by examination with a ×100 objective fitted with direct-interference contrast (DIC). Oocysts appear as spheres, 4–7 μm in diameter, which are occasionally crumpled and deformed and sometimes a fold or suture line is visible. Oocysts have a doughnut appearance when examined with DIC and one to four sporozoites or a residuum is visible internally. Cysts are oval to spherical and 6–16 μm in diameter. When examined with DIC, a distinct cyst wall and nuclei, axonemes and median may be

TABLE 8.2. *Morphology of* Cryptosporidium *oocysts and* Giardia *cysts*

	Giardia cysts	*Cryptosporidium* oocysts
Size	6–16 μm	4–7 μm
Shape	Oval to spherical	Spherical, often crumpled or deformed
Appearance with DIC	Highly refractile; distinct and raised cyst wall	Highly refractile; doughnut-shaped
Internal structure visible with DIC	Nuclei, median bodies and axonemes	Up to four banana-shaped sporozoites, cytoplasm or a residuum

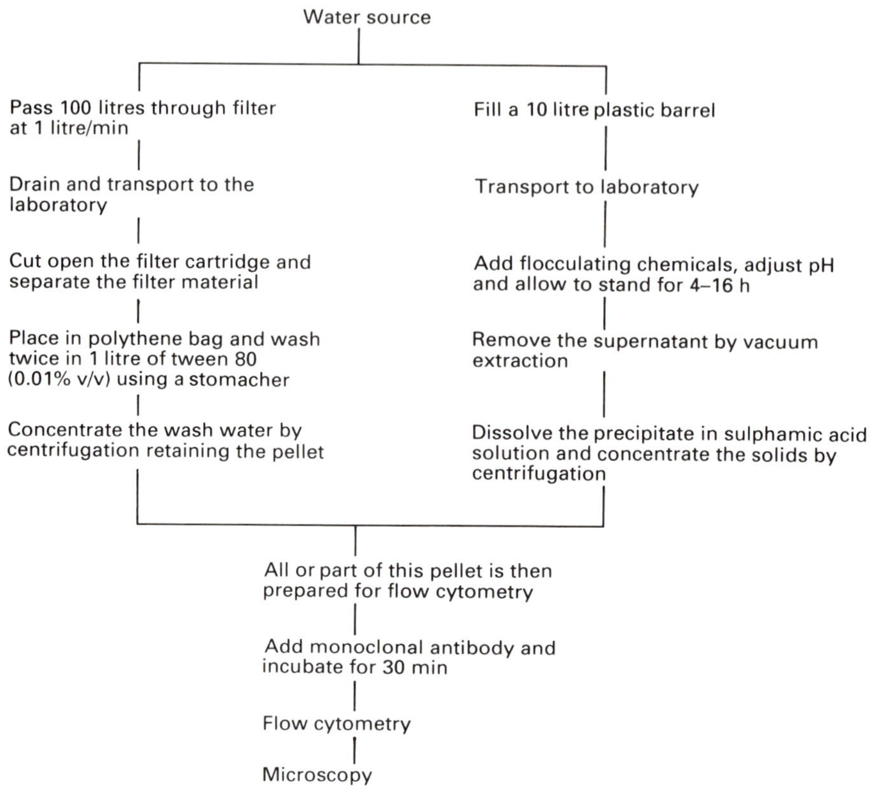

FIG. 8.3.Protocol for detecting *Cryptosporiclium* and *Giardia* in environmental samples.

visible and cysts, to be confirmed as such, should have at least two of these morphological characteristics.

The procedures used in our laboratory are outlined in Fig. 8.3.

Method Development

Flow cytometry

Flow cytometry has the potential to solve many of the problems associated with the detection of low numbers of specific pathogens in environmental samples. Flow cytometers are at present designed and developed for use in haematology laboratories to analyse samples that contain high numbers of the cells of interest. The application of flow cytometry to environmental pathogens is a new research area and is limited by the available instrumentation. Research is, however, in progress to develop instruments specifically for such applications. Recent developments in laser technology have made possible the use of small, compact and robust pulse lasers in place of the existing laser systems. As well as significantly reducing the cost of flow cytometers, this technology has the potential to increase the sensitivity of instruments and thus permit analysis of samples without the need to sort, collect and confirm suspect particles.

Pulse laser flow cytometry will facilitate the use of 'time-resolved fluorescence', which uses fluorochromes with a long fluorescence decay time to label target organisms and then excites these fluorochromes with a short pulse of laser light. Once the autofluorescence background of the debris in environmental samples has died away, which takes only 4–5 ns, the cytometer can look for longer-lived fluorescence. As the fluorescence detection is performed in total darkness, this method also allows the use of highly sensitive photon-counting techniques to detect very small particles.

Array detection systems are another recent development with the potential for increasing the sensitivity of flow cytometers for use in environmental microbiology. This technology can produce an image from light scatter and fluorescence detectors, rather than merely the light intensity measurement used at present, and would aid considerably with the differentiation of the target organism from other debris. Array detection can also be used to produce a spectral fingerprint of the fluorochrome used for labelling. By passing the fluorescence signal through a prism before measuring it on an array of detectors, a unique signal for the fluorochrome can be determined, which permits its differentiation from autofluorescence signals.

The increase in sensitivity achieved by these techniques should allow labelling techniques to be extended from immunological methods to molecular techniques. *In situ* hybridization fluorescence techniques are already widely used in clinical flow cytometry. The application of these methods to environmental microbiology would facilitate the use of flow cytometry.

Although the techniques discussed here are at an early stage of development, they should make it possible to develop flow cytometry methods based on small and relatively inexpensive instruments that require little or no calibration and have the sensitivity to detect a range of specific microorganisms in environmental samples.

Viability

A major limitation of the methods described here is that the immunofluorescence staining techniques do not distinguish between living and dead organisms. It is important to be able to distinguish between cysts and oocysts that have been in the environment for some time and as a result have become non-viable, and those that are fresh and viable.

Methods to determine viability, such as excystation and animal infectivity, are not applicable to environmental samples because of the very low numbers of cysts and oocysts normally present. It has been reported that the fluorochromes propidium iodide (PI) and fluorescein diacetate (FDA) are useful for determining *Giardia* cyst viability (Schupp & Erlandsen 1987a). Cysts that stain with PI are regarded as non-viable and those that stain with FDA are viable. There are, however, problems with this technique. Sauch *et al.* (1991) have reported that it does not work with cysts that have been exposed to chlorine and, in our experience, a high proportion of some cyst strains do not stain with either fluorochrome.

An alternative method to the use of fluorochromes for viability assessment is to examine cysts carefully with DIC microscopy (Schupp & Erlandsen 1987b) for the presence of degraded internal organelles or for a space between the cell wall and the internal organelles. Both of these are said to indicate cyst non-viability.

Definitive methods have not as yet been published for determination of *Cryptosporidium* oocyst viability. However, initial investigations with fluorochrome staining appear promising (Smith *et al.* 1991) and such techniques have already proved useful for identifying oocysts.

Molecular techniques

Abbaszadegan *et al.* (1991) reported the successful use of a cDNA gene probe for the detection of *Giardia* cysts in environmental samples. The method appears to be as sensitive as immunofluorescence staining and epifluorescence microscopy. However, only two sewage effluents and two surface waters were examined and a far wider range of samples will have to be examined before its use and potential can be fully evaluated. The technique depends on the use of a radioactive label, which may limit its application in the water industry.

An mRNA polymerase chain reaction (PCR) protocol for the detection of *Giardia* has been reported (Mahbubani *et al.* 1991). By splitting the sample, carrying out an excystation procedure on one half and comparing the results with those of PCR, the viability of *Giardia* cysts could be measured, as based on the fact that more mRNA was detected in excysted trophozoites than in cysts. This report should be interpreted with care, since the experiments were carried out under ideal conditions without any of the debris found in environmental samples. Furthermore, tests were not carried out on cysts that had been exposed to environmental conditions, such as storage in water or chlorination.

Molecular techniques for the detection of specific microorganisms in environmental samples are still at an early stage of development. Various problems, such as interference from toxic components in the concentrated samples, will have to be overcome before such methods can be used routinely. Numerous publications on this subject give the impression that these problems do not exist, but few if any laboratories have been able to use these methods for routine monitoring.

Note. The views expressed are those of the authors and are not necessarily those of Thames Water Utilities Ltd.

References

ABBASZADEGAN, M., GERBA, C.P. & ROSE, J.B. 1991 Detection of *Giardia* cysts with a cDNA probe and applications to water samples. *Applied and Environmental Microbiology* **57**, 927–931.

ADAM, D.A. 1991 The biology of *Giardia* spp. *Microbiological Reviews* **55**, 706–732.

BADENOCH, J. 1990 Cryptosporidium *in Water Supplies*. London: HMSO.

CURRENT, W.L. 1986 *Cryptosporidium*: its biology and potential for environmental transmission. *Critical Reviews in Environmental Control* **17**, 21–51.

FAYER, R. & UNGAR, B.L.P. 1986 *Cryptosporidium* spp. and cryptosporidiosis. *Microbiological Reviews* **50**, 453–458.

ISAAC-RENTON, J.L., FUNG, C.P.J. & LOCHAN, A. 1986 Evaluation of a tangential-flow multiple-filter technique for the detection of *Giardia lamblia* cysts in water. *Applied and Environmental Microbiology* **52**, 400–402.

KORICH, D.G., MEAD, J.R., MADORE, M.S., SINCLAIR, M.A. & STERLING, C.R. 1990 Effects of ozone, chlorine dioxide, chlorine, and monochloramine on *Cryptosporidium parvum* oocyst viability. *Applied and Environmental Microbiology* **57**, 2610–2616.

LECHEVALLIER, M.W., NORTON, W.D. & LEE, R.G. 1991 Evaluation of a method to detect *Giardia* and *Cryptosporidium* in water. *Monitoring Water in the 1990s: Meeting New Challenges*, eds Hall, J.R. & Glysson, D., pp. 483–498. Philadelphia: American Society for Testing and Materials.

MAHBUBANI, M.H., BEJ, A.K., PERLIN, M., SCHAEFER, F.W., JAKUBOWSKI, W. & ATLAS, R.M. 1991 Detection of *Giardia* cysts by using the polymerase chain recation and distinguishing live from dead cysts. *Applied and Environmental Microbiology* **57**, 3456–3461.

MILLER, R.A., BRONSDON, M.A. & MORTON, W.R. 1986 Determination of the infectious dose of *Cryptosporidium* and the influence of inoculum size on disease severity in a primate model. *Abstracts of the Annual Meeting of the American Society of Microbiology*, Washington, 23–28 March 1986, p. 48. Washington, DC: American Society of Microbiology.

ONGERTH, J.E. & STIBBS, H.H. 1987 Identification of *Cryptosporidium* oocysts in river water. *Applied and Environmental Microbiology* **53**, 672–676.

ROSE, J.B., MADORE, M.S., RIGGS, J.L. & GERBA, C.P. 1986 Detection of *Cryptosporidium* and *Giardia* in environmental waters. *Proceedings of the AWWA Water Quality Technology Conference*, pp. 417–424. Portland, OR: American Water Works Association.

SAUCH, J.F., FLANIGAN, D., GALVIN, M.L., BERMAN, D. & JAKUBOWSKI, W. 1991 Propidium iodide as an indicator of *Giardia* cyst viability. *Applied and Environmental Microbiology* **57**, 3243–3247.

SCHUPP, D.G. & ERLANDSEN, L.S. 1987a A new method to determine *Giardia* cyst viability: correlation of fluorescein diacetate and propidium iodide staining with animal infectivity. *Applied and Environmental Microbiology* **53**, 704–707.

SCHUPP, D.G. & ERLANDSEN, L.S. 1987b Determination of *Giardia* cyst viability by differential contrast, phase, or brightfield microscopy. *Journal of Parasitology* **73**, 723–729.

SHAPIRO, H.M. 1988 *Practical Flow Cytometry*. New York: Alan R. Liss.

SHAW, P.K., BRODSKY, R.E., LYMAN, D.O., WOOD, B.T., HIBLER, C.P., HEALY, G.R., MACLEOD, K.I.E., STAHL, W. & SCHULTZ, M.G. 1977 A community-wide outbreak of giardiasis with evidence of transmission by a municipal water system. *Annals of Internal Medicine* **87**, 426–432.

SMITH, H.V., GRIMASON, A.M., BENTON, C. & BARKER, J.F.W. 1991 The occurrence of *Cryptosporidium* spp. oocysts in Scottish waters, and the development of a fluorogenic viability assay for individual *Cryptosporidium* spp. oocysts. *Water Science and Technology* **24**, 169–172.

VESEY, G. & SLADE, J.S. & FRICKER, C.R. 1991 Taking the eye strain out of environmental *Cryptosporidium* analysis. *Letters in Applied Microbiology* **13**, 62–65.

VESEY, G., SLADE, J.S., BYRNE, M., SHEPHARD, K. & FRICKER, C.R. 1993a A new method for concentrating *Cryptosporidium* oocysts from water. *Journal of Applied Bacteriology* In press.

VESEY, G., SLADE, J.S., BYRNE, M., SHEPHARD, K., DENNIS, P.J. & FRICKER, C.R. 1993b Routine monitoring of *Cryptosporidium* in water using flow cytometry. *Journal of Applied Bacteriology* In press.

9: Indirect Impedance for Microbial Screening in the Food and Beverage Industries

P.DRUGGAN,[1] S.J.FORSYTHE[2] AND P.SILLEY[3]
[1]*Life Technologies Ltd, PO Box 35, Trident House, Renfrew Road, Paisley PA3 4EF, UK;*
[2]*Department of Life Sciences, Nottingham Polytechnic, Clifton Lane, Nottingham NG11 8NS, UK;*
and [3]*Don Whitley Scientific Ltd, 14 Otley Road, Shipley, West Yorkshire BD17 7SE, UK*

The use of indirect impedance was first proposed by Owens *et al.* (1989) as a method to overcome the disadvantages of the direct technique. These are the need to use media in which conductance changes vary as a function of microbial growth and which have initial conductance readings within the measuring range of the available instrumentation. The principle of the indirect technique is based on monitoring CO_2 production as an indicator of microbial growth. The CO_2 produced is absorbed under alkaline conditions to give an electrical signal not directly dependent on the nature of the growth medium. This immediately offers the advantage of the use of selective media and test samples with high salt concentrations. Furthermore, the selected growth medium need not necessarily be optimized for electrical response, which allows the use of media previously considered unsuitable for impedance applications (Bolton 1990). This chapter describes working protocols that demonstrate the suitability of indirect impedance in the food and beverage industries.

Medium Design

Bridson & Brecker (1970) pointed out that the design of many culture media is based on previous formulations by one-step changes that improve the isolation or provide greater biomass for a given target organism. The following provides a guide for the adaptation of formulations for use in impedance microbiology.

New Techniques in
Food and Beverage Microbiology

0-632-03755-5

Peptones are the most common nitrogen source in media, and yet they remain a source of confusion. While amino acid profiles of individual peptones can be useful, more relevant information can be obtained from their amino nitrogen (AN) and total nitrogen (TN) content. The AN/TN ratio is a measure of the degradation products of a peptone, and is conventionally expressed as a percentage. By altering the enzyme-to-substrate ratio and the duration of hydrolysis during peptone manufacture, it is possible to vary the polypeptide chain length and the AN value. Although the AN/TN value is convenient for the characterization of protein hydrolysates, it does not give an indication of the molecular weight distribution of the various amino acid, peptide and polypeptide fractions. The choice of peptone for a culture medium is empirical, but we have found that peptones with AN/TN values of *c.* 45–50% are best for conductimetric work. We have also shown that acid hydrolysates with an AN/TN value of 60% do not perform as well in the direct method. This may be due to nutrient depletion that results from the method of digestion, but we have not pursued this further. Unfortunately, the AN/TN ratio cannot be chosen by the principle of 'the higher, the better'. A number of studies (Payne 1976) have shown that peptides can be superior to free amino acids in bacterial nutrition. This is partly due to protection against degradation and reduced antagonism in utilization. There is also a limit on the useful size for peptides and polypeptides. Organisms such as *Escherichia coli*, which only possess intracellular peptidases, are limited because of steric hindrance in the size of polypeptide they can transport across the cell membrane.

It was originally suggested that in the indirect method the maximum electrical response was determined by the amount of nutrient available to organisms and by the concentration and volume of absorbent (Owens *et al.* 1989). Since the latter are standardized in a given system, it is the availability of nutrients that is the most significant factor in the performance of an assay system. It is this that makes the choice of nitrogen source so significant. In his assessment of the indirect method, Bolton (1990) obtained a significant increase in the rate of evolution of CO_2 of a number of organisms by adding glucose to Whitley impedance broth (WIB). This is a fairly simple but effective method of medium development. It was also suggested that the indirect method might lend itself to the use of traditional diagnostic selective media that contained high concentrations of salts and that were not usable for impedance studies in the past because of high initial medium conductivity.

Most traditional media rely on the increase in biomass for detection, either as an enrichment step in an extended diagnostic procedure, or in the production of indicator compounds from the fermentation of selected carbohydrates, alcohols and organic acids. In the indirect method, however, it is metabolic activity, not biomass, that determines the detection time. Most facultative microbes will respire aerobically if sufficient oxygen is present. In this situation, carbohydrates such as glucose will be catabolized through the tricarboxylic acid (TCA) cycle, to generate energy, carbon dioxide and water. This metabolism, which is common in the early stages of growth and generates large quantities of carbon dioxide, is later replaced by fermentation. It is interesting that, of the four strains of bacteria that failed significantly to increase their rate of carbon dioxide production when WIB was supplemented with glucose (Bolton 1990), three were strict aerobes. The other organism, *Enterococcus faecalis*, lacks a conventional cytochrome system and TCA cycle.

When the choice of medium formulation for the indirect method is to be made, it should be remembered that high concentrations of carbohydrates in traditional media may not be necessary. When carbohydrate utilization is a diagnostic feature, pH indicators need not be relied on, because there are a number of 4-methylumbelliferyl- and indoxyl-linked substrates which, when cleaved, release fluorogenic and chromogenic moieties respectively. This allows media to be held at constant pH by buffering and still indicate carbohydrate utilization. We have recently exploited chemically defined media for the indirect method. This has initially been for the Enterobacteriaceae, which can grow on media that contain a single inorganic nitrogen source and a single carbon source. Many can also utilize single amino acids as sole carbon/nitrogen sources. Pseudomonads can readily be detected with a medium based on a modification of SGAP-10C medium (Huguet & Ribas 1991). This contains the single carbon/nitrogen source, monosodium glutamate. Glutamate is converted to 2-oxoglutarate, an intermediate of the TCA cycle, by a transamination step. It should be remembered that inhibition may play a more significant part in indirect impedance, because the maximum change is determined only by the concentration of available nutrients.

Electrical Detection System

Experiments were carried out with the rapid automated bacterial impedance technique (RABIT) system (Don Whitley Scientific, Shipley, UK). The detection routine depends on definition by the user of suitable detection criteria. Microbial growth is then recorded when the change, in microsiemens (μS), for three consecutive readings is equal to or greater than the predetermined value. It is usually necessary to carry out preliminary experiments to optimize the detection parameters.

Indirect Impedance

Preparation of test cell

Carbon dioxide is trapped in an alkaline agar absorbent that bridges the cell electrodes (Owens *et al.* 1989). The alkaline solution is prepared by dissolving 1 g agar (Oxoid Agar No. 1) in 50 ml water and autoclaving at 121 °C for

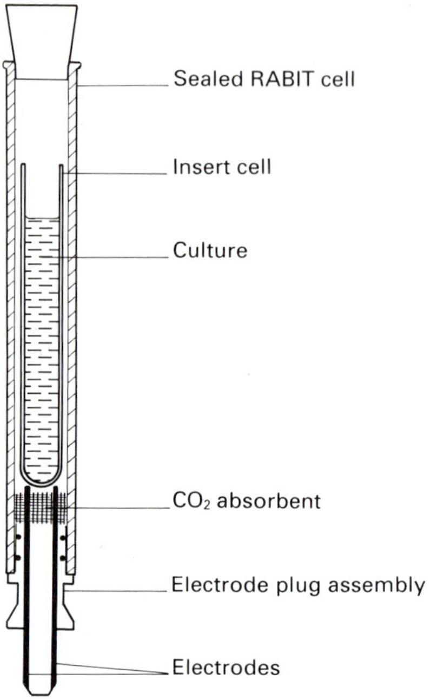

FIG. 9.1. The indirect cell for the rapid automated bacterial impedence technique (RABIT).

15 min. When cooled to about 45 °C this is added to 50 ml of a 0.7% KOH solution. While still warm, 0.7 ml of the resulting solution is added to each dry RABIT cell. Cells are tightly stoppered and stored in the vertical position for a minimum of 3 h before use. In all cases test samples are added to a 12 × 75 mm insert cell (Fig. 9.1) that contains the appropriate media.

Test protocols

The principles of indirect impedance, which have been presented, can be applied to a wide range of applications in various industries. Applications for the detection of yeasts, *Clostridium tyrobutyricum*, coliforms and *Escherichia coli* and the use of the technique for the study of biofilms are described as representative protocols. The standard RABIT detection criteria for these protocols are: time resolution 6 min, detection threshold −10 μS, temperature 37 °C and test duration 24 h.

Detection of yeasts in fruit juice

Yeast growth, in the most commonly used culture media, tends to produce relatively small changes in net charge, presumably because the major meta-

TABLE 9.1. *Detection of yeasts in RABIT with the indirect impedance technique*

Organism	Plate count (cfu/ml)	TTD (h) Wort	TTD (h) CBAT	Mean total change (μS)
Zygosaccharomyces bailii	>1.0exp4	20.5	25.9	
Zygosaccharomyces bailii	1.9exp3	31.9	31.9	
Zygosaccharomyces bailii	1.8exp2	63.3	46.5	
Rhodotorula rubra 004	6.2exp5	NT	13.3	2396
Rhodotorula rubra 004	6.2exp4	NT	17.2	2653
Rhodotorula rubra 008	5.8exp5	NT	13.6	2718
Rhodotorula rubra 008	5.8exp4	NT	16.7	2200
Rhodotorula rubra 010	3.5exp5	NT	21.3	2493
Rhodotorula rubra 010	3.5exp4	NT	24.6	1892

NT, not tested.

bolic end-product, carbon dioxide, is not ionized in solution at the pH values of most available culture media (Owens 1985). The indirect impedance technique allows the screening for and quantification of yeast populations as a function of the time for detection of measurable CO_2 production. The data presented here are taken from a number of studies carried out with European fruit juice manufacturers, who provided samples contaminated with *Zygosaccharomyces bailii* and with *Rhodotorula rubra.*

Samples (1 ml) are aseptically added to 4 ml of the respective media (wort broth (LabM) and carbon-base ammonium tartrate (CBAT; Connolly *et al.* 1988)) in a RABIT insert cell and immediately placed in the incubator module. All tests are run in duplicate according to the standard criteria but using a temperature of 30 °C and test durations from 24 to 72 h.

The technique can detect the presence of yeasts in fruit juices and avoids the need for the growth medium to be optimized with respect to production of an electrical signal (Table 9.1). It also emphasizes the importance of selecting the correct growth medium and relating this choice to both resuscitation and enumeration of the target organism. While high numbers of *Z. bailii* isolated from the fruit juice are detected more rapidly in wort broth than in the CBAT medium, the reverse is true for a product containing low numbers.

Subsequent work has shown that the time to detection (TTD) can be considerably reduced at a lower incubation temperature.

Detection of Clostridium tyrobutyricum

Clostridium tyrobutyricum is of considerable importance as a spoilage organism in high-pH cheeses. It is particularly prevalent in some cheeses from France,

TABLE 9.2. *Comparison of media for detection of* Clostridium *spp. in RABIT with the indirect technique*

	WAB	WAB + 1% SMP	WAB + neomycin
Cl. tyrobutyricum			
TTD (h)	3.7	3.9	15.9
Total change (μS)	11 334	10 302	3263
Rate of change (μS/6min)	−157	−215	−68
Cl. butyricum			
TTD (h)	6.9	5.9	>24
Total change (μS)	3432	10 720	839
Rate of change (μS/6min)	−31	−101	−8
Cl. sporogenes			
TTD (h)	5.1	5.2	16.1
Total change (μS)	9293	10 944	5321
Rate of change (μS/6min)	−62	−217	−83

Italy, Spain and Switzerland. *Cl. tyrobutyricum* ferments lactic acid to produce carbon dioxide, hydrogen and butyric acid. The indirect technique is particularly suitable for the detection of this organism, because carbon dioxide, which is the cause of 'butyric blowing', is readily detected.

Clostridium tyrobutyricum 0088 isolated from blown cheeses in Italy, *Clostridium butyricum* 0087 from an Italian dairy and *Clostridium sporogenes* NCTC 532 are grown in Whitley anaerobe broth (WAB) in a Compact anaerobic workstation (Don Whitley Scientific Limited) for 24 h at 37 °C and subsequently serially diluted in WAB to give approximately 10^5 cfu/ml. The working cultures are removed from the anaerobic workstation and used to inoculate the respective media (WAB, WAB plus 1% skimmed milk powder (SMP) and WAB plus 100 mg/ml neomycin). All experiments are carried out in triplicate with a 0.1-ml inoculum and 4.9 ml of the respective media, using the standard RABIT test criteria. The concentration of KOH in the alkaline agar bridge is increased from 0.7% to 1.0%, as preliminary work had shown this was necessary to absorb the amount of CO_2 produced by *Cl. tyrobutyricum*.

With WAB alone, *Cl. tyrobutyricum* is detected in 3.7 h while *Cl. butyricum* is detected in 7 h. The greatest difference, however, is in terms of the total change, where *Cl. tyrobutyricum* not only produces the greatest change but also the greatest rate of change (Table 9.2), which is often a useful parameter for data evaluation. Addition of SMP promotes gas production for both *Cl. butyricum* and *Cl. sporogenes*, so reducing the differential effects seen with WAB. Addition of neomycin both increases TTD and reduces gas production, when compared with WAB alone, which clearly shows that the neomycin was inhibitory to all three organisms. The most significant finding is, however, the

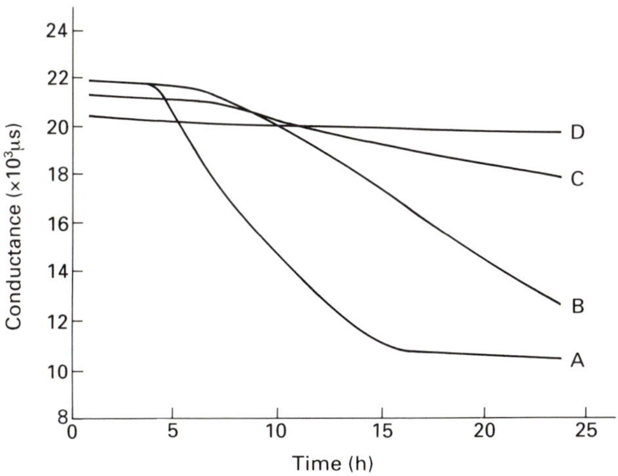

FIG. 9.2.Indirect RABIT response curves for *Clostridium* spp. grown in Whitley anaerobe broth: A, *Cl. tyrobutyricum*; B, *Cl. sporogenes*; C, *Cl. butyricum* and D, control medium blank.

shape of the *Cl. tyrobutyricum* curve (Fig. 9.2), which has a flat baseline and a sharp and characteristic inflection point. This is consistent throughout all tests and is, therefore, considered to have the potential to act as a selective indicator for the presence of *Cl. tyrobutyricum* in samples.

Simultaneous detection of coliforms and Escherichia coli

A novel formulation for the simultaneous detection of coliforms and *E. coli* has been developed. The formulation was based on the ability of coliform bacteria to ferment lactose and produce acid and carbon dioxide, indicated by a change in broth colour from red to yellow and a negative conductance change in the indirect RABIT cell. Furthermore, presence of *E. coli* in the sample was detected by its ability to cleave the substrate methylumbelliferyl-β-D-glucuronide (MUG) to yield fluorescent methylumbelliferone. Fluorescence was determined at the end of a RABIT test by exposing the conductance cell to ultraviolet light at 366 nm.

Batches of rehydrated coliform medium (Don Whitley Scientific Limited) are stored at 0–4 °C after autoclaving and discarded after 1 month. Initial experiments are carried out with 0.1-ml samples inoculated into 4.9 ml medium with the standard RABIT detection criteria. These studies show the protocol to differentiate clearly between *E. coli* ATCC 25922 and *Klebsiella aerogenes* DWS 0056, a food isolate. Total conductance changes and TTDs are reproducible and both organisms produce acid in sufficient quantity to elicit a

TABLE 9.3. *Detection and enumeration of* E. coli *in the presence of high numbers of* K. aerogenes *in a novel coliform broth with the RABIT system and indirect impedance*

Plate count (cfu/ml)				
E. coli	*K. aerogenes*	Mean TTD (h)	AP	F
3.6exp7	1.2exp5	4.0 (0.45)	3/3	3/3
3.6exp6	1.0exp5	6.5 (0.09)	3/3	3/3
3.6exp5	1.1exp5	8.4 (0.17)	3/3	3/3
3.3exp4	1.2exp5	8.7 (0.57)	3/3	3/3
3.4exp3	1.0exp5	8.7 (0.19)	3/3	3/3
3.4exp2	1.1exp5	9.5 (0.09)	3/3	3/3^{w}
3.4exp1	1.0exp5	9.3 (0.09)	3/3	0/3

AP, number of tubes showing acid production; F, number of tubes showing fluorescence; ^{w}, weak fluorescence. Standard deviation ($\pm$) of mean values is shown in parentheses.

complete colour change in the broth, from red to yellow. The distinction is seen when exposed to long-wave ultraviolet light: strong fluorescence is evident in all cells inoculated with *E. coli*, while *K. aerogenes* does not produce fluorescence at any dilutions.

Further development assessed both the relationship between TTD and the standard plate count and the ability of the broth to recover low numbers of *E. coli* in the absence and presence of *K. aerogenes*. A lower limit of 1 cfu/ml was detected in RABIT with the new broth and the relationship between TTD and plate count was similar for both the organisms used, with the lowest inocula detected in around 18 h. Furthermore, an inoculum of *E. coli* at this level still produced strong fluorescence. The efficiency of the coliform broth in recovering low numbers of *E. coli* when heavily outnumbered by *K. aerogenes* is shown in Table 9.3. Fluorescence, due solely to *E. coli*, was strong at all dilutions down to 10^{-5}, weak at 10^{-6} and absent at 10^{-7}. Taking the 10^{-6} dilution as the limit of *E. coli* detection, it can be calculated that about 30 cfu of *E. coli* were detected in the presence of 1.1×10^{4} *K. aerogenes*, in 0.1 ml of inoculum.

Screening foods for coliforms and E. coli

The above method was then used to screen frozen meat products and bean-sprouts for coliforms and *E. coli* and was compared with the traditional pour-plate method with violet red bile agar (VRBA). Portions (10 g) of sausage rolls, beefburgers and cottage pie are aseptically weighed into sterile stomacher

TABLE 9.4. *Detection of coliforms and* E. coli *in food products*

Product	Dilution	VRBA* count (cfu/ml)	TTD† (h)	Total change (μS)	AP	F
Sausage roll	10^{-1}	68	10.9	3732	3/3	1/3
	10^{-2}	<10	15.8	2192	3/3	0/3
	10^{-3}	<1	>24	1032	0/3	0/3
Beefburger	10^{-1}	87	9.6	4430	3/3	3/3
	10^{-2}	<10	12.1	3071	3/3	1/3
	10^{-3}	<1	>24	1163	0/3	1/3
Cottage pie	10^{-1}	10	11.2	3488	3/3	0/3
	10^{-2}	<10	19.9	1649	0/3	0/3
	10^{-3}	<1	>24	950	0/3	0/3
Beansprouts	10^{-1}	>1000	9.2	3325	3/3	0/3
	10^{-2}	110	12.9	3254	3/3	0/3
	10^{-3}	15	16.0	3360	3/3	0/3
	10^{-4}	<10	17.9	1881	3/3	0/3

* Results expressed as mean of two plate counts.
† Results expressed as mean of three replicates.
AP, number of tubes showing acid production; F, number of tubes showing fluorescence.

bags with 90 ml of MRD. After stomaching for 1 min the homogenate is further diluted in MRD and 1-ml volumes are withdrawn and added to 4 ml coliform broth in indirect RABIT cells. Duplicate 1-ml volumes are also transferred to sterile Petri dishes and mixed with molten VRBA. These plates are overlaid and incubated at 37 °C for 24 h. In a similar manner, the coliform broth is used to screen fresh stir-fry beansprouts.

Comparative results in the recovery of coliforms from food products are shown in Table 9.4. The highest numbers of coliforms were found in frozen beefburger, in which a VRBA coliform count of 87 cfu/ml corresponded to a mean TTD of 9.6 h with the coliform broth. Fluorescence was observed in all three cells at this dilution, indicating the presence of *E. coli* in the sample. At the dilutions of 10^{-2}, mean TTD was extended to 12 h and fluorescence was seen in only one of the three cells, suggesting the presence of very low numbers of *E. coli*, as confirmed by the standard coliform plate count. Lower numbers of coliforms were found in the other two meat products, but the count of only 10 cfu/ml in the 10^{-1} dilution of cottage pie corresponded with a mean TTD of 11.2 h. Results from the analysis of fresh beansprouts showed that the coliform count corresponded to a mean TTD of 12.9 h, with the three replicates showing good reproducibility. At the lowest dilution of 10^{-4}, a total count of 1.5×10^3 was obtained, of which less than 10 were coliforms, detected in 17.9 h. Fluorescence was not seen at any dilution, suggesting that *E. coli* was absent from the microbial flora of this product.

Collation of the results of the above experiments allowed construction of a calibration curve. The finding that coliforms and *E. coli* in food products and pure culture showed a significant correlation ($r = 0.92$) between plate count and TTD is of assistance in producing a protocol for the screening of samples for these organisms. While MUG has been used extensively in the food industry to enumerate *E. coli* (Moberg 1985), concerns have been expressed about the incidence of false negative results. Because of this, we recommend a test duration of 24 h, so that the increased incubation time will allow for expression of β-glucuronidase.

Indirect Impedance in the Study of Microbial Biofilms

Microorganisms in biofilms are 10–100-fold more resistant to biocides than when tested as a suspension (Holah *et al.* 1990; Keevil *et al.* 1990). This apparent resistance is probably due to the slow diffusion of the biocide through the polymeric matrix in which the adherent bacteria are embedded. Methods of assessing biocide activity against biofilms are, therefore, required for the determination of effective hygiene standards in the food industry. The indirect impedance technique offers the facility rapidly to determine the microbial activity in biofilms *in situ*, because the biofilm can be placed in the impedance cell. The nature of the inanimate surface, whether glass, plastic or metal, etc., is irrelevant, so offering a versatile method for studying biofilms and biocides. The antimicrobial effect of a biocide depends upon various extrinsic factors, such as biocide concentration, temperature, pH and the presence of inactivating substances. The effect of concentration on biocide efficacy depends on the biocide in question. It is normally defined as the gradient of the plot of the $\log_{10}$ of bacterial number surviving against the $\log_{10}$ of biocide concentration, and is referred to as the concentration exponent or dilution coefficient, n. The relationship, however, depends on the uniform mixing of the biocide with the microorganism, which may not be the case with biofilms (Keevil *et al.* 1990).

Method

Microbial biofilms are produced on 1-cm stainless steel discs (1.57 cm^2) as a simple model for demonstrating the application of indirect impedance. Since there is, at present, no standard protocol for assessing biocide activity against biofilms, the European suspension test (EST; Anon. 1987) is used, with appropriate modifications. Sodium hypochlorite is used as the model for chlorine-based disinfectants. Sodium thiosulphate (5 g/l) is used as the inactivation solution, as recommended in the EST, despite reports that *Staphylococcus aureus* is sensitive to it (Bloomfield 1990). Since many biocides, especially sodium hypochlorite, are inactivated by organic matter, bovine serum albumin (BSA) is incorporated as the internal organic matter standard. As recommended in the EST, a concentration of 0.03% w/v BSA is used to represent a 'clean' sample and 0.1% w/v an 'unclean' sample.

Biofilm formation

Fresh tryptone soya broth (TSB) (20 ml) is inoculated with an overnight culture of *Pseudomonas aeruginosa* NCIB 10421, *Proteus mirabilis* NCIB 12596, *Staphylococcus aureus* NCTC 10788 and *Saccharomyces cerevisiae* ATCC 9763. The inoculated medium is poured into a sterile petri dish containing the designated number of steel discs and incubated at 37 °C. It is necessary to ensure that the discs do not overlap. After overnight incubation, the growth medium is discarded and the discs are washed three times in 10 ml sterile saline. The model biofilms are stored in sterile saline until required, but for not more than 30 min.

Biocide activity assessment

The discs are immersed in 1.5 ml BSA at a concentration of 0.03% or 1% for clean or unclean conditions respectively. The latter is diluted to 0.03% by adding 0.5 ml stock plus 1 ml sterile distilled water. Appropriate double-strength biocide (1.5 ml) is added for a known time interval, for example 5 min for EST. Control discs are prepared by adding 1.5 ml water of standard hardness. Exposure to biocide is terminated by carefully decanting the biocide solution and replacing with 5 ml inactivation solution. This is shaken gently for 1 min, carefully decanted and replaced with a further 5 ml of inactivation solution. The discs are gently transferred to the inner RABIT cell, containing 2 ml of appropriate growth medium, and tests are set up according to the standard RABIT detection criteria. The surviving biofilm viability is determined according to appropriate impedance calibration curves, constructed from decimal dilutions from 10^{-2} to 10^{-8} of respective cultures.

Required cell density for 'time to detection'

The cell density required to produce sufficient carbon dioxide for the TTD is determined by removing 1 ml of the inner tube contents as soon as the TTD is recorded by the RABIT system. The cell suspension is subsequently diluted in sterile saline from 10^{-1} to 10^{-8}, followed by inoculation of TSA plates in triplicate.

Enumeration of biofilm cell numbers by plate count

The biofilm disc is placed in a universal container containing 10 ml sodium phosphate buffer (50 mmol/l, pH 7) and 2 g sand is added. The mixture is shaken on a wrist-action shaker for 5 min and diluted 10-fold, after which TSA plates are inoculated in triplicate with the diluted cells. The cell count is expressed as colony-forming units per disc (cfu/disc).

Biofilms were produced by the three bacterial strains tested but not the yeast *S. cerevisiae*. This was in agreement with the results of Holah *et al.* (1990) and is probably due to the absence of extracellular polysaccharide production by *S. cerevisiae*. *Staph. aureus* produced the most dense biofilm, followed by *Ps. aeruginosa* and *P. mirabilis*.

In spite of the higher-cell-density biofilm produced by *Staph. aureus*, the detection time was longer than that of the other less dense biofilms (Table 9.5). This is due to differences in carbon dioxide production rate and demonstrates that the growth medium influences the TTD, because shorter detection times were obtained with TSB in the inner tube than is the case with WIB. This was not surprising, because WIB was originally designed for specific use in the direct impedance technique. The cell densities required to cause TTD showed that a cell density of $4-10 \times 10^8$ cfu/ml was necessary to produce sufficient carbon dioxide for a detectable impedance change. Since the cell densities were similar, regardless of whether the inoculum was a suspension or a biofilm immobilized on disc, the use of a standard calibration curve generated by suspended cells was justifiable. In an extreme case, it is considered that certain organisms may not be amenable to indirect impedance, because insufficient carbon dioxide is produced. This possibility should be investigated when studying environmental biofilms by isolating individual

TABLE 9.5. *Comparison of biofilm cell densities and detection times for three European suspension test bacterial strains*

Organism	Plate count (cfu/disc)*	Detection time (h)	
		WIB	TSB
Ps. aeruginosa	5.0×10^6	4.0	4.5
P. mirabilis	5.0×10^6	4.5	3.2
Staph. aureus	2.0×10^7	6.5	5.0

* Disc area equals 1.57 cm^2.

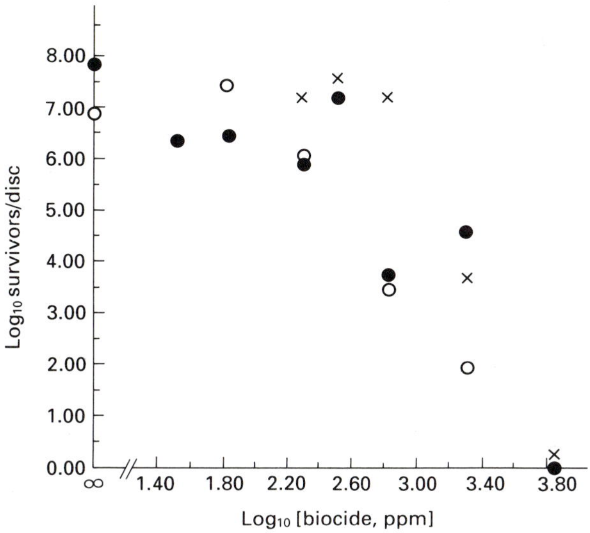

FIG. 9.3.●, Survival of *Ps. aeruginosa*; ○, *P. mirabilis* and ×, *Staph. aureus* at different sodium hypochlorite concentrations. Data points are averages of three biofilm discs. Standard errors were less than 0.2 units.

members of the consortium and testing for carbon dioxide production in TSB, or some other suitable medium.

Relationship between biocide concentration and survivor number

Comparison of the three EST bacterial strains showed that *Ps. aeruginosa* was the most resistant to sodium hypochlorite, with a dilution coefficient, that is, the ratio of $\log_{10}$ surviving population to $\log_{10}$ sodium hypochlorite concentration, of −2.4 as compared with −3.7 for *Staph. aureus* and −4.8 for *P. mirabilis*. The survival rates for exposure to different concentrations of sodium hypochlorite are shown in Fig. 9.3. Timed exposure to 370 ppm sodium hypochlorite gave death indices of −0.11, −0.25 and −0.25 $\log_{10}$ survivors/disc/min exposure for *Ps. aeruginosa*, *P. mirabilis* and *Staph. aureus* respectively (see Fig. 9.4 for *Ps. aeruginosa* and *Staph. aureus* plots). The EST requires the biocide to reduce the microbial number of five strains by 5 logs_{10} in 5 min (5–5–5 test; Mossel 1963). As reported by Holah *et al.* (1990), *Ps. aeruginosa*, *P. mirabilis* and *Staph. aureus* are 10- to 100-fold more resistant to biocides when grown as biofilms, as compared with suspensions. Data obtained for biocide efficiency against microorganisms in suspension should, therefore, not be used where biofilm formation has taken place.

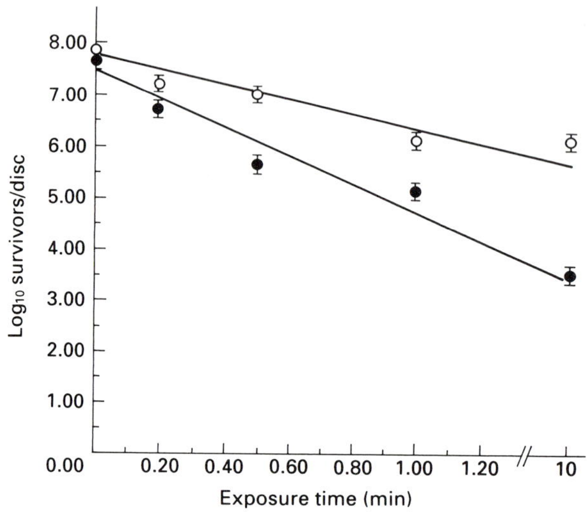

FIG. 9.4.○, Survival of *Ps. aeruginosa* and ●, *Staph. aureus* after timed exposure to 370 ppm sodium hypochlorite.

Skimmed milk powder biofilm

A 2-day biofilm of organisms from skimmed milk powder, resuspended to 20% w/v at 25 °C, was analysed by indirect impedance. The cell density in the biofilm was 4×10^5 cfu/disc (equivalent to 2×10^5 cfu/cm^2) on TSA after mechanical removal of the cells with sand and agitation. Seven colony types were isolated for further study. The biofilm gave a TTD of approximately 5 h. After treatment with 1% sodium hypochlorite for 5 min, the TTD was increased to almost 7 h and no viable cells were detected (<500 cells/disc). Of the seven colony types isolated, only three gave positive conductance changes by the direct impedance method with WIB as the detection medium. This demonstrates the usefulness of indirect impedance for measuring the bacterial activity of a biofilm.

Justification of Indirect Impedance

Indirect impedance offers a non-disruptive means for measuring microbial activity in a biofilm, irrespective of the support material. The microbial activity is measured by the production of carbon dioxide and is, therefore, applicable to most microorganisms. Since impedance measurement is semi-automated, large numbers of samples can be analysed and the results stored for subsequent data retrieval and analysis. Although direct impedance can be used in many circumstances, the indirect method offers a number of advantages. First, not all microbes produce well-defined impedance curves. Notably in this study, *Ps. aeruginosa* NCIB 10421, which is the specified strain for the European suspension test protocol, gave poor direct impedance curves as compared with indirect responses. Second, a reason for considering the indirect technique is that the organisms in the biofilm need not be subjected to mechanical removal, that is, sand plus agitation, which leads to increased sampling errors due to treatments that are not reproducible, uneven biofilm disruption with surviving cell clumps, and difficulty in ensuring complete cell removal. Because indirect impedance measures the production of carbon dioxide from a microflora, there are many other biofilms that can be studied. Complex biofilms amenable to indirect impedance include dental plaques, catheters, contact lenses and microalgal biofouling of water conduits.

Acknowledgements

The technical support from Luci Mansfield and Richard Beadham is gratefully acknowledged for the biofilm work and from Gillian Iredale and Andrew Pridmore for the other studies. Thanks are also given to the fruit juice manufacturers who supplied contaminated samples, various companies for providing the frozen food samples and Gibco Life Sciences for provision of media.

References

ANON. 1987 *Method of Test for the Antimicrobial Activity of Disinfectants in Food Hygiene*. Strasbourg: Council of Europe.

BLOOMFIELD, S.F. 1990 Mechanisms of action of chemical biocides. *Methods for Assessing Antimicrobial Activity*, Society for Applied Bacteriology, Technical Series 27, eds Denyer, S.P. & Hugo, W.B., pp. 77–83. Oxford: Blackwell Scientific Publications.

BOLTON, F.J. 1990 An investigation of indirect conductimetry for detection of some food-borne bacteria. *Journal of Applied Bacteriology* **69**, 655–661.

BRIDSON, E.Y. & BRECKER, A. 1970 Design and formulation of microbial culture media. *Methods in Microbiology*, Vol. 3A, eds Norris, J.R. & Ribbons, D.W., pp. 229–295. London: Academic Press.

CONNOLLY, P., LEWIS, S.J. & CORRY, J.E.L. 1988 A medium for the detection of yeasts using a conductimetric method. *International Journal of Food Microbiology* **7**, 31–40.

HOLAH, J.T., HIGGS, C., ROBINSON, S., WORTHINGTON, D. & SPENCELEY, H. 1990 A conductance based surface disinfection test for food hygiene. *Letters in Applied Microbiology* **11**, 255–259.

HUGUET, J.M. & RIBAS, F. 1991 SGAP-10C agar for the isolation and quantification of *Aeromonas* from water. *Journal of Applied Bacteriology* **70**, 81–88.

KEEVIL, C.W., MACKERNESS, C.W. & COLBOURNE, J.S. 1990 Biocide treatment of biofilms. *International Biodegradation* **26**, 169–179.

MOBERG, L.J. 1985 Fluorogenic assay for rapid detection of *Escherichia coli* in food. *Applied and Environmental Microbiology* **50**, 1383–1387.

MOSSEL, D.A.A. 1963 The rapid evaluation of disinfectants for use in the food processing plants. *Laboratory Practice* **12**, 898–899.

OWENS, J.D. 1985 Formulation of culture media for conductimetric assays: theoretical considerations. *Journal of General Microbiology* **131**, 3055–3076.

OWENS, J.D., THOMAS, D.S., THOMPSON, P.S. & TIMMERMAN, J.W. 1989 Indirect conductimetry: a novel approach to the conductimetric enumeration of microbial populations. *Letters in Applied Microbiology* **9**, 245–249.

PAYNE, J.W. 1976 Peptides and microorganisms. *Advances in Microbial Physiology* **13**, 55–113.

10: Monoclonal Antibody-based ELISAS

H.J.BALL, S.KERR AND D.P.MACKIE
Veterinary Sciences Division, Department of Agriculture for Northern Ireland, Stoney Road, Stormont, Belfast BT4 3SD, UK

Monoclonal antibody (MAb) and enzyme-linked immunosorbent assay (ELISA) techniques have provided a new diagnostic approach in microbiology. ELISA for the detection of microbial antigens takes various forms but most commonly involves their capture from samples in MAb-coated microtitre wells. The captured antigen is detected with another layer of specific antibody, with or without various amplification steps. One of these secondary reagents is conjugated to an enzyme, and substrate digestion is the final demonstration of the presence of an antigen. The technique is known as antigen-capture or sandwich ELISA.

In bacteriology, sandwich ELISAs have been developed for the detection of defined virulence factors such as toxins, and for the detection of specific bacteria such as *Salmonella* and *Listeria* (reviewed by Notermans & Wernars 1991). Affinity-purified polyclonal antiserum (Prusak-Sochaczewski & Luong 1989) and MAb (Mattingly 1984) have been used in these ELISAs for the detection of *Salmonella*. Such studies have been directed towards the development of genus-specific assays (Choi *et al.* 1992) and assays with a narrower specificity for the detection of serovars of greater importance in food poisoning, such as *Salmonella typhimurium* (Lee *et al.* 1990) and *Salmonella enteritidis* (Brigmon *et al.* 1992).

Sandwich ELISAs for the detection of *Salmonella* in food samples are now commercially available. These include Tecra (Bioenterprises Ltd, Roseville, NSW, Australia), Locate (Rhone-Poulenc, Glasgow, UK) *Salmonella* TEK (Organon Teknika Corp., NC, USA) and one distributed by Kirkegaard Perry Laboratories Inc. (MD, USA). These kits use affinity-purified polyclonal or monoclonal antibodies and have a similar sensitivity. None is completely *Salmonella*-specific. Bioenterprises Ltd have also recently introduced a novel *Salmonella* kit (Immunocapture) involving capture of *Salmonella* from preenriched test samples on dipsticks, which are then transferred to a second

New Techniques in
Food and Beverage Microbiology

0-632-03755-5

enrichment medium. After incubation, the second-medium samples are tested by sandwich ELISA.

The advantages of MAb ELISA in terms of specificity are well recognized, as is the potential for automation. Although the sensitivity of sandwich ELISA for toxins is lower than that of cell culture and animal toxicity tests, it is generally adequate for the diagnosis of pathologically significant concentrations. For the detection of low numbers of whole organisms in food, however, they are not sufficiently sensitive without preenrichment. Thus these tests cannot, at present, replace culture methods, but they yield results at an earlier stage of a culture protocol.

This chapter describes the development of genus-specific MAbs for *Salmonella*, their use in a sandwich ELISA and the application of sandwich ELISAs for the detection of subclinical bovine mastitis and *Escherichia coli* verotoxin 1.

Salmonella Sandwich ELISA

Selection of antigen

The Enterobacteriaceae have variable numbers of shared antigens (Brenner 1984). The lipopolysaccharide (LPS) somatic, capsular and flagellar antigens allow the genera to be divided into numerous serovars. For organisms such as *Salmonella*, the search for genus-specific antigens amongst the somatic and flagellar antigens is, in many ways, a reversal of the emphasis on subdivision of pathogenic species into strains, largely for epidemiological purposes.

Since the outer membrane proteins (OMP) of these organisms are immunologically less variable, and probably more conserved genetically, than the LPS antigens and, because they are exposed on the outer cell surface, they have potential as detectable genus-specific antigens. The outer membrane (OM) can be purified by selective solubilization of the inner membrane after sonic disruption of washed concentrated cells. The greater part of the immunodominant LPS fraction of the OM is removed by dialysis after solubilization in the chaotropic agent guanidine thiocyanate (Moldow *et al.* 1972).

Solubilization of outer membrane proteins

Reagents

The reagents used are as follows:

1 Standard buffer: 10 mol/l Tris (1.21 g/l), 5 mol/l EDTA (1.86 g/l), pH 7.8.

2 Sarcosine: 2% sodium *N*-lauroylsarcosine in standard buffer.

3 Guanidine thiocyanate (6 mol/l) in standard buffer (7.09 g/10 ml).
4 Urea (6 mol/l) in standard buffer (360 g/l).

Method

1 Prepare 50–100-fold concentrated, washed cells from an overnight broth culture in standard buffer containing 0.1 mmol/l of the protease inhibitor phenylmethylsulfonyl fluoride.
2 Cool in crushed ice while sonicating for 4 × 1-min pulses at 20 kilocycles/s with 1-min intervals between pulses.
3 Centrifuge the sonicate at 15 000 *g* to sediment intact cells.
4 Remove supernatant, mix with a quarter-volume of 2% sarcosine, and leave at room temperature for 30 min.
5 Centrifuge the detergent-treated supernatant at 300 000 *g* for 1 h.
6 Resuspend the pellet in standard buffer, mix with an equal volume of 2% sarcosine solution and leave for 1 h at room temperature.
7 Centrifuge at 300 000 *g* for 1 h and wash the pellet by resuspending in standard buffer and centrifuging again at 300 000 *g*.
8 Resuspend the washed pellet of OM in 6 mol/l guanidine thiocyanate solution, and leave at room temperature for 1 h.
9 Dialyse the solubilized OMP against 100 volumes of 6 mol/l urea.
10 Centrifuge the dialysate at 300 000 *g* for 1 h to remove insoluble material.

The largely hydrophobic OMP remains soluble in 6 mol/l urea and provides an antigen preparation for immunization of BALB/c mice and for coating microtitre trays for ELISA. Further fractionation of these solubilized proteins by standard methods can improve the purity of selected antigens, but the immunogenic potential of OMPs is largely unknown.

Hybridoma selection

MAb can be selected by screening of the hybridoma supernatants by ELISA with a variety of OMP preparations. To detect a *Salmonella*-specific MAb, hybridomas must be screened with OMP preparations from more than one *Salmonella* somatic antigen group, and also from closely related genera such as *Citrobacter*, *Escherichia*, *Klebsiella* and *Serratia*. The OMP for use in screening hybridomas must first be titrated to obtain the optimal coating dilution on microtitre trays. Serum from the immunized BALB/c mouse can be used for this standardization.

Method

The ELISA reagents are listed in Table 10.1. All reagents are added at 100 μl per well except the H_2SO_4 stopper of the substrate reaction, which is added at 50 μl per well.

TABLE 10.1. *ELISA reagents*

Reagent	Ingredients*	Final composition*
Coating buffer (0.05 mol/l carbonate buffer)	Solution A: 2.1 g Na_2CO_3 made up to 100 ml Solution B: 8.4 g $NaHCO_3$ made up to 500 ml	10 ml A + 40 ml B, pH adjusted to 9.5 and made up to 200 ml
PBS (0.01 mol/l phosphate-buffered saline)	Solution A: 4.8 g NaH_2PO_4 made up to 200 ml Solution B: 14.2 g Na_2HPO_4 made up to 500 ml NaCl	27 ml A + 73 ml B + 17 g NaCl, pH adjusted to 7.2 and made up to 2000 ml
Wash fluid	Solution A and B of PBS, NaCl, Tween 20	2000 ml PBS + 5 ml 20% (v/v) Tween 20
PTN (phosphate buffer, Tween and NaCl)	Solution A and B of PBS NaCl Tween 80	200 ml PBS + 4 g NaCl + 400 µl 20% (v/v) Tween 80
Substrate buffer (0.1 mol/l citrate/phosphate buffer)	Citric acid: 1.92 g made up to 100 ml Solution B of PBS	24 ml citric acid solution + 26 ml B, pH adjusted to 5.0 and made up to 100 ml
Substrate	Substrate buffer 10 mg 3,3′,5,5′-tetramethyl-benzidine (TMB) in 1 ml dimethyl sulphoxide (DMS) H_2O_2	10 ml substrate buffer + 100 µl TMB/DMS + 10 µl H_2O_2

* Solutions in distilled water unless otherwise stated.

1 Coat microtitre wells with doubling dilutions of the OMP solution in coating buffer at 4 °C overnight or at 37 °C for 1 h.
2 Wash with six changes of wash buffer.
3 Add a 1:1000 dilution of mouse serum in PTN (phosphate buffer, Tween and NaCl), and incubate at 37 °C for 1 h.
4 Wash with six changes of wash buffer.
5 Add a dilution of goat antimouse peroxidase conjugate in PTN. The optimum dilution of the conjugate is calculated by titration. Incubate at 37 °C for 1 h.
6 Wash with six changes of wash buffer.
7 Add substrate and incubate at 37 °C for 10–20 min.
8 Stop the reaction by addition of 2.5 mol/l H_2SO_4, and read the reaction on an ELISA reader.
9 Controls: omit mouse antiserum to detect non-specific reactions between antigen and goat antimouse conjugate and omit coating antigen to detect non-specific coating by mouse antisera.

The dilution of OMP that gives the highest reading in comparison with the controls is selected for hybridoma screening. The initial screening should include OMP from a *Salmonella* serovar different from that used in the immunization, and from at least two representatives of closely related genera. Selected hybridomas should then be tested against a wide range of *Salmonella* and other enterobacterial OMPs. Hybridoma selection is based on the highest ELISA titre against *Salmonella* antigen, with the lowest cross-reactivity with the other enterobacterial antigens. The most promising hybridomas are cloned twice by limiting dilution and used to produce MAbs in the ascites and/or dialysis sac fluid systems (Sjorgren-Jansson & Jeansson 1985).

Purification of MAb

Before use as a sandwich ELISA reagent, IgG in ascites must be purified in order to limit the non-specific effects of the other serum proteins. Caprylic acid precipitation of extraneous serum proteins is a simple, effective method of purification (McKinney & Parkinson 1987).

Reagents

The reagents used are as follows:
1 Acetate buffer (0.06 mol/l): 0.817 g sodium acetate solution in distilled water, adjusted to pH 4.3 with 20% glacial acetic acid and made up to 100 ml.
2 Caprylic acid.
3 PBS (0.01 mol/l), pH 7.2.

Method

1 Add 2 volumes of acetate buffer slowly (1–2 min) to 1 volume of ascites fluid while stirring and, if necessary, adjust pH to 4.6 with 0.1 mol/l NaOH.
2 Add 25 μl caprylic acid per ml of diluted ascites fluid slowly over 5 min, while stirring rapidly.
3 Leave to stir more slowly for a further 30 min.
4 Centrifuge at 10 000 *g* for 30 min and discard precipitate. Ignore floating material, which normally settles out after dialysis.
5 Dialyse overnight in 0.01 mol/l PBS (pH 7.2). Discard any precipitate left after centrifugation at 5000 *g* for 15 min.

The caprylic acid-purified IgG can, if necessary, be concentrated by precipitation by the dropwise addition of an equal volume of saturated ammonium sulphate solution in chilled conditions. The purified MAb is used to coat microtitre plate wells to capture specific antigens (Fig. 10.1). After incubation of test samples, capture is demonstrated by the addition of another, biotinylated, MAb (Hofmann *et al.* 1982). The biotin is the link to the next reagent, streptavidin (Sigma), which is conjugated to an enzyme. With a chromogenic substrate, the enzyme produces a measurable reaction that is quantitatively related to the captured antigen.

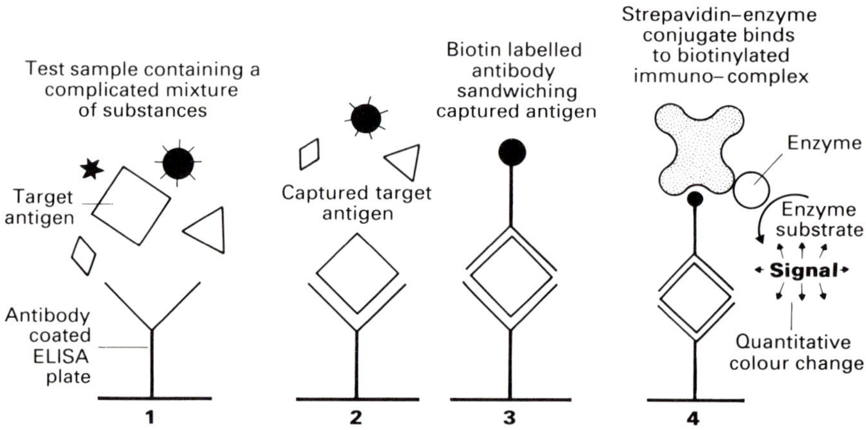

Fig. 10.1. Schematic representation of sandwich or antigen-capture ELISA.

Biotinylation

Reagents

The reagents used are as follows:
1 Carbonate/bicarbonate buffer (0.1 mol/l), pH 8.4.
2 Biotin: aminohexanoyl-biotin-*N*-hydroxysuccinimide in dimethyl formamide (004302, Cambridge Bioscience, Cambridge, UK), 10 mg/ml (w/v).
3 PBS (0.01 mol/l), pH 7.2 (Table 10.1).

Method

1 Dialyse the caprylic acid-purified IgG against 0.1 mol/l carbonate/bicarbonate buffer overnight.
2 Measure the protein level of the solution spectrophotometrically at 280 nm and adjust to between 5 and 10 mg/ml with carbonate/bicarbonate buffer.
3 Add (v/v) 1 part biotin to 25 parts antibody solution. If the eventual test sensitivity is poor, a range of biotin:antibody ratios between 1:5 and 1:50 should be used. Excessive biotinylation may result in a high level of non-specific binding of the biotinylated antibody. Low biotinylation may give reduced sensitivity because of the presence of non-biotinylated antibody, which competes for epitopes (Gretch *et al.* 1987).
4 Place on mixing wheel for 1 h at room temperature.
5 Dialyse against two changes of 100 volumes of 0.1 mol/l PBS over at least 24 h. Biotin is slow to dialyse.

ELISA procedure

Optimization of the reagents

MAbs are stored at −70 °C. The purified reagents can withstand freezing and thawing several times without loss of titre, but are best stored in small volumes. The biotinylated MAb is conveniently stored at 4 °C for several months diluted in PBS with 1% bovine serum albumin and a preservative (0.02% sodium azide).
1 Carry out a chequerboard titration in which the coating MAb is diluted in one direction, against the biotinylated MAb in the other (see under 'Method below'). MAb dilutions should start from 1:500.
2 Dilutions of an antigen reactive in hybridoma screening should be used as capture antigen, and a different microtitre plate used for each antigen dilution.
3 Reconstitute 1 mg streptavidin peroxidase (Sigma S5512) in 5 ml distilled water and store in small volumes at −70 °C. Use at a dilution of 1:2000 until

the optimum MAb capture and biotinylated reagent dilutions have been established, and then titrate further to establish the optimum streptavidin peroxidase dilution.

4 Controls must be included to detect non-specific activity of the reagents. Essential controls are the omission of coating MAb, coating MAb plus test sample and test sample alone. When a reagent is omitted, a corresponding volume of PTN is added in its place.

Conditions that give maximum sensitivity with minimum background are selected for further use.

Method

The ELISA reagents are listed in Table 10.1. All microtitre well additions are 100 µl, except for the final H_2SO_4 addition, which is 50 µl.

A 96-well microtitre plate can be used to test 46 broth samples in duplicate. The four remaining wells can be used for duplicate negative and positive controls that consist of an uninoculated broth and a *Salmonella* culture respectively. Before testing, care must be taken to ensure that the test sample is sterile.

1 Coat microtitre wells with optimal dilution of the purified MAb in 0.05 mol/l carbonate buffer (pH 9.5) at 37 °C for 1 h or at 4 °C for overnight.

2 Wash with six changes of wash fluid.

3 Add test sample and incubate at 37 °C for 1–2 h.

4 Wash with six changes of wash fluid.

5 Add optimal dilution of biotinylated MAb in PTN and incubate at 37 °C for 1 h.

6 Wash with six changes of wash fluid.

7 Add optimal dilution of streptavidin peroxidase in PTN and incubate at 37 °C for 1 h.

8 Wash with six changes of wash fluid.

9 Add substrate and incubate at 37 °C for 10–20 min.

10 Stop the reaction with 2.5 mol/l H_2SO_4.

11 Read the reaction on an ELISA reader.

Absorbance values more than three times the average negative control are regarded as positive.

Ascites fluid contains serum proteins, including host immunoglobulins, some of which may cross-react with enterobacterial antigens. To test samples liable to contain such antigens, it is advisable to use MAb from cells grown in dialysis sac culture (Sjorgren-Jansson & Jeansson 1985) for at least one side of the sandwich.

Polyclonal serum is often used as the second antibody reagent in sandwich ELISA. Since this contains antibody to more than the single antigen epitope that attaches to the MAb, it can increase the sensitivity of the assay. Polyclonal sera, however, often display non-specific activity in ELISA, which results in higher negative control background. In addition, the presence of antibody to enterobacterial antigen in normal serum and the recognized cross-reactivity amongst the Enterobacteriaceae makes the use of polyclonal sera in the *Salmonella* sandwich ELISA inadvisable.

In a test such as the *Salmonella* ELISA, where whole cells or fragments of OM are captured, the same MAb can be used on both sides of the sandwich. Theoretically, it should be possible to enhance sensitivity by using MAbs to different epitopes on either side of the sandwich, but the optimum sensitivity of different combinations can only be established by experiment. The end result depends on a number of factors, including the size of the captured molecule and the number of epitopes.

Interpretation

Several *Salmonella* sandwich ELISA surveys have been carried out (Table 10.2) with MAbs to OMPs developed by the methods described (Kerr *et al.* 1992). The most promising preenrichment method for veterinary diagnostic samples, for testing by ELISA, appeared to be 24-h culture in selenite broth. Although false negative ELISA results in relation to culture were obtained, these were from tissue samples obtained from individual animals in which *Salmonella* had been detected by ELISA of other samples.

False negative results were also obtained from samples expected to contain low levels of *Salmonella* (e.g. environmental swabs and poultry meat samples (see Table 10.2). Improved preculture methods might enhance the *Salmonella* detection in such samples.

The false positive ELISA results investigated were all due to *Enterobacter* spp. This genus is distantly related to *Salmonella* and had not been used for the original hybridoma screening. This illustrates the complex nature of enterobacterial interrelationships.

Several *Salmonella* serovars from various somatic O-groups, including *S. arizonae*, were detected in these surveys, indicating the largely *Salmonella*-specific nature of the detected antigen. So far, 86 different *Salmonella* serovars have been detected by the sandwich ELISA and only one, a strain of *S. harmelen*, has been missed.

Sandwich ELISA for Detection of Subclinical Mastitis

A sandwich ELISA to demonstrate a host antigen associated with an inflammatory response in milk has been used to detect infectious subclinical mastitis in cattle (Ball *et al.* 1991a). The MAb used in this test came from a mouse

TABLE 10.2. *Summary of results obtained with monoclonal antibody sandwich ELISA to detect* Salmonella

Experimental samples	No.	Preculture	Salmonella culture-positive	False positives in relation to culture	False negatives in relation to culture
1 Veterinary diagnostic	2100	Selenite broth (24 h)	58 (2.8%)	65 (3%)	3 (0.14%)
2 Veterinary diagnostic	1186	Selenite broth (24 h)	7 (0.6%)	88 (7.4%)	1 (0.08%)
3 Veterinary diagnostic	284	Selenite broth (24 h)	80 (28%)	11 (3.8%)	32 (11.3%)
4 Veterinary diagnostic	250	Selenite broth (24 h)	8 (3.2%)	0	0
5 Veterinary diagnostic	840	Peptone broth (24 h)	82 (9.7%)	85 (10%)	39 (4.6%)
6 Medical diagnostic	870	Selenite broth (24 h)	11 (1.3%)	108 (12.4%)	5 (0.6%)
7 Food diagnostic	487	Peptone broth (24 h) followed by impedance media (24 h)	47 (9.6%)	21 (4.3%)	13 (2.6%)
8 Mixed*	286	Peptone broth (24 h) followed by selenite broth (24 h)	10 (3.5%)	18 (6.3%)	6 (2.1%)
9 Mixed*	437	Peptone broth (24 h) followed by Rappaport–Vassiliadis broth (24 h)	13 (3%)	2 (0.5%)	9 (2%)
10 Mixed*	71	Peptone broth (24 h)	2 (2.8%)	3 (4.2%)	0

* Mixed poultry-associated samples, i.e. cloacal swabs, meal ingredients, environmental swabs and poultry meat.

immunized with a double-immunodiffusion immunoprecipitin line developed between *Staphylococcus aureus*-infected mastitic milk and mouse hyperimmune serum to whole cells of *Staph. aureus*. The antigen detected has been characterized to only a limited extent and has a molecular mass of *c.* 23.5 kDa.

Although such a test is not pathogen-specific, since antigen is detected only in active bacterial infection it has the advantage of distinguishing between infectious and non-infectious mastitis (Ball *et al.* 1991b).

Test protocol

The same MAb is used for both the capture and the biotinylated stages of the sandwich ELISA, and milk samples are tested directly without pretreatment.

Interpretation

Ten herd surveys were carried out with the sandwich ELISA to detect the inflammatory response antigen in milk (Table 10.3). The results were analysed in comparison with the milk cell count (MCC), a standard means of grading milk quality and detecting subclinical mastitis. The positive results of the MCC and ELISA were similar for each herd. Bacteriological results were not included in the analysis of the results, because the microbiological quality of milk samples in these surveys was variable. A test to screen the samples before bacterial culture would reduce this problem. Since the numbers of infected

TABLE 10.3. *Summary of results obtained with monoclonal antibody sandwich ELISA to detect subclinical mastitis in ten herds*

Herd no.	No. of samples	No. with MCCs* >400 000 cells/ml	No. of ELISA positives†	False positive in relation to MCC	False negative in relation to MCC
1	109	19 (17.5%)	19 (17.5%)	6 (5.5%)	6 (5.5%)
2	132	32 (24.2%)	26 (19.7%)	5 (3.8%)	11 (8.3%)
3	158	20 (12.6%)	27 (17.1%)	18 (11.4%)	11 (7%)
4	113	34 (30%)	38 (33.6%)	8 (7%)	4 (3.5%)
5	64	17 (26.6%)	21 (32.8%)	5 (7.8%)	1 (1.6%)
6	83	16 (19.3%)	14 (16.9%)	1 (1.2%)	3 (3.6%)
7	56	15 (26.8%)	9 (16%)	1 (1.8%)	7 (12.5%)
8	93	7 (7.5%)	5 (5.4%)	0	2 (2.1%)
9	77	30 (39%)	32 (41.5%)	6 (7.8%)	4 (5.2%)
10	75	21 (28%)	20 (26.7%)	1 (1.3%)	2 (2.7%)
Totals	960	211 (22%)	211 (22%)	51 (5.3%)	51 (5.3%)

* Milk cell count.

† ELISA levels greater than 0.100 absorption units were considered positive.

cows in many herds is low, a screening test would also substantially reduce the bacteriological workload. The MCC can be used as a screening test but only the more expensive Coulter counters can carry out counts on fresh milk. In the case of Coulter counters that require overnight fixation of the milk cells, the bacterial culture is most conveniently carried out on milk samples before addition of the fixative.

The sandwich ELISA can be carried out on fresh milk, and the ELISA antigen appears to be a better indicator of active infection than the MCC. In a study with experimentally infected animals, analysis of the MCC and ELISA results, in relation to bacterial isolation, demonstrated that, after the resolution of infection, ELISA levels returned to normal more rapidly than the MCC (Ball *et al.* 1991b). In the same study, it was shown that fewer false negative results were obtained with the ELISA than with the MCC.

Sandwich ELISA for Verotoxin 1

Several sandwich ELISAs have been developed for bacterial toxins. One method for the detection of *Escherichia coli* verotoxin 1 (VT1) used a combination of MAb for capture and polyclonal antibody for development (Downes *et al.* 1989). An alternative method used the natural VT1 cell receptor, glycolipid globotriosyl ceramide, for capture and MAb or polyclonal antibody for detection (Ashkenazi & Cleary 1989; Basta *et al.* 1989). The disadvantages of polyclonal sera in sandwich ELISAs have already been noted, and the VT1 cell receptor, although commercially available, is expensive.

A MAb hybridoma (cell line 13C4; Strockbine *et al.* 1985) against an epitope on the B subunit of VT1 is available from the American Type Culture Collection (Rockville, MD, USA). The B subunit is the non-enzymic binding part of the toxin. It is an oligomer of five identical units (Stein *et al.* 1992), each of which contains the 13C4 epitope, which makes it possible to use the MAb on either side of a sandwich ELISA.

Test protocol

The same MAb is used on both sides of the sandwich ELISA. Clinical enteritis samples can be tested directly or after overnight enrichment on solid media. Bacteria from a colony sweep into polymyxin solution (2 mg/ml) and, incubated for 30 min at 37 °C, release more toxin than untreated cells.

Interpretation

The VT1 sandwich ELISA was used to test animal enteritis samples directly and after 24 h culture on agar medium (Table 10.4). Only six of 14 VT1-

TABLE 10.4. *Summary of results obtained with monoclonal antibody sandwich ELISA to detect* E. coli *VTI toxin*

No. of samples	VTI detected in: Faeces only	Faeces and culture	Culture only	Total VTI detected
240	3	6	5	14

positive reactions occurred with both test samples. This probably indicates the presence of low levels of VT1-positive strains in the faeces after antibiotic treatment, natural recovery from infection, or enzymatic degredation of the toxin. Alternatively, VT1-negative strains may grow better during enrichment. Although direct demonstration of toxin in clinical samples provides earlier diagnosis, and may suggest a greater pathological significance for the toxin, the results of this small survey indicate that the testing of both direct and enriched samples gives the best result.

Conclusions

Monoclonal antibody-based sandwich ELISAs are sensitive, specific and cost-effective. They therefore readily lend themselves to the examination of large numbers of samples on a routine basis. Another important consideration of the tests described above is that they detect specific antigens that are indicative of a disease or a potential disease state.

With the emphasis on quality control testing for potential human pathogens in the food industry, this technique should find an application, provided that the test is sufficiently sensitive and gives an earlier result than conventional culture. The time taken to obtain results from the *Salmonella* sandwich ELISA described depends largely on the preenrichment method, and the optimum culture protocol is liable to vary for different types of samples. For veterinary diagnostic samples (Table 10.2), preculture for 24 h in selenite broth was considered to give adequate enrichment for ELISA. However, the numbers of *Salmonella* in clinical samples is liable to be greater than in most contaminated food samples and should therefore be more easily detectable. It can be postulated that, if all the bacterial virulence factors involved in human food poisoning are ever fully characterized, specific sandwich ELISAs to detect them in precultured food samples might provide the most effective control standard.

The aims of the three MAb sandwich ELISAs described in this chapter are the diagnostic detection of a specific bacterial genus, a recognized bacterial virulence factor and a host response to subclinical bovine mastitis. There is potential for a wider diagnostic application of MAb-based ELISA technology, not only in the detection of well-characterized factors, but also for the detection of as yet unrecognized antigens.

References

Ashkenazi, S. & Cleary, T.G. 1989 Rapid method to detect Shiga toxin and Shiga-like toxin I based on binding to globotriosyl ceramide (Gb_3), their natural receptor. *Journal of Clinical Microbiology* **27**, 1145–1150.

Ball, H.J., Finlay, D., Mackie, D.P., Greer, D., Pollock, D. & McNair, J. 1991a Application of a monoclonal antibody-based enzyme-linked immunosorbent assay for detection of an inflammatory response antigen in subclinical mastitic milk samples. *Journal of Clinical Microbiology* **29**, 1625–1628.

Ball, H.J., Greer, D., Finlay, D. & Mackie, D.P. 1991b Examination of milk samples from experimentally infected heifers by an *N*-acetyl-β-D-glucosaminidase test and an antigen-capture ELISA. *Research in Veterinary Science* **51**, 299–305.

Basta, M., Karmali, M. & Lingwood, C. 1989 Sensitive receptor-specific enzyme-linked immunosorbent assay for *Escherichia coli* verocytotoxin. *Journal of Clinical Microbiology* **27**, 1617–1622.

Brenner, D.J. 1984 Enterobacteriaceae. *Bergey's Manual of Systematic Bacteriology*, Vol. 1, eds Krieg, N.R. & Holt, J.G., pp. 408–413. Baltimore: Williams and Wilkins.

Brigmon, R.L., Bitton, G. & Farrah, S.R. 1992 Detection of *Salmonella enteritidis* in environmental samples by monoclonal antibody based ELISA. *Journal of Immunological Methods* **152**, 135–142.

Choi, D., Tsang, R.S.W. & Ng, M.H. 1992 Sandwich capture ELISA by a murine monoclonal antibody against a genus-specific LPS epitope for the detection of different common serotypes of salmonellas. *Journal of Applied Bacteriology* **72**, 134–138.

Downes, F.P., Green, J.H., Greene, K., Strockbine, N., Wells, J.G. & Wachsmuth, I.K. 1989 Development and evaluation of enzyme-linked immunosorbent assays for detection of Shiga-like toxin I and Shiga-like toxin II. *Journal of Clinical Microbiology* **27**, 1292–1297.

Gretch, D.R., Suter, M. & Stinski, M.F. 1987 The use of biotinylated monoclonal antibodies and streptavidin affinity chromatography to isolate herpes virus hydrophobic proteins or glycoproteins. *Analytical Biochemistry* **163**, 270–277.

Hofmann, K., Titus, G., Montibeller, J. & Finn, F.M. 1982 Avidin binding of carboxyl-substituted biotin and analogues. *Biochemistry* **21**, 978–984.

Kerr, S., Ball, H.J., Mackie, D.P., Pollock, D.A. & Finlay, D.A. 1992 Diagnostic application of monoclonal antibodies to outer membrane protein for rapid detection of salmonella. *Journal of Applied Bacteriology* **72**, 302–308.

Lee, H.A., Wyatt, G.M., Bramham, S. & Morgan, M. 1990 Enzyme-linked immunosorbent assay for *Salmonella typhimurium* in food. *Applied and Environmental Microbiology* **56**, 1541–1546.

McKinney, M.M. & Parkinson, A. 1987 A simple, non-chromatographic procedure to purify immunoglobulins from serum and ascites fluid. *Journal of Immunological Methods* **96**, 271–278.

Mattingly, J.A. 1984 An enzyme immunoassay for the detection of all *Salmonella* using a combination of a myeloma protein and a hybridoma antibody. *Journal of Immunological Methods* **73**, 147–156.

MOLDOW, C., ROBERTSON, J. & ROTHFIELD, L. 1972 Purification of bacterial membrane proteins, the use of guanidinium thiocyanate and urea. *Journal of Membrane Biology* **10**, 137–152.

NOTERMANS, S. & WERNARS, K. 1991 Evaluation and interpretation of data obtained with immunoassays and DNA–DNA hybridation techniques. *International Journal of Food Microbiology* **11**, 35–50.

PRUSAK-SOCHAZHEWSKI, E. & LUONG, J.H.T. 1989 An improved ELISA method for the detection of *Salmonella typhimurium*. *Journal of Applied Bacteriology* **66**, 127–135.

SJORGREN-JANSSON, E. & JEANSSON, S. 1985 Large scale production of monoclonal antibodies in dialysis tubing. *Journal of Immunological Methods* **84**, 359–364.

STEIN, P.E., BOODHOO, A., TYRRELL, G.J., BRUNTON, J.L. & REID, R.J. 1992 Crystal structure of the cell-binding B oligomer of verotoxin-1 from *E. coli*. *Nature* **355**, 748–750.

STROCKBINE, N.A., MARQUES, L.R.M., NEWLAND, J.W., SMITH, H.W., HOLMES, R.K. & O'BRIEN, A.D. 1985 Characterization of monoclonal antibodies against Shiga-like toxin from *Escherichia coli*. *Infection and Immunity* **50**, 695–700.

11: Specific and Rapid Detection of Foodborne Bacteria with rRNA Sequences and the Polymerase Chain Reaction

KATHLEEN A.GRANT,[1] JOANNE H.DICKINSON[1] AND R.G.KROLL[2]

[1]*AFRC Institute of Food Research, Reading Laboratory, Earley Gate, Whiteknights Road, Reading RG6 2EF, UK; and* [2]*MAFF, Torry Research Station, PO Box 31, 135 Abbey Road, Aberdeen AB9 8DG, UK*

One of the main aims in several areas of applied microbiology has been the development of rapid, sensitive and specific methods for the detection and identification of bacteria. Methods based on enrichment, isolation of colonies on plates and subsequent biochemical and serological tests have been the mainstay of microbiology for many years. These have generally proved inadequate for many types of microbiological analysis of perishable goods, because of the delay in obtaining results. It is also clear that the use of physiological traits for bacterial identification and classification is not always reliable, because they have little relationship to the genealogy and hence real identity of a particular organism. For instance, the ability to utilize a particular sugar can be found in a wide range of unrelated organisms. Similarly, loss of this ability by a simple mutation or by its non-expression or its gain by genetic transfer does not substantially alter the identity of the organism or its spoilage or pathogenic potential, but it can significantly change its taxonomic classification as determined by physiologically based tests.

Microbial taxonomy has made great strides in the development of alternative methods of classification and identification based on the analysis of key chemical constituents, such as quinones and fatty acids (Collins & Jones 1981). It is, however, clear that, to get stable and comprehensive taxonomic groupings, and hence identification methods, it is necessary to look to the nucleic acids of bacteria. The base sequence of DNA contains all the information that specifies the characteristics and potential activities of an organism. Phenotypic responses to environmental changes occur widely but only within

New Techniques in
Food and Beverage Microbiology

0-632-03755-5

the capacity encoded in the DNA, unless more DNA is acquired from elsewhere. Indeed, whole genome DNA–DNA homologies have been used for many years to determine relatedness among microorganisms, although much more detailed information may be obtainable from the entire genome sequence. Molecular biological techniques are advancing rapidly, but it is clearly not feasible, at present, to sequence the entire genome of every organism.

Several new approaches are emerging to exploit DNA information. 'Snapshots' of the arrangement of the entire genome can be made by restriction fragment length polymorphisms, rRNA gene restriction patterns or pulsed-field gel electrophoresis (Owen 1989). Another approach is to find unique but unspecified stretches of DNA for use as DNA probes (Fitts *et al.* 1983). Alternatively, particular gene(s) known to be characteristic of the organism, or important in its pathogenicity, such as haemolysin, toxin production and others, can be sequenced and complementary probes for them designed (Datta *et al.* 1990; Neill *et al.* 1990).

It is also possible to use a molecule which, because of constraints on its ability to mutate throughout evolution by the need to maintain the functionality of the gene product, can act as a molecular chronometer. The sequence of such a molecule will, therefore, relate to the phylogeny of the organism. Of several candidate molecules, 16S rRNA has proved the most reliable (Woese 1987).

The rRNA molecule is attractive, because it can now readily be sequenced and can be considered a mosaic of conserved or variable and hypervariable regions, in which some bases have changed throughout evolution. Oligonucleotide probes can be designed against the variable regions, automatically synthesized and non-radioactively labelled. Indeed, several ingenious, commercially available, kits specifically to identify a range of organisms have been successfully developed (Klinger & Johnson 1988). These probes should be highly sensitive, because there are many thousands of copies of the rRNA molecule in each cell. In practice, more than 10^5 cfu/ml are necessary and samples have still to be incubated to increase the number of target organisms to detectable levels. Fluorescent rRNA probes can also be made, where the bacteria can be specifically detected by fluorescence microscopy (Amann *et al.* 1990).

Another significant factor has been the development of the polymerase chain reaction (PCR), in which DNA sequences can be amplified specifically with oligonucleotide primers, to give over 10^6-fold amplification of the selected region of DNA in a few hours. The PCR has been used to amplify selected genes, including pathogenicity factors, to give sensitive and specific detection methods for a variety of organisms, such as staphylococcal enterotoxin genes (Wilson *et al.* 1991) and *Listeria* haemolysin genes (Furrer *et al.* 1991). The

TABLE 11.1. *Examples of rDNA PCR methods*

Organism	Primer sequences	Sensitivity of detection	Substrate	Reference
Brochothrix spp.	16S rDNA	$1-10^4$ cfu	Pure culture	Grant *et al.* 1993
Aerococcus viridans	16S rDNA	13 cfu	Pure culture	Grant *et al.* 1992
E. coli	16S rDNA	500 cells/g	Soil	Tsai & Olson 1992
Mycobacterium paratuberculosis	16S rDNA	10^7 bacteria/g	Faeces	Giessen *et al.* 1992
Chlamydia pneumoniae	16S rDNA	0.4–4 inclusion-forming units	Pure culture	Gaydos *et al.* 1992
Chlamydia pneumoniae	16S rDNA	Not done	Throat specimens	Gaydos *et al.* 1992
Carnobacterium	16S rDNA	$<10-10^4$ cfu/ml	Pure culture	Brooks *et al.* 1992
Carnobacterium	16S rDNA	$<10-10^4$ cfu/ml (PCR + hybridization)	Meat homogenate	Brooks *et al.* 1992
Legionella	5S rDNA	Not done	Pure culture	Mahbubani *et al.* 1990

PCR has also been used to amplify rDNA sequences from a range of different bacteria and low numbers of cells can be specifically and rapidly detected (Table 11.1). This chapter examines the practical aspects of this approach and gives examples of the results obtained. Methods for rRNA sequencing and sequence alignment can be found elsewhere (Woese 1987; Embley 1991).

DNA Extraction

Although PCR is usually performed with purified DNA, relatively impure samples of DNA, such as those from lysed bacterial cells, can also be used. However, when the PCR is performed directly on food samples, the reaction is generally inhibited by as yet unknown components of the food, and some processing of the sample before execution of the PCR is usually required. Experimental PCR assays are usually first set up with purified DNA and, once established, the assay can be used to assess how effective the purification/extraction methods are when whole bacterial cells, bacteria in food or other kinds of samples are used.

The particular method used for DNA extraction will depend on the starting material, whether a pure culture, a food sample or soil, etc. When

TABLE 11.2. *Rapid method of bacterial genomic DNA extraction*

1 Resuspend cells from a 10-ml overnight broth culture in 100 μl of 50 mg/ml lysozyme, transfer to 1.5 ml microfuge tube and incubate at 37 °C for 30 min
2 Add 0.5 ml GES (0.5 mol/l guanidium thiocyanate, 100 mmol/l EDTA and 0.5% (v/v) sarkosyl) and incubate cells for 1 h at 50 °C to lyse bacterial cells
3 Cool the lysate on ice, add 0.25 ml of cold 7.5 mol/l ammonium acetate and leave on ice for 10 min
4 Add 0.5 ml chloroform and 2-pentanol (24 : 1) mixture. Mix phases thoroughly and spin at high speed in microfuge for 10 min
5 Transfer the aqueous layer to a fresh tube and add 0.54 vols of ice-cold isopropanol. Invert the tubes to mix and to precipitate strands of DNA. Spin tubes at low speed in microcentrifuge for 20 s, to deposit DNA
6 Wash the DNA three times with 70% ethanol
7 Air-dry and resuspend in 100 μl sterile distilled water or TE (10 mmol/l Tris–HCl, 1 mmol/l EDTA pH 8)

starting with a pure culture, many DNA extraction methods will give purified DNA suitable for the PCR. The method shown in Table 11.2 is a modification of that described by Pitcher *et al.* (1989). It is rapid, gives a high yield of good-quality DNA and can be used to extract DNA from a wide range of bacteria. Staphylococci require treatment with lysostaphin (50 mg/ml) instead of lysozyme. Gram-negative bacteria do not require lysozyme treatment and the 1-h incubation at 50 °C, after addition of GES, can be omitted. The purity and quantity of DNA can be determined by spectrophotometry at 260–280 nm or by agarose gel electrophoresis.

Whole bacterial cells must be lysed to release DNA prior to use in the PCR assay. Often boiling for 5–10 min is sufficient for cell lysis, but with some bacteria, particularly Gram-positive microorganisms, more rigorous methods of lysis will be required. Table 11.3 lists a variety of lysis methods that can be used.

TABLE 11.3. *Rapid methods for releasing bacterial DNA from whole cells*

Boil cells for 5–10 min and then place immediately on ice

Boil cells for 5 min with 1% Triton X-100 and then place immediately on ice

Incubate cells at 60 °C for 30 min with 100 μg/ml proteinase K. Boil cells for 5 min to destroy proteinase K

Treat cells with lysozyme at 1 mg/ml, incubate at 37 °C for 15 min. Treat with 100 μg/ml proteinase K at 60 °C for 30 min. Boil for 5 min to destroy proteinase K

Treat cell pellet with 0.5 ml ice-cold acetone and hold on ice for 10 min. Decant acetone after 5 min spin. Treat with 0.5 ml of 20 μg/ml mutanolysin in Tris–HCl pH 6.5 and incubate at 37 °C for 30 min

Detection of bacteria in food samples by PCR is more problematic, and there are two main considerations. First, there are often low numbers of the bacteria of interest in a large sample of food and, second, components of the food may inhibit the PCR. One solution is to use enrichment culture to allow the bacteria of interest to multiply to a detectable level before PCR and this

TABLE 11.4. *Methods used to extract bacterial DNA from food samples*

Food sample	Brief description of method	Organism detected and detection level	Reference
Pasteurized milk	Bacterial cells recovered by centrifugation, washed and lysed with lysozyme and proteinase K	*Listeria monocytogenes* 10 cfu/10 ml	Furrer *et al.* 1991
Dried skimmed milk	Milk solubilized in sodium hydroxide. Bacterial cells recovered by centrifugation and lysed by heating at 99 °C	*Staphylococcus aureus* 10^5–10^6 cfu/ml	Wilson *et al.* 1991
Soft cheese	1 Sample homogenized, lysis buffer added and heated to 40 °C. After centrifugation aqueous phase applied to DNA affinity purification column	*Listeria monocytogenes*, detection level varied depending on type of cheese, 10^3–$>10^8$ cfu/ml	Wernars *et al.* 1991b
	2 Soft cheese homogenized with digestion buffer, centrifuged and treated with pronase. Bacterial cells recovered by centrifugation and lysed with lysozyme and proteinase K or by boiling	*Escherichia coli* 10^3 cfu/g	Meyer *et al.* 1991
Meat homogenate	1 Filtering of homogenized sample. Succession of centrifugation steps to recover cells. Lysis by heating with Triton X-100	*Listeria monocytogenes* 8×10^5 cfu/ml	Wang *et al.* 1992
	2 Bacteria recovered with lectin-coated magnetic beads. Bacteria eluted from beads and lysed by boiling	*Brochothrix thermosphacta* 10^5 cfu/ml	Grant *et al.* 1993

will also dilute out any inhibitory effects of the food components. The major drawback to this approach is the overall loss of rapidity. Processing samples, to eliminate the inhibitory components and to concentrate bacterial nucleic acids in the sample, is often necessary. Some of the procedures that have been used depend upon the nature of the food sample (Table 11.4). Although conventional DNA extraction methods have been used with a degree of success and although they help to concentrate bacterial nucleic acids from the food sample, the use of organic solvents in a routine procedure with a large number of samples may be impractical. More research in this area is required if PCR is to become accepted as a routine method for the direct detection of small numbers of bacteria in food.

Primer Design

The specificity of a PCR assay is determined by the sequences of the oligonucleotide primers. A number of workers have used the variable and conserved regions of the 16S rRNA sequences to design primers for use in PCR detection assays (Table 11.1). Primers based on the variable regions amplify rDNA specifically from a particular genus or species. Primers based on the conserved regions can be used as controls, because in this case the same sized PCR product should be produced from a wide range of bacteria. Use of control primers ensures that the PCR is working and that DNA is present in the samples. Examples of 16S rDNA primers are shown in Table 11.5. Some species have similar variable regions in their rRNA sequences, and it may be difficult to select species-specific primers for use in PCR assays. However, this problem can usually be overcome by hybridizing PCR products with a specific probe. In PCR, primer binding depends on the annealing temperature, while in hybridization both salt concentration and temperature can be manipulated to give highly stringent conditions under which it is possible to distinguish between sequences that differ by only one or two bases.

General rules govern primer selection with the aim of minimizing secondary structure in each primer and preventing the formation of dimers by the primers (Table 11.6). Either of these possibilities can result in a failure of the PCR to amplify specific sequences. However, adherence to these guidelines does not automatically guarantee the success of PCR. Some primer pairs fail to work, while some produce stronger PCR bands than others. There is, at present, no obvious explanation for this, but secondary structure of the DNA may be important. Some primer pairs may cross-react with other bacteria. In view of this it can be beneficial to design several different primer pairs, all of which can be tried out at an early stage, the most effective pair being selected for further use.

TABLE 11.5. *Examples of 16S rDNA primers*

Brochothrix spp. (Grant *et al.* 1993)

A 5′-CACAGCTGGGGATAACATCGA
5′-CGAAGGTTACCAAACCGGCT

These primers give a 1324-bp product with DNA from *Brochothrix* spp.

Aerococcus spp. (Grant *et al.* 1992)

AC_2 5′-GTGCTTGCACTTCTGACGTTAGC
AC_4 5′-TGAGCCGTGGGCTTTCACAT

These primers give a 540-bp product with DNA from *Aerococcus* spp.

Escherichia coli (Tsai & Olson 1992)

RW01 5′-AACTGGAGGAAGGTGGGGAT
DG74 5′-ACGGGCGGTGTGTAC

These primers give a 237-bp product with *Escherichia coli* DNA

Chlamydia pneumoniae (Gaydos *et al.* 1992)

CpnA 5′-TGACAACTGTAGAAATACAGC
CpnB 5′-ATTTATAGGAGAGAGGCG

These primers give a 463-bp product with DNA from *Chlamydia pneumoniae*

Universal primers for Gram-positive bacteria

AR1 5′-GAGAGTTTGATCCTGGCTCAGGA
KK 5′-TTACCGCGGCTGCTGGCACGT

These primers give a 520-bp product with DNA from Gram-positive bacteria

TABLE 11.6. *General guidelines for primer design*

The GC content of each primer should not be too high (50–60%). The GC content of a primer pair should be similar

Long stretches of polypurines and polypyrimidines should be avoided

The melting temperature (T_m) of each primer should be similar (T_m calculated using Wallace formula T_m (°C) = 2(A + T) + 4(C + G)

There should be no complementarity between primers, particularly at the 3′ end

Primers should not be self-complementary

TABLE 11.7. *PCR components and concentration*

PCR buffer	50 mmol/l KC1; 10 mmol/l Tris–HCl pH 8.3; 3 mmol/l $MgCl_2$; 0.0001% gelatin
dNTPs	100–200 μmol/l each
Primers	100 nmol/l each
Taq polymerase	0.5 units/reaction
Sterile distilled water	Up to a total volume of 25 μl

Polymerase Chain Reaction

The PCR is usually performed with total volumes of between 25 and 100 μl, but we generally perform PCR assays in a total reaction volume of 25 μl, adding 2.5 μl of DNA sample material. It is convenient to keep all stock reagents as 10× concentrates and then to prepare a mixture of all reagents, excluding sample DNA, sufficient for the total number of samples. The mix can then be dispensed (22.5 μl) into the appropriate number of tubes. One drop of mineral oil is added to prevent evaporation during heating, followed by the DNA sample. This minimizes the amount of pipetting and helps to prevent sample contamination.

A standard PCR buffer is suitable for most PCR assays but we have obtained better results with higher magnesium concentrations (final concentration 3 mmol/l) than are generally found in commercially prepared buffers (1.5 mmol/l). It should be stressed that the magnesium requirement depends on the primers used and on the dNTP concentration and should be titrated for each primer pair. A deoxyribonucleotide concentration of 100 μmol/l and primer concentrations of 100 nmol/l have been found to work well for 16S rDNA primers. Concentrations of dNTPs and primers should be kept as low as possible to minimize misincorporation and mispriming errors. Table 11.7 gives the composition of a typical PCR reaction mixture.

One of the most important considerations is to minimize the risk of contamination with DNA that could give a false positive result. To this end, PCR assays should be set up with positive displacement pipettes and in a room separate from the areas where PCR products are analysed. It is important to include negative control samples without DNA in the PCR assays. This should not give a PCR product; a positive result from this tube is indicative of DNA contamination and none of the positive results from other reaction tubes can be relied upon.

Reaction conditions

An example of some PCR conditions are given in Table 11.8. The time, temperature and number of cycles will depend on the DNA sample and the

TABLE 11.8. *Conditions for PCR*

Initial denaturing of DNA	94 °C	4 min	
Denaturing of DNA	94 °C	30 s	30 cycles
Annealing of primers	55 °C	30 s	
Extension of primers	72 °C	2 min	
Final extension	72 °C	10 min	
Samples held at 4 °C until analysed			

primers used. The annealing temperature is determined from the melting temperature of the primers; 5 °C below the T_m is generally satisfactory, but high annealing temperatures may be helpful if the primers cross-react with other bacteria. The higher the annealing temperature, the greater the stringency, so minimizing non-specific annealing of primers and misincorporation of dNTPs at the 3′ end.

The times given for each step of the cycle work well for 16S rDNA primers. Short steps also help to eliminate non-specific annealing and extension of primers. We have found 30 cycles of amplification sufficient to generate PCR products from as little as 100 μg of purified genomic DNA.

Detection of PCR products

PCR products can be visualized by agarose gel electrophoresis. A portion (e.g. 5–10 μl) of the PCR products is mixed with DNA loading buffer and run on a 1.4% agarose gel containing 0.25 μg/ml ethidium bromide along with suitable DNA size markers. Electrophoresis is performed in TAE buffer (40 mmol/l Tris-acetate, 1 mmol/l EDTA) at 90 V for 1–1.5 h. The gel is then examined on a UV transilluminator at 312 nm and photographed. The size of the expected fragment can be calculated from the position of the primers on the 16S rRNA sequence. Any bands of a different size are the result of non-specific amplification and can usually be eliminated by altering the PCR conditions, such as concentration of primers, dNTPs or magnesium, or by changing the annealing temperature. Also, if the expected product is present, but only as a faint band, alteration of the assay conditions may increase the amount of PCR product.

Probing with PCR Products

Hybridization of PCR products to an internal probe is performed for three main reasons: first, to confirm that the PCR product has been amplified from the gene from which the primers were designed; second, to improve sensitivity of detection (probing PCR products can result in 10–100-fold increase in sensitivity); and, finally, to distinguish between PCR products that have been amplified from closely related bacteria. A probe is used that binds to a particular sequence of the DNA within the PCR product. Several non-radioactive labelling kits are available. We routinely use a non-radioactive labelling and detection method – the ECL 3′ oligo labelling kit (Amersham International). In this method the probe is labelled with fluorescein and hybridized with the PCR product. The hybrid is incubated with antifluorescein antibody conjugated to horseradish peroxidase. The antibody binds to the probe and enzyme activity is detected with the chemiluminescent substrate luminol, the oxidation of which, in the presence of an enhancer, emits sufficient energy for detection on X-ray film. This non-radioactive method works particularly well for detecting PCR products and exposure times of as little as 2 min are required. With the use of the appropriate probe under stringent hybridization conditions, it is possible to distinguish PCR products from closely related bacteria.

Typical results obtained with 16S rDNA primers in PCR assays are shown in Figs 11.1 and 11.2. Primers specific to the 16S rDNA of *Brochothrix* spp. amplify purified DNA from *Brochothrix*, but not that from other bacteria (Fig. 11.1(a)). The universal primers give a 520-bp fragment with DNA from all strains tested (Fig. 11.1(b)). The *Brochothrix*-specific primers are sensitive and can detect 100 μg purified DNA and 30 cfu when cell lysates are used (Fig. 11.1(c) & (d)).

16S rDNA primers for *Aerococcus* amplify DNA from colonies of *Aerococcus viridans* but not DNA from closely related bacteria (Fig. 11.2(a)). The universal primers give a 520-bp fragment with DNA from all the bacteria tested (Fig. 11.2(b)). Although a number of strains of staphylococci give a similar sized product with the *Aerococcus* 16S rDNA primers, hybridization of the PCR product with an *A. viridans*-specific internal probe eliminates problems of cross-reactivity (Fig. 11.2(c)). The *Aerococcus* 16S rDNA primers are very sensitive and can detect 13 cfu when cell lysates are the PCR starting material (Fig. 11.2(d)).

Detection of Bacteria in Food Samples with 16S rDNA Primers

In common with others, we have experienced difficulties with the use of PCR directly on foods (Wernars *et al.* 1991a, b; Brooks *et al.* 1992). However, some success has been achieved with two different approaches. In the first, bacteria were separated from meat homogenates with lectin-coated magnetic particles and then specifically detected by PCR. With this method, detection levels of 1.2×10^5 cfu/ml of *Brochothrix thermosphacta* in meat homogenates have been

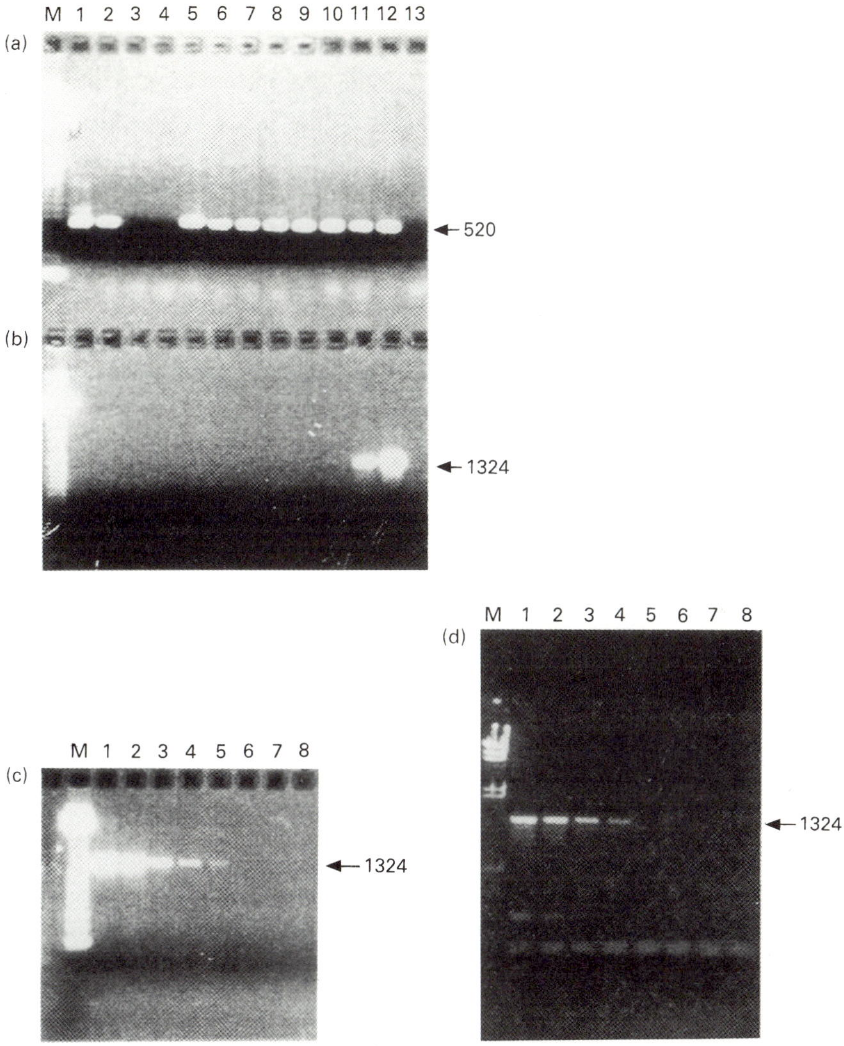

FIG. 11.1. Agarose gel electrophoresis of PCR products obtained with purified DNA and (a) universal 16S rDNA primers and (b), (c) and (d) *Brochothrix*-specific 16S rDNA primers. (a) and (b) Lane 1, *Carnobacterium divergens*; Lane 2, *C. mobile*; Lanes 3 and 4, sterile distilled water; Lane 5, *Lactobacillus fermentum*; Lane 6, *Streptococcus porcinus*; Lane 7, *Staphylococcus aureus*; Lane 8, *Vagococcus salmoninarum*; Lane 9, *Bacillus thuringiensis*; Lane 10, *Listeria seeligeri*; Lane 11, *Brochothrix campestris*; Lane 12, *Brochothrix thermosphacta*; Lane 13, sterile distilled water; Lane M, 123-base pair (bp) ladder (Gibco BRL). (c) PCR products obtained from different concentrations of *Brochothrix thermosphacta* DNA. Lane 1, 100 pg; Lane 2, 10 pg, Lane 3, 1 pg; Lane 4, 100 fg; Lane 5, 10 fg; Lane 6, 1 fg; Lane 7, 100 ag; Lane 8, sterile distilled water; Lane M, 123-bp ladder. (d) Cell lysates of *Brochothrix thermosphacta*. Lane 1, 3×10^4 cfu; Lane 2, 3×10^3 cfu; Lane 3, 3×10^2; Lane 4, 30 cfu; Lane 5, 3 cfu; Lane 6, 10-fold dilution of Lane 5; Lane 7, 10-fold dilution of Lane 6; Lane 8, sterile distilled water; Lane M, $\lambda \times$ *Hind*III markers (Gibco BRL). (a), (b) and (c) are reproduced with permission from Grant *et al.* (1993).

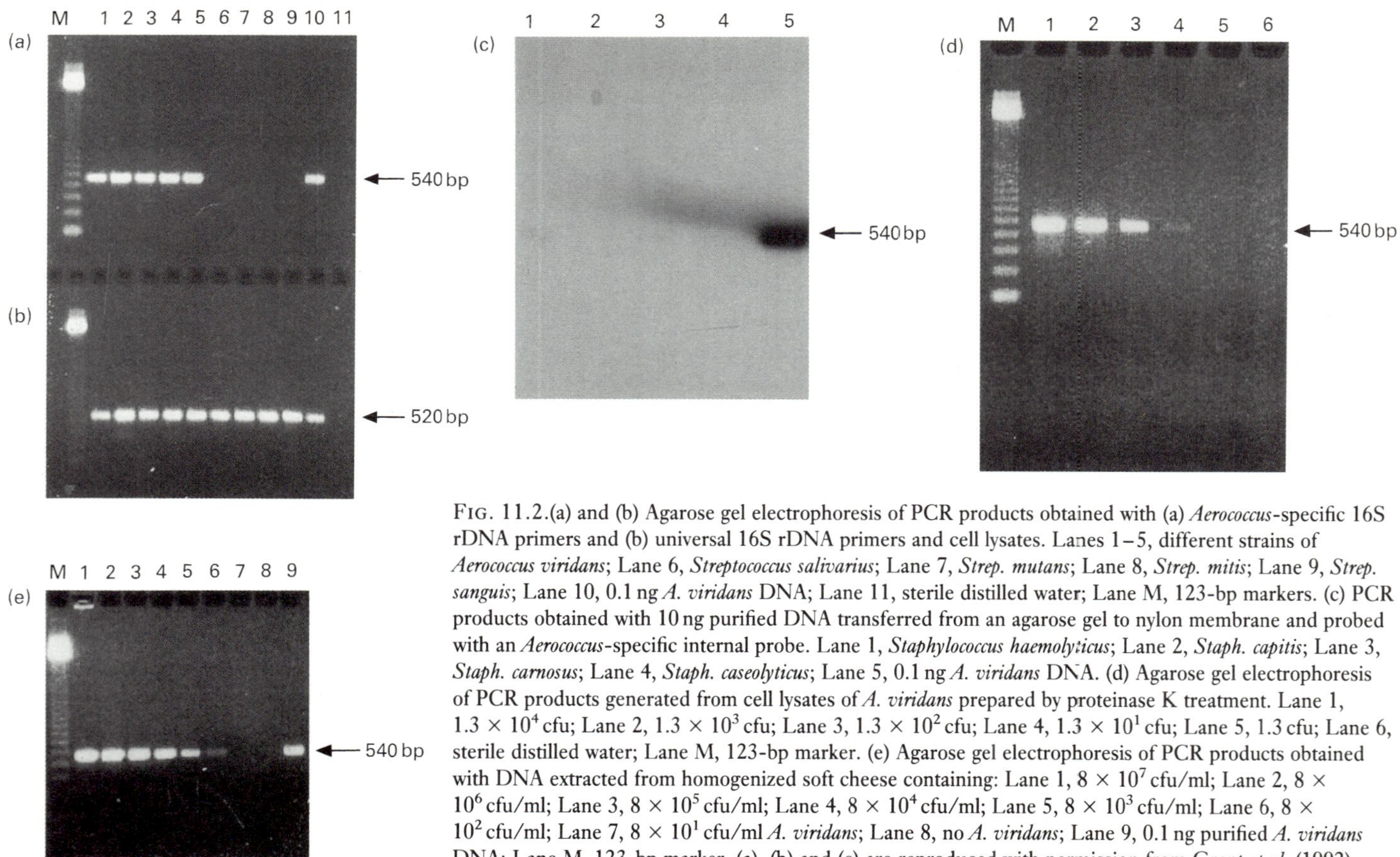

FIG. 11.2.(a) and (b) Agarose gel electrophoresis of PCR products obtained with (a) *Aerococcus*-specific 16S rDNA primers and (b) universal 16S rDNA primers and cell lysates. Lanes 1–5, different strains of *Aerococcus viridans*; Lane 6, *Streptococcus salivarius*; Lane 7, *Strep. mutans*; Lane 8, *Strep. mitis*; Lane 9, *Strep. sanguis*; Lane 10, 0.1 ng *A. viridans* DNA; Lane 11, sterile distilled water; Lane M, 123-bp markers. (c) PCR products obtained with 10 ng purified DNA transferred from an agarose gel to nylon membrane and probed with an *Aerococcus*-specific internal probe. Lane 1, *Staphylococcus haemolyticus*; Lane 2, *Staph. capitis*; Lane 3, *Staph. carnosus*; Lane 4, *Staph. caseolyticus*; Lane 5, 0.1 ng *A. viridans* DNA. (d) Agarose gel electrophoresis of PCR products generated from cell lysates of *A. viridans* prepared by proteinase K treatment. Lane 1, 1.3×10^4 cfu; Lane 2, 1.3×10^3 cfu; Lane 3, 1.3×10^2 cfu; Lane 4, 1.3×10^1 cfu; Lane 5, 1.3 cfu; Lane 6, sterile distilled water; Lane M, 123-bp marker. (e) Agarose gel electrophoresis of PCR products obtained with DNA extracted from homogenized soft cheese containing: Lane 1, 8×10^7 cfu/ml; Lane 2, 8×10^6 cfu/ml; Lane 3, 8×10^5 cfu/ml; Lane 4, 8×10^4 cfu/ml; Lane 5, 8×10^3 cfu/ml; Lane 6, 8×10^2 cfu/ml; Lane 7, 8×10^1 cfu/ml *A. viridans*; Lane 8, no *A. viridans*; Lane 9, 0.1 ng purified *A. viridans* DNA; Lane M, 123-bp marker. (a), (b) and (c) are reproduced with permission from Grant *et al.* (1992).

achieved (Grant *et al.* 1993). The second approach used rapid nucleic acid extraction from soft cheese followed by PCR. With soft cheese homogenate inoculated with *A. viridans*, 8×10^2 cfu/ml have been detected (Fig. 11.2(e)).

Conclusions

To date, PCR primer/probe systems based on rRNA sequences have been designed for a restricted number of organisms. As rRNA databases are expanded, it is probable that many more will be described. However, it seems likely that these systems will not be applied to every organism but only to key organisms for a particular situation. As with PCR systems which employ DNA from other target genes, very sensitive and specific detection of isolated DNA or pure cultures can be achieved. It is, however, important to optimize the DNA extraction method, the primer sequences and the PCR conditions and to include both positive and negative controls. As with other gene targets, it is possible to look only for the production of PCR products of the correct size. This is not difficult and gives results within a few hours, but it relies on having sufficient unique regions for which primers can be designed to give specific PCR product formation, and this may not always be possible. Use of rRNA sequences has many advantages but the amount of sequence variation within and between species and genera is not always sufficient to give truly specific PCR product formation. The degree of stringency that can be applied during the PCR can be limited as compared with probe hybridizations, and PCR products of the correct size can be produced from rDNA with several base mismatches. With the use of oligonucleotide probes, single base differences can sometimes be detected under high-stringency conditions. It is sometimes necessary to use a probe specific for the PCR product, internal to the primers, to confirm specificity (Grant *et al.* 1992). However, this clearly reduces rapidity and increases the amount of experimental work.

When PCR systems have been applied to mixed cultures and foods, problems have been encountered. The presence of high numbers of lysed bacteria can prevent the formation of PCR products, presumably by inhibiting the *Taq* polymerase and/or by non-specific binding of the PCR primers. With rDNA as the target, PCR products from non-target organisms can also be formed in mixed cultures, but these problems can be eliminated usually by the use of an internal probe. When applied to food, the PCR is usually strongly inhibited. Several sample clean-up methods have been tried, with some success, such as centrifugation, affinity chromatography, solvent extraction and magnetic particles, and the detection of low numbers of bacteria has been demonstrated (Meyer *et al.* 1991; Wernars *et al.* 1991b; Wang *et al.* 1992; Grant *et al.* 1993). Although in one case, in which the PCR was strongly inhibited, the PCR products could still be detected by use of a secondary probe (Brooks *et al.* 1992), it is obvious that this area deserves further effort.

If these problems can be addressed, a truly rapid, sensitive and specific method will become available for use directly on samples to detect bacteria. However, for this to make a practical impact, two other problems must be addressed. Since DNA is heat-stable, the DNA from organisms killed by heat treatment would still be amplified by the PCR. Thus, heat-treated foods could appear to contain a particular organism, though the organism was non-viable. One approach is to include a short enrichment step in which only viable cells can multiply (Golsteyn Thomas *et al.* 1991; Wernars *et al.* 1991a). However, it has recently been suggested that mRNA/PCR systems could be used, because mRNA has a short half-life (Bej *et al.* 1991). In some circumstances, the detection of DNA from apparently non-viable cells could be advantageous, as where cells are difficult to cultivate in the laboratory or in the so-called 'somni-stage'. At present single cells can in some circumstances be detected from food or clinical samples. However, it should be pointed out that typical sample volumes in PCR reactions are 2–10 µl. Thus, if one cell can be detected, the overall sensitivity is only 100–500 cells/ml of extract. This may not be a problem with clinical samples, but in foods, the presence of one cell in 25 g of food may have to be detected. Indeed, any new method would have to assure the absence of one cell in 25 g. Thus, significant improvements in sample preparation are necessary or a short enrichment phase is still required.

It seems probable that PCR/rDNA systems may become widely used for rapid identification of isolated colonies, but that direct application to foods and clinical samples requires further work. It also seems likely that PCR/rDNA systems can be run together with other PCR systems to detect specific genes. The rDNA system may be used to confirm the genus or species and a specific gene system used to confirm that a particular species-specific trait was present.

References

AMANN, R.I., BINDER, B.J., OLSON, R.J., CHISOLM, S.W., DEVEREUX, R. & STAHL, D.A. 1990 Combination of 16S rRNA-targeted oligonucleotide probes with flow cytometry for analyzing mixed microbial populations. *Applied and Environmental Microbiology* **56**, 1919–1925.

BEJ, A.K., MAHBUBANI, M.H. & ATLAS, R.M. 1991 Detection of viable *Legionella pneumophila* in water by polymerase chain reaction and gene probe methods. *Applied and Environmental Microbiology* **57**, 597–600.

BROOKS, J.L., MOORE, A.S., PATCHETT, R.A., COLLINS, M.D. & KROLL, R.G. 1992 Use of the polymerase chain reaction and oligonucleotide probes for the rapid detection and identification of *Carnobacterium* spp. from meat. *Journal of Applied Bacteriology* **70**, 121–126.

COLLINS, M.D. & JONES, D. 1981 The distribution of isoprenoid quinone structural types in bacteria and their taxonomic implications. *Microbiological Reviews* **45**, 316–354.

DATTA, A.R., WENTZ, B.A. & RUSSEL, J. 1990 Cloning of the listeriolysin O gene and development of specific gene probes for *Listeria monocytogenes*. *Applied and Environmental Microbiology* **12**, 3874–3877.

EMBLEY, T.M. 1991 The linear PCR reaction: a simple robust method for sequencing amplified rRNA genes. *Letters in Applied Microbiology* **13**, 171–174.

FITTS, R., DIAMOND, M., HAMILTON, C. & NERI, M. 1983 DNA–DNA hybridisation assay for detection of *Salmonella* spp. in foods. *Applied and Environmental Microbiology* **46**, 1146–1151.

FURRER, B., CANDRIAN, U., HOEFELEIN, C. & LUETHLY, J. 1991 Detection and identification of *Listeria monocytogenes* in cooked sausage products and in milk by *in vitro* amplification of haemolysin gene fragments. *Journal of Applied Bacteriology* **70**, 372–379.

GAYDOS, C.A., QUINN, T.C. & EIDEN, J.J. 1992 Identification of *Chlamydia pneumoniae* by DNA amplification of the 16S rRNA gene. *Journal of Clinical Microbiology* **58**, 754–757.

GIESSEN, J.W.B. VAN DER, HARING, R.M., VANCLAVE, E., EGER, A., HAAGSMAR, J. & ZEIJST, B.A.M. VAN DER 1992 Evaluation of the abilities of three diagnostic tests based on the polymerase chain reactions to detect *Mycobacterium paratuberculosis* in cattle: application in a control programme. *Journal of Clinical Microbiology* **30**, 1216–1219.

GOLSTEYN THOMAS, E.J., KING, R.K., BURCHAK, J. & GANNON, V.P.J. 1991 Sensitive and specific detection of *Listeria monocytogenes* in milk and ground beef with the polymerase chain reaction. *Applied and Environmental Microbiology* **57**, 2576–2580.

GRANT, K.A., DICKINSON, J.H., COLLINS, M.D. & KROLL, R.G. 1992 Rapid identification of *Aerococcus viridans* using the polymerase chain reaction and an oligonucleotide probe. *FEMS Microbiology Letters* **95**, 63–68.

GRANT, K.A., DICKINSON, J.H., PAYNE, M.J., CAMPBELL, S., COLLINS, M.D. & KROLL, R.G. 1993 Use of the polymerase chain reaction and 16S rRNA sequences for the rapid detection of *Brochothrix* spp. in foods. *Journal of Applied Bacteriology* **74**, 260–267.

KLINGER, J.D. & JOHNSON, A.R. 1988 A rapid nucleic acid hybridisation assay for *Listeria* in foods. *Food Technology* **42**, 66–70.

MAHBUBANI, M.H., BEJ, A.K., MILLER, R., HAFF, L., DICESARE, J. & ATLAS, R.M. 1990 Detection of *Legionnella* with polymerase chain reaction and gene probe methods. *Molecular and Cellular Probes* **4**, 175–187.

MEYER, R., LUTIG, J. & CANDRIAN, U. 1991 Direct detection by polymerase chain reaction (PCR) of *Escherichia coli* in water and soft cheese and identification of enterotoxigenic strains. *Letters in Applied Microbiology* **13**, 268–271.

NEILL, R.J., FANNING, G.R., DELARHOZ, F., WOLFF, R. & GEMSKI, P. 1990 Oligonucleotide probes for the detection and differentiation of *Staphylococcus aureus* strains containing genes for enterotoxins A, B and C and toxic shock syndrome. *Journal of Clinical Microbiology* **28**, 1514–1518.

OWEN, R.J. 1989 Chromosomal DNA fingerprinting – a new method of species and strain identification applicable to microbial pathogens. *Journal of Medical Microbiology* **30**, 89–99.

PITCHER, D.G., SAUNDERS, N.A. & OWEN, R.J. 1989 Rapid extraction of bacterial genomic DNA with guanidium thiocyanate. *Letters in Applied Microbiology* **8**, 151–156.

TSAI, Y.-L. & OLSON, B. 1992 Detection of low numbers of bacterial cells in soils and sediments by polymerase chain reaction. *Applied and Environmental Microbiology* **58**, 754–757.

WANG, R.F., CAO, W.-W. & JOHNSON, M.G. 1992 Development of cell surface protein associated gene probe specific for *Listeria monocytogenes* and detection of the bacteria in food by PCR. *Molecular and Cellular Probes* **6**, 119–129.

WERNARS, K., DELFGOU, E., SOETORO, P.S. & NOTERMANS, S. 1991a A successful approach for detection of low numbers of enterotoxigenic *Escherichia coli* in minced meat by using the polymerase chain reaction. *Applied and Environmental Microbiology* **57**, 1914–1919.

WERNARS, K., HENVELMAN, C.J., CHAKRABORTY, T. & NOTERMANS, S.H.W. 1991b Use of polymerase chain reaction for direct detection of *Listeria monocytogenes* in soft cheese. *Journal of Applied Bacteriology* **70**, 121–126.

WILSON, I.G., COOPER, J.E. & GILMOUR, A. 1991 Detection of enterotoxigenic *Staphylococcus aureus* in dried skimmed milk: use of the polymerase chain reaction for amplification and detection of staphylococcal enterotoxin genes *entB* and *entC*I and the thermonuclease gene *nuc. Applied and Environmental Microbiology* **57**, 1793–1798.

WOESE, C.R. 1987 Bacterial evolution. *Microbiological Reviews* **51**, 221–271.

12: Detection of Toxigenic Microorganisms in Foods by PCR

I.G.Wilson,[1] A.Gilmour[2] and J.E.Cooper[2]
[1]*Public Health Laboratory, Belfast City Hospital, Lisburn Road, Belfast BT9 7AD, UK;*
and [2]*Agriculture and Food Science Centre, Queen's University of Belfast, Newforge Lane, Belfast BT9 5PX, UK*

The polymerase chain reaction (PCR) has been used for the detection and identification of a wide range of pathogenic microorganisms (Hill & Keasler 1991). This has been possible because of the relative simplicity and robustness of the method and the continued expansion in the availability of DNA sequence information.

The requirements of food microbiology are, however, somewhat different to those of medical or veterinary microbiology in that bacteria are more often present in mixed culture and in the presence of complex food components which may interfere with tests. The numbers of bacteria present are frequently of great significance and, in addition, sublethal injury of pathogens may give rise to misleading negative results on culture. These factors have been discussed by Sharpe (1991), Wilson *et al.* (1991) and Wolcott (1991).

Testing for Enterotoxin Genes in Foods

Culture of bacteria followed by immunoassay of the culture supernatant is the method of choice for detection of toxin-producing organisms in foods. Immunoassay is also used for the detection of enterotoxins in foods. Although they are the most widely used detection methods, occasions may arise when they are inadequate. Heat-stable enterotoxins may survive extreme processing conditions which injure or kill the cells that produce them and therefore appear culture-negative. Injured cells may be resuscitated and produce enterotoxins, and thermally inactivated enterotoxins may reactivate spontaneously (Bennett 1992). False negative reactions may occur in serological tests because of interference from food components. False positive reactions

New Techniques in
Food and Beverage Microbiology

0-632-03755-5

may be caused by non-specific interactions with food components. Some strains may produce different enterotoxins under different conditions of growth. Unidentified new enterotoxins may be produced by some strains and supposedly non-enterotoxigenic species or strains may be found to produce enterotoxins. Finally, enterotoxins may be produced by strains that are atypical in diagnostic reactions (e.g. coagulase-negative *Staphylococcus aureus*; Wilson *et al.* 1991).

Bennett (1992) has described the factors affecting the serological inactivation of staphylococcal enterotoxins by heat which leaves the biological activity of the toxins undiminished. This has led to false negative serological results, while the toxins show pathogenic effects which have actually become greater than those of the unheated toxins in humans and in animal models.

Toxin gene detection is complementary to immunological testing and allows the identification of strains capable of toxin production, even under conditions in which toxins have not been produced or these molecules are not recognized immunologically. Either technique alone may, in some cases, produce misleading results. The usefulness of genotypic testing, rather than immunological assay, is in distinguishing the potential for toxin production from the actual production of toxins in a given situation.

Sensitivity

It has often been demonstrated that PCR is able to detect a single copy of a gene. Frequently, this sensitivity has been achieved in experiments involving mammalian tissues (Higuchi *et al.* 1988; Saiki *et al.* 1988) or environmental samples (Tsai & Olson 1992). In practice, the detection of bacteria in foods is often about 1000-fold less sensitive (Meyer *et al.* 1991; Wernars *et al.* 1991; Wilson *et al.* 1991) and may vary with the type of food examined. Factors responsible for this attenuation of sensitivity must now be considered.

Cell debris

Components of the bacterial cell and extracellular substances may interfere with the availability or integrity of target DNA. It should also be remembered that the bacterium sought will often be greatly outnumbered by the rest of the microflora.

Nucleases

Staphylococcus aureus produces an extremely heat-stable nuclease which, depending on the conditions of DNA extraction, may not be neutralized and may degrade target DNA and primers.

Food components

Foods may contain a large variety of compounds that are able to interfere with the amplification reaction in a number of ways. These may include binding the DNA and rendering it unavailable for enzymic amplification, and inactivating *Taq* polymerase, perhaps by sequestration of Mg^{2+}, which is essential for the activity of this enzyme.

Sampling errors

Uneven distribution of bacteria throughout some foods, especially non-homogeneous foods, and the small sample size (10 μl) used in the amplification reaction make the potential sampling errors large. If DNA is to be extracted from intact cells in the PCR reaction tube, the highest sensitivity that can be expected for a single-copy gene is 100 cfu/ml (1 cfu/10 μl = 100 cfu/ml). Prior extraction of target DNA will allow greater sensitivity, because a larger quantity of target material can be added to the reaction. Scaling up the reaction to achieve single-cell detection is not economically feasible and is limited by the tube volume accommodated by the thermal cycling instrument.

Sample preparation

The preparation of food samples for PCR may, in some cases, be a critical factor in the usefulness of the technique. Ideally, sample preparation should ensure high recovery of cells and DNA and should separate the target DNA from any interfering compounds, while allowing rapid examination of a reasonable number of samples and the economical examination of small numbers of samples. Any method adopted is likely to compromise one or more of these factors.

Sample preparation strategies

For some foods it may be possible to remove bacteria by swabbing or washing (Lampel *et al.* 1990). An important area for future research is the separation of bacterial cells from interfering food debris. DNA extraction by the methods discussed below may prove effective for many combinations of bacteria and foods.

1 Boil cells in the reaction tube to release DNA. This method is the easiest and most rapid; it works well for pure cultures and many cells in aqueous suspensions. It should be tried before more complex methods are adopted.

2 Boil cells, spin down cell debris and transfer an aliquot of released DNA to the reaction tube. This method is only slightly slower than that above, and may be all that is necessary to remove interfering compounds and so allow efficient amplification.

3 Purify sample DNA by phenol/chloroform extraction, guanidium isothiocyanate, powdered glass solution, ion exchange chromatography, or other solid-phase method, quantify, if required, and then amplify. A wide range of DNA extraction kits based on these and other methods is available. Many of these allow good DNA recoveries in less than an hour. Choice will often be governed by whichever method is most efficient for the number of samples to be processed.

4 Concentrate cells and/or purify DNA by membrane filtration. A number of methods to effect physical separation of bacterial cells from foods are currently under investigation. Membrane filter devices are available that allow the recovery of DNA by centrifugation in a few minutes.

The choice of protocol will depend on factors such as the organism being detected. Factors for consideration include the inactivation of nuclease, the detection of low or high numbers of cells, the speed with which results are required and the importance of this relative to sensitivity. The presence of factors inhibitory to PCR in the growth menstruum of the organisms, the identity of inhibitors and their removal should also be considered. The quantity of DNA required may affect the choice of method, since the testing of a single food sample requires much less DNA than will be needed for a stock solution of target DNA used to spike samples in ongoing quantitative experiments. A number of different thermostable polymerases are now available and one of these may be more suitable for a particular detection system than another.

Detection in Skimmed Milk of the *Staphylococcus aureus sec*$^+$ Enterotoxin C_1-encoding Gene

Enterotoxin C is the staphylococcal enterotoxin most commonly associated with bovine, ovine and caprine sources. Subclinical mastitis may be present in the animal and give rise to infected milk. *Staph. aureus* cells present in milk may colonize the skimmed milk production plant and lead to a contaminated product (Wilson *et al.* 1991).

DNA extraction from skimmed milk

Target DNA for PCR can be obtained from bacteria that contaminate skimmed milk by the following method (Huhtanen & Jones 1989; Wilson *et al.* 1991).

Dried skimmed milk (1 g) is weighed aseptically into sterile 50-ml centrifuge tubes and the milk sample is dissolved by adding 35 ml of 0.015 mol/l NaOH and incubating on a roller (Denley Spiramix) at 37 °C for 1 h before

FIG. 12.1. Concentrated bacterial cells (not visible) added to PCR reaction mix.

centrifugation at 14 000 *g* (20 °C) for 20 min. The pellet is resuspended in 10 ml 0.015 mol/l NaOH and centrifuged as before, so that any milk powder not dissolved in the first treatment can be decanted. The cell pellet is dissolved in 50 μl sterile TE8 buffer (10 mmol/l Tris, 1 mmol/l EDTA, pH 8.0) and heated at 99 °C for 15 min. Ten μl is transferred to a PCR tube for amplification (Fig. 12.1).

Primers

Two sets of 25-mer primers are used, one of which amplifies a 631-bp region nested within the 801-bp region (Bohach & Schlievert 1987) of the sec^+ gene that encodes SEC_1 (Betley *et al.* 1990) and which is amplified by the first set of primers.

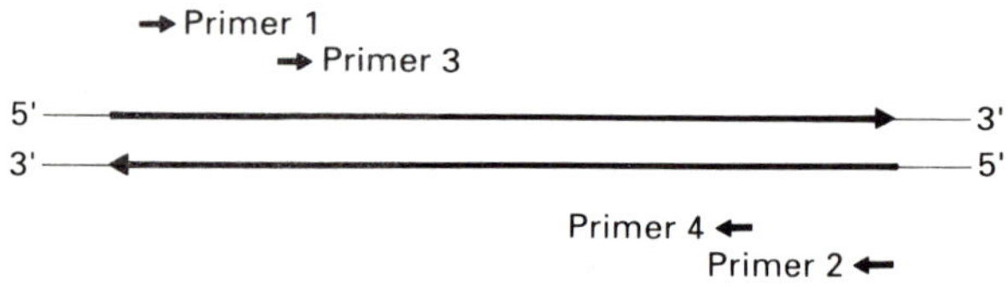

FIG. 12.2.Arrangement of nested primer pairs for amplification of *sec*$^+$ gene. Primers 3 and 4 are nested within the sequence to be amplified by primers 1 and 2. The long arrows represent the 801 bp *sec*$^+$ target sequence.

The primer sequences are:

801-bp product:
Primer 1: 5′-ATG AAT AAG AGT CGA TTT ATT TCA T-3′
Primer 2: 5′-TTA TTC ATT CTT TGT TGT AAG GTG G-3′

631-bp product:
Primer 3: 5′-ACA CCC AAC GTA TTA GCA GAG AGC C-3′
Primer 4: 5′-CCT GGT GCA GGC ATC ATA TCA TAC C-3′

Nesting of primers (Fig. 12.2) improves both sensitivity and specificity, since the 801-bp product provides a further template for amplification of the 631-bp product whose sequence it contains (Mullis & Faloona 1987; Wilson *et al.* 1991). Alternative primers for staphylococcal enterotoxin genes have been described by Johnson *et al.* (1991).

PCR reaction

PCR reaction mixtures can be made efficiently from master mixtures of components from the GeneAmp kit (Perkin-Elmer, Beaconsfield, England), as described in the instruction sheet. Each reaction tube is filled with a mixture of 53.5 μl double-distilled sterile water, 10 μl of 10× reaction buffer (100 mmol/l Tris, pH 8.3, 500 mmol/l KCl, 15 mmol/l $MgCl_2$, 0.1% gelatin), 16 μl of dinucleoside triphosphate mixture (each dinucleoside triphosphate at 1.25 mol/l), 5 μl of primer 1 (20 mol/l), 5 μl of primer 2 (20 mol/l), 10 μl of target DNA and 0.5 μl of AmpliTaq recombinant DNA polymerase. The total volume of the mixture is 100 μl. As a positive control, DNA extracted from *Staph. aureus* NCTC 10655 will give a positive signal with these primers.

Evaporation in the tube during thermal cycling can be prevented by the addition of an overlay of 100 μl liquid paraffin. Tubes should be heated to 95 °C for 5 min before cycling to inactivate any proteases which might affect the activity of AmpliTaq polymerase.

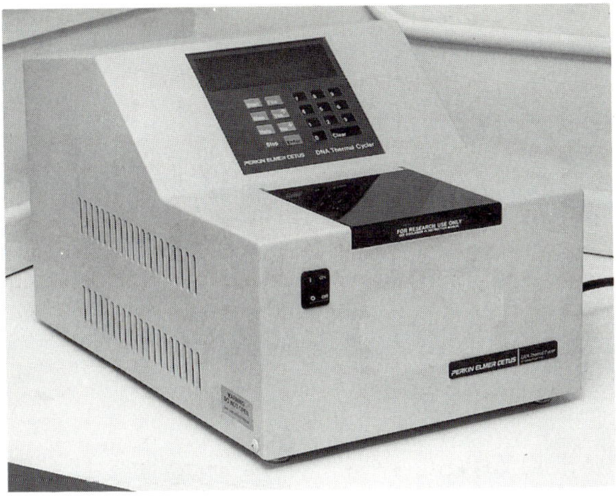

FIG. 12.3. Thermal cycler as used for the amplification reaction.

The reaction conditions used in the Perkin-Elmer thermal cycler (Fig. 12.3) are:

Initial denaturation: 94 °C for 1 min
Subsequent denaturations: 94 °C for 30 s
Primer annealing: 50 °C for 30 s
Polymerization: 72 °C for 30 s

Fifty cycles are completed in under 3 h. Under certain conditions, better sensitivity may be obtained by cycling for 25 cycles with primers 1 and 2 only, removing a 10-μl aliquot of amplified product to another tube containing a reaction mix with primers 3 and 4 only, and amplifying for a further 25 cycles. If a large number of samples is to be analysed, this method adds considerably to the analysis time.

Detection of amplified DNA

A 15-μl aliquot of amplified product is examined after electrophoresis in 1% Ultra-Pure agarose gel (BRL, Cambridge, England) in 1× Tris–borate buffer (0.089 mol/l boric acid, 0.002 mol/l EDTA, pH 8.0 (Kodak BufferEZE formula 3, Eastman Kodak, Liverpool, UK)). After thorough mixing with tracking dye (40% (w/v) sucrose, 0.025% (w/v) bromophenol blue, Sigma), each sample is loaded into one well of the gel.

A DNA-size ladder with fragments of known length should be run in each gel to facilitate positive identification of amplification products by size.

Electrophoresis is carried out for about 1 h at 70 V. The gel is then stained

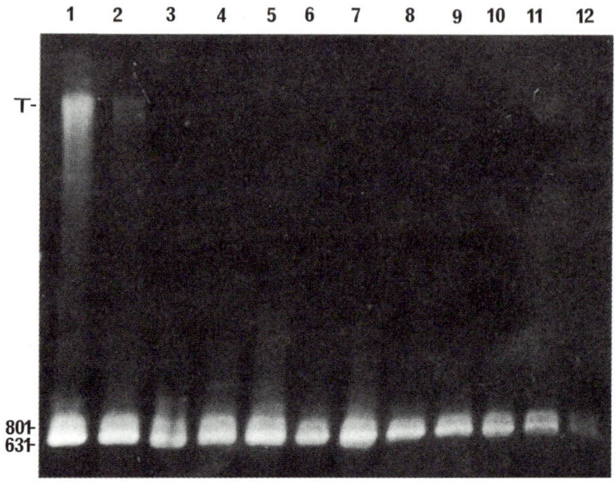

FIG. 12.4. Amplified DNA fragments detected by agarose gel electrophoresis. PCR products (801 and 631 bp) from amplification of decimal dilutions of target DNA (increasing dilution of target DNA from left to right, 1 mg–10 fg). Note the genomic target DNA bands (T) at the top of lanes 1 and 2, which are diluted out at higher dilutions. PCR product bands are still produced from low numbers of template DNA strands (lane 12).

with an aqueous solution of 0.5 μg/ml ethidium bromide (Sigma) for 20 min, and washed briefly in tap water (CAUTION: ethidium bromide is a powerful mutagen; appropriate protective clothing should be worn). The gel can be examined visually, with an ultraviolet transilluminator, and photographed (Fig. 12.4). This process may be completed in 1.5 h.

If speed is not critical and more certain identification of bands is required, gels can be blotted on to membranes and these can be hybridized with a probe of sequence internal to the region to be amplified (Notermans *et al.* 1988; Jaulhac *et al.* 1992). This provides evidence that bands are not merely of the size expected, but that they also have the correct sequence.

Conclusion

The polymerase chain reaction provides an alternative to phenotypic methods of toxin detection. It is of particular value in situations where test methods for the presence of toxins require supplementation by methods capable of showing the genetic ability to produce toxins.

It may be possible to detect as yet unknown toxins in families of toxins where a conserved polypeptide structure is recognized as responsible for the pathogenic effects (Marrack & Kappler 1990). The structurally and functionally important disulphide bridge region of the staphylococcal enterotoxins is an example (Bergdoll *et al.* 1974). Primers that correspond to the conserved region of the toxin may permit the detection of toxins which can only be recognized by animal feeding tests. Degenerate primers, that is, a mixture of closely related primers containing a combination of slightly different sequences because of variations in bases at certain positions in the polymer, may be of particular use in screening for likely sequences.

PCR may be of use in determining whether an organism is capable of producing a given toxin, irrespective of whether that toxin has been produced under laboratory conditions. This may be an advantage or a disadvantage depending on the circumstances. For most investigations, immunological testing will remain the method of choice, but the possibilities and drawbacks of genotypic and phenotypic testing should be considered in each situation.

References

BENNETT, R.W. 1992 The biomolecular temperament of staphylococcal enterotoxin in thermally processed foods. *Journal of the Association of Official Analytical Chemists International* **75**, 6–12.

BERGDOLL, M.S., HUANG, I.-Y. & SCHANTZ, E.J. 1974 Chemistry of the staphylococcal enterotoxins. *Journal of Agricultural and Food Chemistry* **22**, 9–13.

BETLEY, M.J., SCHLIEVERT, P.M., BERGDOLL, M.S., BOHACH, G.A., IANDOLO, J.J., KHAN, S.A., PATTEE, P.A. & REISER, R.R. 1990 Staphylococcal gene nomenclature. *ASM News* **56**, 182.

BOHACH, G.A. & SCHLIEVERT, P.M. 1987 Nucleotide sequence of staphylococcal enterotoxin C_1 gene and relatedness to other pyrogenic toxins. *Molecular and General Genetics* **209**, 15–20.

HIGUCHI, R., VON BEROLDINGEN, C.H., SENSABOUGH, G.F. & ERLICH, H.A. 1988 DNA typing from single hairs. *Nature* **332**, 543–546.

HILL, W.E. & KEASLER, S.P. 1991 Identification of foodborne pathogens by nucleic acid hybridization. *International Journal of Food Microbiology* **12**, 67–76.

HUHTANEN, C.N. & JONES, C.O. 1989 A procedure for the direct microscopic count of bacteria in non-fat dry milk. *Journal of Food Protection* **52**, 404–411.

JAULHAC, B., BES, M., BORNSTEIN, N., PIÉMONT, Y., BRUN, Y. & FLEURETTE, J. 1992 Synthetic DNA probes for detection of genes for enterotoxins A, B, C, D, E and TSST-1 in staphylococcal strains. *Journal of Applied Bacteriology* **72**, 386–392.

JOHNSON, W.M., TYLER, S.D., EWAN, E.P., ASHTON, F.E., POLLARD, D.R. & ROZEE, K.R. 1991 Detection of genes for enterotoxins, exfoliative toxins, and toxic shock syndrome toxin 1 in *Staphylococcus aureus* by the polymerase chain reaction. *Journal of Clinical Microbiology* **29**, 426–430.

LAMPEL, K.A., JAGOW, J.A., TRUCKNERS, M. & HILL, W.E. 1990 Polymerase chain reaction for detection of invasive *Shigella flexneri* in food. *Applied and Environmental Microbiology* **56**, 1536–1540.

MARRACK, P. & KAPPLER, J. 1990 The staphylococcal enterotoxins and their relatives. *Science* **248**, 705–711.

MEYER, R., LÜTHY, J. & CANDRIAN, U. 1991 Direct detection by polymerase chain reaction (PCR) of *Escherichia coli* in water and soft cheese and identification of enterotoxigenic strains. *Letters in Applied Microbiology* **13**, 268–271.

MULLIS, K.P. & FALOONA, F.F. 1987 Specific synthesis of DNA *in vitro* via a polymerase-catalysed chain reaction. *Methods in Enzymology* **155**, *Recombinant DNA* ed. Wu, R., pp. 335–351. London: Academic Press.

NOTERMANS, S., HEUVELMAN, K.J. & WERNARS, K. 1988 Synthetic enterotoxin B DNA probes for detection of enterotoxigenic *Staphylococcus aureus* strains. *Applied and Environmental Microbiology* **54**, 531–533.

SAIKI, R.K., GELFAND, D.H., STEOFFEL, S., SCHARF, S.J., HIGUCHI, R., HORN, G.T., MULLIS, K.B. & ERLICH, H.A. 1988 Primer-directed enzymic amplification of DNA with a thermostable DNA polymerase. *Science* **239**, 478–491.

SHARPE, A.N. 1991 On multiplexed separations in quick detection of micro-organisms in foods. *Food Microbiology* **8**, 167–170.

TSAI, Y.-L. & OLSON, B.H. 1992 Detection of low numbers of bacterial cells in soils and sediments by polymerase chain reaction. *Applied und Environmental Microbiology* **58**, 754–757.

WERNARS, K., HEUVELMAN, C.J., CHAKRABORTY, T. & NOTERMANS, S.H.W. 1991 Use of the polymerase chain reaction for direct detection of *Listeria monocytogenes* in soft cheese. *Journal of Applied Bacteriology* **70**, 121–126.

WILSON, I.G., COOPER, J.E. & GILMOUR, A. 1991 Detection of enterotoxigenic *Staphylococcus aureus* in dried skimmed milk: use of the polymerase chain reaction for amplification and detection of staphylococcal enterotoxin genes *ent*B and *ent*C_1 and the thermonuclease gene *nuc*. *Applied and Environmental Microbiology* **57**, 1793–1798.

WOLCOTT, M.J. 1991 DNA-based rapid methods for the detection of foodborne pathogens. *Journal of Food Protection* **54**, 387–401.

13: Combined Subtraction Hybridization and PCR Amplification for Generating High-specificity Bacterial DNA Probes

A.J.BJOURSON[1] AND J.E.COOPER[1,2]
[1]*Food Microbiology Division, Department of Agriculture for Northern Ireland, Newforge Lane, Belfast BT9 5PX, UK;*
and [2]*Agriculture and Food Science Centre, Queen's University of Belfast, Newforge Lane, Belfast BT9 5PX, UK*

The basic feature of subtraction hybridization protocols is the removal from one cell type of nucleic acid sequences that are shared with other cell types (sources of *subtracter sequences*), to leave only sequences unique to the cell type or organism in question.

Subtracter sequences may be from one or several related organisms and can be modified in a variety of ways to permit separation of unwanted hybrids from the cell-specific sequences in the hybridization mixture. These include immobilization on solid supports to facilitate removal from the mobile phase by centrifugation (Scott *et al.* 1983; Bjourson & Cooper 1988) and biotinylation to allow capture of hybrids by affinity chromatography with streptavidin (Bjourson & Cooper 1988), by binding to avidin-coated beads (Straus & Ausubel 1990; Sun *et al.* 1992) or by streptavidin–phenol–chloroform extraction (Sive & St John 1988).

These techniques are limited in their capacities to generate a pool of highly enriched cell-specific sequences, either by reliance on a single, partially efficient subtraction/separation system, or by an inability to amplify the small quantities of cell-specific nucleic acid generated by each round of subtraction.

We describe here a combined subtraction hybridization and PCR amplification procedure that overcomes these problems. It was originally designed to isolate unique DNA sequences from *Rhizobium* spp. (Bjourson *et al.* 1992) but is applicable to any group of bacteria to generate species- or strain-specific DNA probes.

New Techniques in
Food and Beverage Microbiology

0-632-03755-5

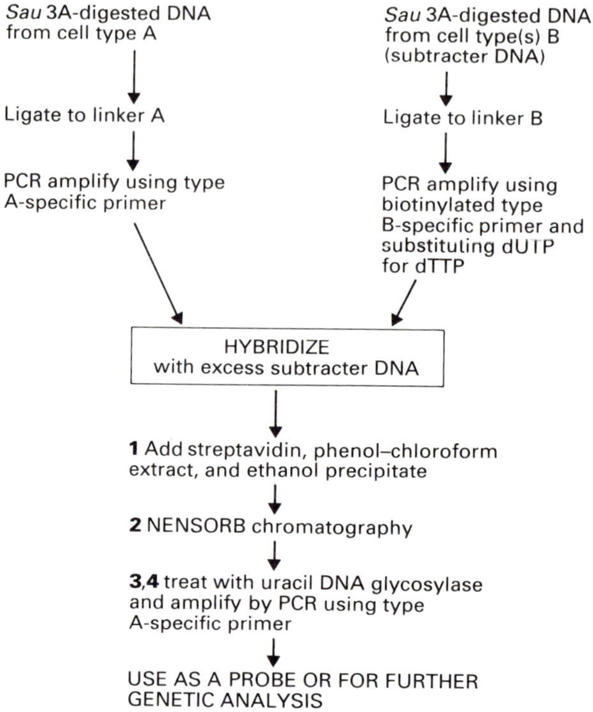

FIG. 13.1.Scheme showing the four strategies used to isolate unique nucleic acid sequences from cell type 'A' by a combined subtraction hybridization and PCR amplification.

Outline of Procedure

A combination of four distinct separation strategies can be used to isolate unique DNA sequences from the genome of cell type 'A' after its hybridization with total genomic DNA from related cell types, 'B'. The steps involved are shown diagrammatically in Fig. 13.1. Cell type 'A' in this scheme is the organism for which a DNA probe is to be constructed, and cell type 'B' refers to a group of related strains or species of bacteria that collectively provide the source of subtracter DNA.

*Sau*3A-digested DNA from cell type 'A' is ligated to a linker, denatured to the single-stranded form and hybridized in solution with a vast excess of subtracter DNA from cell type 'B', which has been restricted, ligated to a subtracter-specific biotinylated linker and amplified by PCR to incorporate dUTP, and has similarly been denatured. Subtracter DNA and 'A–B' hybrids are then removed by phenol–chloroform extraction of a streptavidin–biotin–

DNA complex. NENSORB chromatography of the sequences remaining in the aqueous layer captures biotinylated subtracter DNA that may have escaped removal by the phenol–chloroform treatment. Traces of contaminating subtracter DNA are removed by digestion with uracil DNA glycosylase. Finally, remaining sequences are amplified by PCR with the type 'A'-specific primer, labelled and tested for specificity in dot-blot hybridizations against total genomic target DNA from cell type 'B'. Removal of cross-hybridizing sequences is normally achieved after one or two rounds of subtraction/amplification.

Probe Generation

The information given below is generally applicable to probe generation in bacteria. The organism for which a probe is required is termed the *probe strain* and the group of related organisms, which share some DNA homology with the probe strain, are termed *subtracter strains*.

Isolation of DNA

Pure cultures of bacteria are grown in 50-ml conical flasks containing 25 ml of appropriate growth medium. Of each broth culture 5 ml are pelleted in a bench centrifuge and the resuspended pellets (50 μl) transferred to 1.5-ml centrifuge tubes, followed by washing three times with 500 μl TE buffer (10 mmol/l Tris–HCl, 1 mmol/l EDTA, pH 8.0). The final pellet is resuspended in 100 μl of a solution containing 25% sucrose, 1 mg/ml lysozyme, 10 mmol/l Tris–HCl, pH 8.0, at 37 °C for 15 min. Cells are lysed by the addition of 200 μl of lysis solution (5 mol/l guanidine isothiocyanate, 0.1 mol/l EDTA, pH 7.0). The lysate is mixed gently with 150 μl of 7.5 mol/l ammonium acetate and the mixture is extracted with 500 μl of chloroform/isoamyl alcohol (24 : 1 v/v) by mixing and centrifuging in a microcentrifuge for 10 min. The aqueous layer is transferred to a clean 1.5-ml centrifuge tube and the DNA is precipitated by the addition of 0.54 vol 2-propanol. DNA is collected by spooling on a pipette tip, washed twice in 100 μl of 70% ethanol and then resuspended in 20 μl of TE buffer. The concentration is estimated by gel electrophoresis of 5-μl volumes. Approximately 1 μg DNA from each strain is digested with restriction endonuclease *Sau*3A, and the restriction fragments are purified with PREP-A-GENE matrix (BioRad) as described by the manufacturer, and resuspended in 20 μl of TE buffer.

Synthesis of oligonucleotides and preparation of linkers

Oligonucleotides with the following base sequences are synthesized:

TB7006 5′ HO-AGCGGATAACAATTTCACACAGGA-OH 3′
TB7007 5′ BIOTIN-CGCCAGGGUUUUCCCAGUCACGAC-OH 3′
TB7008 5′ P-GATCTCCTGTGTGAAATTGTTATCCGCT-OH 3′
TB7009 5′ P-GAUCGUCGUGACUGGGAAAACCCUGGCG-OH 3′

Oligonucleotides are resuspended in sterile TE buffer at a final concentration of 200 μmol/l and stored in aliquots at −20 °C. Each of the appropriate oligonucleotides (5 μg) is combined to produce the linkers L-P (TB7006 and TB7008) and L-S (TB7007 and TB7009). The mixtures are heated to 65 °C and cooled slowly at room temperature to produce double-stranded linkers containing 5′-phosphorylated *Sau*3A-compatible overhangs at one end. Linker L-P is ligated to *Sau*3A-digested probe strain DNA and linker L-S is ligated to similarly digested subtracter DNA. In each case *Sau*3A-digested DNA (200 ng) is mixed with 600 ng of the appropriate linker and the mixture is ligated with DNA ligase. Excess linkers are removed with PREP-A-GENE matrix by procedures described by the manufacturer, and linked DNA is eluted in 20 μl of TE buffer at 50 °C.

Preparation of probe strain DNA

Probe strain DNA (1 μg) modified by ligation to linker L-P is amplified with 45 cycles of PCR. Each cycle consists of denaturation at 94 °C (1 min 20 s), annealing at 55 °C (1 min) and DNA polymerization at 72 °C (2 min) in an automated thermal cycler (Perkin Elmer Cetus, model 480). Reactions are performed in sterile 0.5-ml tubes with 100-μl final reaction volumes containing Tris, pH 8.3, 10 mmol/l; KCl, 50 mmol/l; $MgCl_2$, 1.5 mmol/l; gelatin, 0.01% (w/v); dNTPs, 200 μmol/l; primer TB7006, 1 μmol/l; 0.5 units of AmpliTaq DNA polymerase (Perkin Elmer Cetus). Evaporation from the tubes is prevented by addition of a 100-μl mineral oil overlay.

Preparation of subtracter DNA

Subtracter DNA from the required number of subtracter strains, modified by ligation to linker L-S, is amplified by PCR as individual strain DNA according to the amplification conditions described for probe strain DNA, except that primer TB7007 is used instead of TB7006 and dUTP is substituted for dTTP, to give a final dUTP concentration of 300 μmol/l. The efficiency of amplification is examined by submitting 10-μl volumes of each reaction to electrophoresis in composite gels (3% NuSieve agarose, 1% SeaKem agarose (w/v); FMC BioProducts, Sittingbourne, UK). Subtracter DNA PCR products are pooled, the mineral oil is removed by extraction with an equal volume of chloroform/isoamyl alcohol (24:1), the aqueous phase is transferred to a Centricon 30 microconcentrator (Amicon Ltd, UK) and the mixture is concentrated by spin dialysis to yield a final volume of 25–50 µl. The volume is made up to 2 ml with 1 mmol/l EDTA and the spin dialysis step repeated until the residual volume of the mixture is approximately 25 µl. This step is repeated with another 2 ml of 1 mmol/l EDTA.

Subtraction hybridization

PCR-amplified probe strain DNA (1–5 ng) and approximately 20 µg subtracter DNA are mixed in a 0.5-ml microcentrifuge tube containing hybridization solution, to give a total volume of 10 µl containing 50 mmol/l HEPES, pH 7.5; 500 mmol/l NaCl; 1 mmol/l EDTA and 0.1% SDS (final concentrations). The mixture is overlaid with 50 µl mineral oil and the DNA is denatured at 99 °C for 10 min, cooled rapidly on ice and incubated at 64 °C for 48 h to allow subtraction hybridization to take place.

Isolation of probe strain DNA sequences from the subtraction mixture

A solution (100 µl) containing 500 mmol/l NaCl, 1 mmol/l EDTA, and 50 mmol/l HEPES is added and the mixture is briefly centrifuged, excess mineral oil removed and 10 µg of streptavidin is added. The tube is mixed gently at room temperature for 5 min and a further 20 µg of streptavidin is added. The mixture is extracted with an equal volume of phenol/chloroform (50:50, v/v) and centrifuged in a microcentrifuge for 10 min at 15 000 *g*. The aqueous phase is transferred to a fresh tube, SDS is added to 0.1% and the mixture is extracted twice more with phenol–chloroform and once with chloroform, and the DNA remaining in the aqueous phase is precipitated after the addition of 3 mol/l sodium acetate (0.1 vol) and 2 vol 99% ethanol. The pellet, which may not be visible, is washed with 100 µl 70% ethanol and redissolved in 20 µl TE buffer, pH 8.0. The deproteinized DNA is desalted with a NENSORB 20 purification cartridge (Du Pont Ltd, UK), as described by the manufacturers, and finally resuspended in 10 µl TE buffer. This subtracted DNA (5 µl), enriched for probe strain-specific sequences, is used for another subtraction cycle and 1 µl of the remaining 5 µl is used to prepare logarithmic dilutions in 9 µl aliquots of TE buffer. Of each dilution 10 µl is mixed with PCR reagents (Tris, pH 8.3, 10 mmol/l; KCl, 50 mmol/l; $MgCl_2$, 1.5 mmol/l; gelatin, 0.01% (w/v); dNTPs, 200 µmol/l) containing 15 units of uracil DNA glycosylase (Cetus) and incubated at 37 °C for 1–4 h to destroy all traces of dUTP-containing subtracter DNA. Primer TB7006 (5 µl of 200 µmol/l , probe strain-specific) and 0.5 units of *Taq* polymerase are added to a final volume of 100 µl and the reaction mix is amplified for 45 cycles, under the temperature conditions described above, to prepare probe strain

DNA. PCR products are detected by electrophoresis of 10-μl volumes of each PCR mix in composite gels.

^{32}P labelling of subtracted probe strain DNA

Subtracted and PCR-amplified probe strain DNA (5 μl) is radiolabelled with three cycles of amplification. Each cycle consists of denaturation at 95 °C (2 min), renaturation at 55 °C (2 min), and polymerization at 72 °C (10 min). Reactions are performed in 100-μl final volumes containing Tris, pH 8.3, 10 mmol/l; KCl, 50 mmol/l; $MgCl_2$, 1.5 mmol/l; gelatin, 0.01% (w/v); dATP, dTTP, dGTP, 200 μmol/l each; 5 μl [^{32}P]dCTP (400 ci/mmol); primer TB7006, 1 μmol/l, and 0.5 units of AmpliTaq DNA polymerase (Perkin Elmer Cetus). Unincorporated nucleotides are removed by push-column chromatography (Stratagene Ltd, UK).

Preparation of DNA dot blots

Total genomic target DNA (1 μg) from each subtracter strain is denatured at 100 °C for 10 min, snap-cooled on ice and 1 vol of 20 × SSC (1 × SSC is 0.15 mol/l sodium chloride, 0.015 mol/l sodium citrate) is added. Samples are manually spotted on to Hybond-N^+ membranes (Amersham plc) in 5-μl aliquots. Membranes are placed on filter paper soaked in denaturing solution (1.5 mol/l NaCl, 0.5 mol/l NaOH) for 5 min and transferred to another filter paper containing neutralizing solution (1.5 mol/l NaCl, 0.5 mol/l Tris–HCl pH 7.2, 1 mmol/l EDTA) for 1 min. DNA is fixed to the membranes by placing them on filter paper soaked in 0.4 mol/l NaOH (20 min), immersed briefly in 6 × SSC, followed by air-drying. Membranes are stored at 4 °C until used.

Filter hybridizations

Filters are prehybridized for 5 h at 68 °C in bags containing prehybridization solution (4 × SSC; 10 × Denhardt's solution; 100 μg/ml sonicated denatured salmon sperm DNA; 10 mmol/l Tris–HCl, pH 8.0; 0.1% SDS). Hybridization is performed at the same temperature with 2.5 × 10^6 cpm/ml of ^{32}P-labelled subtracted DNA probe for 18 h. Filters are washed in 1 × SSC, 0.5% SDS for 1 h (4 × 15 min), and in 0.1 × SSC, 0.5% SDS at 68 °C for 1 h (4 × 15 min). Filters are subjected to autoradiography, as described by Bjourson & Cooper (1988). An example of probe specificity is shown in Fig. 13.2. Nucleic acid sequences that cross-hybridize with total genomic target DNA from seven strains of *Rhizobium trifolii* (2–8) have been removed from *R. trifolii* strain 1 by two rounds of subtraction/amplification. The former

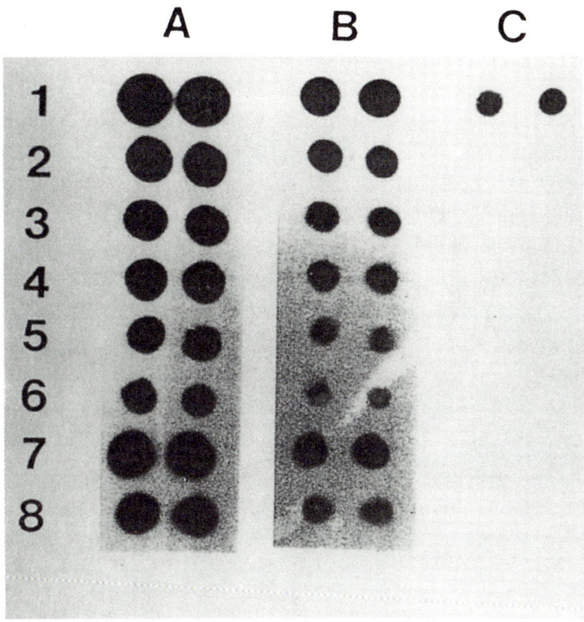

FIG. 13.2. Specificity of DNA sequences from *Rhizobium trifolii* strain 1, generated by subtraction hybridization. DNA from strain 1 was submitted to (a) zero, (b) 1 or (c) 2 rounds of subtraction/amplification against DNA from seven other strains of the same species (2–8). At each stage ^{32}P-labelled strain 1 sequences were hybridized in duplicate to 1 μg of total genomic DNA from the homologous and subtracted strains.

strains were the source of subtracter DNA in this experiment. Although this is an example of a soil bacterium, the principle is equally applicable to any other organism.

Conclusions

This method permits the reliable and rapid isolation of strain- or species-specific bacterial DNA sequences. The isolated fragments can be used directly as probes in colony or dot-blot hybridizations with unknown target DNA, or they can be cloned and further characterized by sequencing or restriction mapping.

The stringency at which the subtraction is performed is an important factor when considering the intended use of the final sequences. Low-stringency conditions will remove some probe strain sequences which have a relatively low base sequence homology with the subtracter DNA, whereas high-stringency subtraction will remove only perfectly matched sequences. Modification of probe and subtracter DNA, by the addition of specific linkers, permits the unbiased amplification of fragments in the 200–2500-bp size range and it is important that the restriction enzyme chosen for the DNA digestion yields fragments within this size range. Despite the need for the large molar excess of subtracter over probe strain DNA, the actual amount required for each round of subtraction is in the 10–20-μg range. Additionally, the subtracter DNA is in an easily replenishable form and this feature would assume special significance for organisms that are difficult to grow or whose DNA is difficult to extract. The method may be applicable to non-culturable organisms.

References

BJOURSON, A.J. & COOPER, J.E. 1988 Isolation of *Rhizobium loti* strain-specific DNA sequences by subtraction hybridization. *Applied and Environmental Microbiology* **53**, 1705–1707.

BJOURSON, A.J., STONE, C.E. & COOPER, J.E. 1992 Combined subtraction hybridization and polymerase chain reaction amplification procedure for the isolation of strain specific *Rhizobium* DNA sequences. *Applied and Environmental Microbiology* **58**, 2296–2301.

SCOTT, M.R.D., WESTPHAL, K.-W. & RIGBY, P.W.C. 1983 Activation of mouse genes in transformed cells. *Cell* **34**, 557–567.

SIVE, H.L. & ST JOHN, T. 1988 A simple subtractive hybridization technique employing photo-active biotin and phenol extraction. *Nucleic Acids Research* **16**, 10937.

STRAUS, D. & AUSUBEL, F.M. 1990 Genomic subtraction for cloning DNA corresponding to deletion mutants. *Proceedings of the National Academy of Sciences* **87**, 1889–1892.

SUN, T.-P., GOODMAN, H.M. & AUSUBEL, F.M. 1992 Cloning of the *Arabidopsis* GA1 locus by genomic subtraction. *Plant Cell* **4**, 119–128.

14: Microbial Detection and Two-dimensional Preservative Mapping Using *lux*-based Bioluminescence

CAROLE L.REID,* W.M.WAITES AND G.S.A.B.STEWART[†]
Department of Applied Biochemistry and Food Science, University of Nottingham, Sutton Bonington, Loughborough, Leicestershire LE12 5RD, UK

Principles of Bioluminescence Mapping

Current methods of microbial detection and enumeration disrupt the food matrix and result in the loss of spatial information. Most foods are composed of complex heterogeneous mixtures of components, so that different locations within the food have variations in levels of nutrients, oxygen, pH, water activity and preservatives. The food matrix, therefore, may be considered as a series of interconnected microenvironments, some of which may be more suitable for microbial growth than others (Dodd *et al.* 1990). Knowledge of sites of microbial growth may be used to extend shelf life by reducing growth-supporting phases. Alternatively, specific targeting of preservatives to these sites may allow reduction in the use of preservative. This approach may lead to the identification of factors that significantly contribute to bacterial growth or survival during processing and assist our understanding of the interactions between microorganisms in foods.

Most methods of microbiological analysis, which involve the familiar techniques of serial dilution and plating for culture, necessarily involve examination of only a proportion of the total population of cells. It is now feasible to examine the functional biochemistry of cells *in situ* in a food as a con-

* Present address: Department of Chemical Engineering, Loughborough University of Technology, Loughborough, Leicestershire LE12 3TU, UK.
[†] To whom correspondence should be addressed.

New Techniques in
Food and Beverage Microbiology

0-632-03755-5

sequence of developments in the fields of bioluminescence and photon counting and imaging.

Bacterial bioluminescence has been demonstrated in *Alteromonas*, *Photobacterium*, *Vibrio* and *Xenorhabdus* species – taxa that also include non-luminous species (Nealson & Hastings 1979). Bioluminescence involves a luciferase-catalysed oxidation of flavin mononucleotide ($FMNH_2$) by molecular oxygen, with the concomitant oxidation of a long-chain aliphatic aldehyde:

$$FMNH_2 + RCHO + O_2 \rightarrow FMN + H_2O + RCOOH + \text{light (490 nm)}$$

For novel bioluminescent assays in microbiology, it is necessary to have instruments capable of detecting low levels of light, and the ability to transfer the genes encoding the enzymic components for light production into normally dark bacteria. These genetically manipulated bacteria must be identical to their parents in all other physiological respects. The *lux* genes from *Vibrio* spp. have been transferred into other species by plasmid transformation, transposition or conjugation (Stewart & Williams 1992).

Bacteria containing the *luxAB* genes only produce the luciferase enzyme and are inherently dark in the absence of exogenous aldehyde. Therefore, there is minimal perturbation of cellular metabolism until aldehyde is added. Determination of bioluminescence is made easily and accurately without damage to the cells, which remain available for subsequent correlations between bioluminescence and colony-forming ability, or for *in vitro* measurements of luciferase, other enzymes or metabolites.

Since the production of light from recombinant organisms depends on a functional intracellular biochemistry, agents that disrupt or kill the *lux*-containing bacteria destroy their metabolism and directly or indirectly affect the intracellular production of $FMNH_2$, so reducing the bioluminescence of the culture (Jassim *et al.* 1990; Meighen 1991). The susceptibility of cells to different antibiotics or toxic materials can, therefore, be determined. This principle can be used to detect inhibitory residues in foods and to evaluate the biological efficacy of food-grade preservatives (Ahmad & Stewart 1991; Stewart *et al.* 1991).

A requirement for the application of bioluminescence is the establishment of a strict correlation between *in vivo* bioluminescence and cell viability, as determined by conventional plate counts. Comparative studies for various biocides and for sublethal cellular injury of *Salmonella typhimurium* and *Listeria monocytogenes* indicate that the correlation is valid (Jassim *et al.* 1990; Ellison *et al.* 1991; Stewart *et al.* 1991; Walker *et al.* 1992a, b).

By means of photon-counting imaging, it is now possible to form images at low photon levels, with technology developed for use in astronomy and spectroscopy. Ultralow-level imaging permits observations of dynamic events in biology and biomedical science and acquires the image as a composite of discrete photon events. Photon-counting imaging devices, with digital image processors, are available and allow visualization of light in two or even three dimensions. The BIQ imaging system (Image Research, Cambridge, UK) can detect single bioluminescent *Escherichia coli* cells (Mitchell 1991). Similar results were obtained by Masuko *et al.* (1991) with the Hamamatsu Argus 100 Vim 3 camera (Hamamatsu Photonics, Enfield, UK). These instruments can give spatial information about bacterial growth and viability in complex food systems in real time and in a non-invasive manner. These cameras can also be used to obtain a conventionally transmitted or reflected light image for comparison with the luminescent image and to localize regions of bioluminescence (Wick 1989).

Detection of microorganisms *in situ* has been achieved by the microscopic examination of cryosections of foods appropriately stained to observe microorganisms. Toluidine blue has been used as a stain to differentiate food components (Flint & Firth 1988). In conjunction with cryosectioning, toluidine blue has made possible the detection of microbial growth sites at or near the surface of spoiling fresh British sausages (Dodd & Waites 1989). With *Lactobacillus plantarum* and cryosectioning model fermentation mixtures, Fox *et al.* (1989) were able to demonstrate uneven distribution of the starter culture in fermented red meats, with bacterial populations aggregated at the interfaces of food components. They also observed similar inconsistencies in commercially produced salami.

Mould ripening is a common feature of dry salamis, such as Italian and Hungarian salamis, giving rise to characteristic flavours and aromas. Moulds are inoculated on to the surface and mycelia penetrate deeply, causing a decrease in lactic acid and a consequent rise in pH (Grazia *et al.* 1986). Duitschaver & Irvine (1971) found that mould growth at the surface of cheese led to a pH gradient, with more alkaline conditions near the surface, which favoured the growth of staphylococci and the production of enterotoxin D. Between 1967 and 1985, six incidents of staphylococcal food poisoning were publicized and traced to fermented meat products, including Genoa and Italian salami, by the United States Department of Agriculture. They involved major producers and large amounts of product, and in each case there was a lack of controls designed to ensure that the final product was safe (Halpin-Dohnalek & Marth 1989).

Since staining techniques do not differentiate clearly between living and dead bacterial cells, *in vivo* bioluminescence, which detects metabolizing cells, is uniquely appropriate as a tool for studies on microbial growth and survival in food systems. To monitor the distribution of preservative concentrations throughout a food matrix, *lux* recombinant indicator bacteria can be incorporated in model foods or, alternatively, they may be applied to the surface of sections of food (Stewart *et al.* 1991). We describe the use of bioluminescent bacteria to map variations in antimicrobial areas within a model meat fermentation and mould-ripened salami.

Mapping Antimicrobial Areas in Foods

Salmonella typhimurium LT2[pSB100], containing the *luxAB* genes from *Vibrio fischeri* (Blisset & Stewart 1989) and selectable with ampicillin, was grown in 5 ml of Luria broth (LB; Maniatis *et al.* 1982), containing 30 μg/ml ampicillin (Sigma), overnight at 30 °C in a tube shaker. A 5% inoculum was made into 50 ml of tryptone soya broth (TSB, Oxoid) and the culture grown in an orbital shaker to an A_{600} of 0.5. This was added to 500 g of raw, minced beef and mixed to disperse as evenly as possible. The inoculated minced beef was firmly packed into a beaker, covered and incubated at 30 °C. At intervals, the surface of the beaker was exposed to aldehyde vapour and the light profile was determined with a Hamamatsu Argus 100 Vim 3 photon-counting camera (Hamamatsu Photonics UK Ltd, Enfield, Middlesex, UK).

Lactic acid bacteria, able to produce bacteriocins (Campden Food and Drink Research Association, Chipping Campden, UK), were added to the inoculated minced beef with glucose (1% w/w): *Pediococcus pentosaceus* strains FBB61 (CRA 509) and the bacteriocin-negative variant (CRA 645), *Leuconostoc mesenteroides* strains NCIB 8189 (CRA 531) and CRA 38, and *Lactobacillus sake* Lb 706 (Schillinger & Lücke 1989). The bacteria were grown in universal bottles in 25 ml of MRS broth (Oxoid) overnight at 30 °C. Cells were harvested by centrifugation, washed and resuspended in phosphate-buffered saline (PBS, Oxoid) and added to the inoculated minced beef, to give approximately 2×10^8 cfu/g. The inoculated beef was packed tightly into sterile 25-ml syringes, with the needle ends removed, loosely covered with a small Petri dish and incubated at 30 °C. The bioluminescence of sections cut from the end and from inside the syringe was observed as above.

The decrease in pH resulting from the lactic fermentation was determined with a stab electrode (Mettler, Leicester, UK). Because of the unevenness of acidification, 10 pH determinations were made at each sampling time and the lowest reading recorded.

In order to quantify light emission and to determine the numbers of indicator bacteria, a second batch of minced beef was prepared as described above. Sterile Ringer's solution (Oxoid) (9 ml) was added to 1 g of meat and the mixture was vortexed. The bioluminescence emitted by 1 ml for 10 s was determined immediately after addition of 10 μl of a 1% (v/v) solution of aldehyde in ethanol, in a scintillation vial placed in a Turner Designs 20e luminometer (Steptech, Herts SG4 7JE, UK). To determine the numbers of *S. typhimurium*, aliquots were serially diluted with Ringer's solution, plated on MacConkey No. 3 (Oxoid) containing 30 µg/ml of ampicillin and incubated overnight at 37 °C.

To determine the profile of antimicrobial activity within ready-to-eat meats, bioluminescent bacteria were evenly incorporated into a simple agarose gel, cast with polyacrylamide gel (PAGE) plates and sufficiently pliable to be laid over the surface of meat slices without air spaces or gaps. The glass plates (10 × 10 cm) from a BioRad Mini-Protean II PAGE system (Hemel Hempstead, UK) were prepared by cleaning thoroughly with ethanol. Spacers (0.75 mm) were inserted and the plates were assembled on the casting stand, to ensure a tight fit with the rubber cushions.

S. typhimurium was grown as described above. *L. monocytogenes* 23074[pSP4] (Walker *et al.* 1992a) was grown in LB containing 7.5 µg/ml chloramphenicol overnight at 30 °C. A 5% inoculum was made into 50 ml of brain–heart infusion broth (BHI, Oxoid) and incubated in a rotary shaker at 30 °C. Cells were harvested at an A_{600} of approximatcly 0.5 by centrifugation for 10 min at 3000 *g* and 4 °C, followed by resuspension in Ringer's solution to one-fifth the initial volume. Concentrated cells (5 ml) were added to 20 ml of sterile Ringer's solution containing 0.75% agar tempered to 50 °C. The bottles were vortexed and immediately poured into the prepared PAGE plates. The gels were allowed to solidify at room temperature and were stored betwecn the glass plates in sealed plastic boxes at 5 °C for up to 10 days. When required, one glass plate was removed by twisting the spacers, and the seeded gel was allowed to slide off the other glass plate and on to the slice of fermented meat. Salami was purchased locally, from a major high street supermarket, and thinly sliced at point of purchase. For storage of the gel-overlaid salami, cling film was wrapped over the top of the container, so as not to touch the gel. To induce bioluminescence, dodecanal (1 µl) was spotted on a large weighing boat and inverted over the salami for 10 s. Light profiles were determined with the Hamamatsu system, as described above, at various times after the overlay had been placed on the salami. The pH of regions of the salami and the overlay were determined with a flat-based electrode (model PHE-2171; Omega, Leicestershire LE9 6XR, UK).

TABLE 14.1. *Bioluminescence profile of* Salmonella typhimurium *growing on minced beef*

Time (h)	Relative bioluminescence	Colony count ($\times 10^8$)
3.5	788	0.6
5	3861	2.59
7.5	8350	8.93
10	8680	17.0

TABLE 14.2. *Effect of lactic acid bacteria in minced beef on the pH, colony count and bioluminescence of* S. typhimurium *LT2[pSB100] after 5 h at 30°C*

Lactic acid bacteria added	pH	Colony count ($\times 10^8$)	Relative bioluminescence
Control	5.80	2.7	3861
P. pentosaceus CRA 509	5.64	2.6	2958
P. pentosaceus CRA 645	5.68	2.2	2160
L. mesenteroides CRA 531	5.58	0.9	301
L. mesenteroides CRA 38	5.60	1.62	1163
L. sake Lb 706	5.63	2.23	1961

In situ *growth of bioluminescent salmonellas in minced beef*

The bioluminescence and the colony count of salmonellas in the minced beef at various intervals are shown in Plate 14.1(a) (*opposite*) and Table 14.1, respectively. Variability in the luminescence is evident in the photon-counting image, signifying the uneven growth of *S. typhimurium* within the minced beef.

Bioluminescence of salmonellas during meat fermentation with lactic acid bacteria

Lactic acid fermentation lowers the pH of the minced beef and reduces the colony count and the bioluminescence of the *S. typhimurium* LT2, as determined by the Turner luminometer (Table 14.2). The image accumulated by the photon-counting camera shows the variation in light emitted after 5 h coculture with the lactic acid bacteria at 30 °C (Plate 14.1(b)). In minced beef coincubated with lactic acid bacteria, less light was detected by the light profile and in the luminometer. *L. mesenteroides* strains CRA 531 and CRA 38 produced the most rapid decrease in pH of the meat and bioluminescence of the salmonellas. There was uneven bioluminescence in the slices, reflecting the uneven distribution of the light-emitting bacteria. This is the net effect of the uneven growth of the salmonellas and inhibitor production by the lactic acid bacteria.

Antimicrobial profiles in slices of fermented meats

The image by reflected light and the pseudocoloured bioluminescence profile of four sliced-meat products, stored at 30 °C for 1 h, are shown in Plate 14.1(c) and (d), respectively. The pH range at the surface of each slice is shown in Table 14.3. It can be seen that less light is emitted by gels overlying salamis with lower pH.

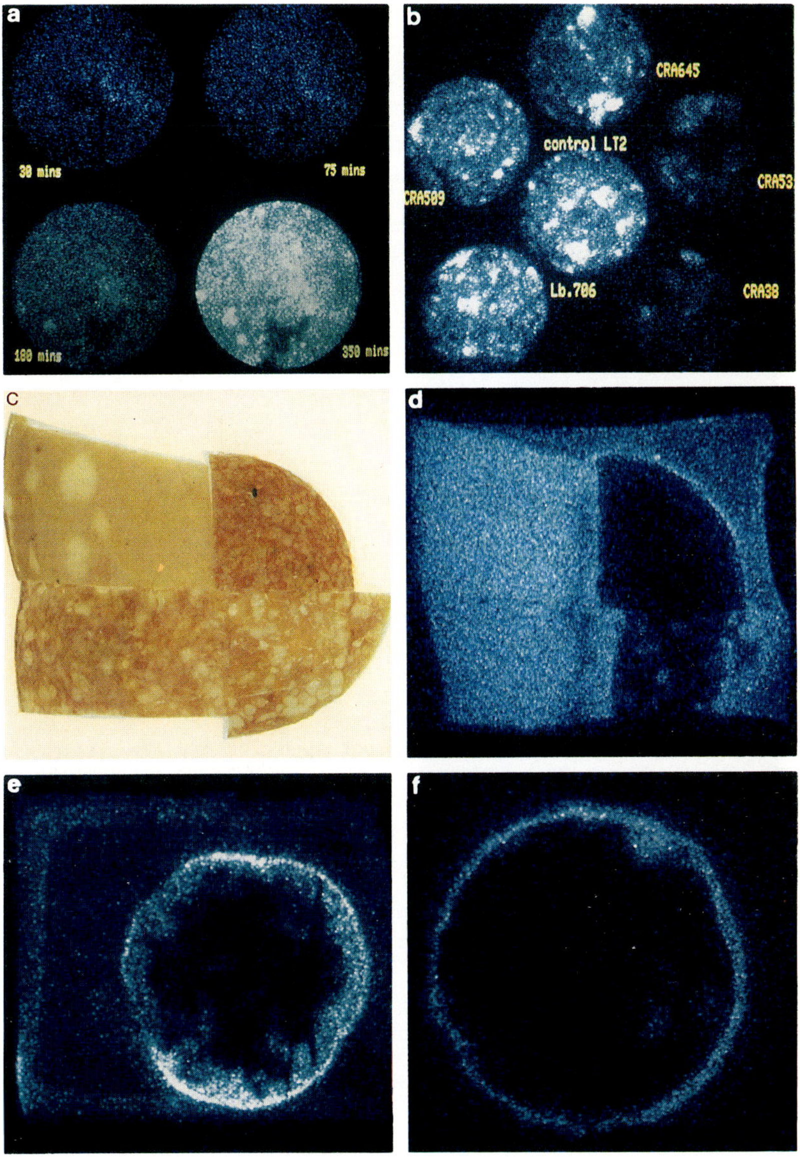

PLATE 14.1. Two-dimensional bioluminescence mapping with *lux* recombinant bacteria and an Argus 100 Vim 3 photon video camera. (a) Bioluminescence of salmonellas in minced beef at various intervals; (b) variation in light emitted after 5-h coculture with lactic acid bacteria; (c) reflected light image of four sliced meat products stored at 30°C for 1 h; top left, Mortadella, top right, German salami, bottom left, Bierwurst, bottom right, Italian salami; (d) pseudocoloured bioluminescence profile of the sliced meats shown in (c); (e) and (f) antimicrobial profile of slices of mould-ripened salami as determined with bioluminescent *S. typhimurium* (e) and biluminescent *L. monocytogenes* (f).

[*facing page 186*]

TABLE 14.3. *pH range of four sliced meat products after storage at 30°C for 1 h*

Meat product	pH range
Mortadella	6.33–6.36
German salami	4.62–4.63
Italian salami	5.36–6.22
Bierwurst	6.34–6.39

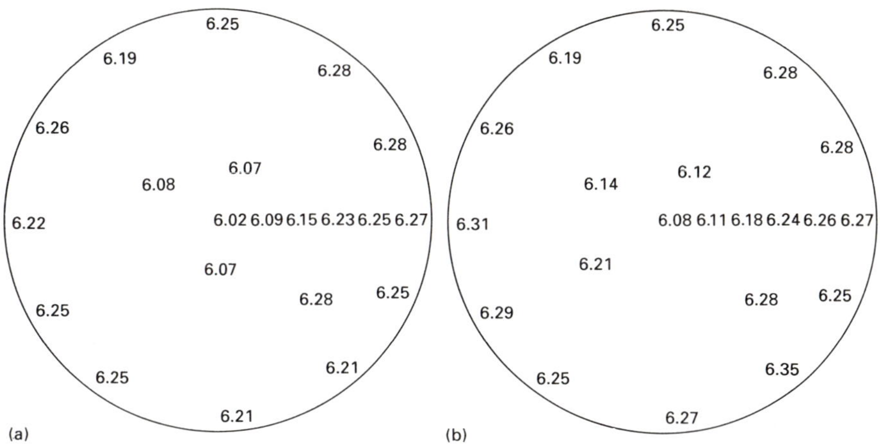

FIG. 14.1. The pH profiles, as determined with a flat-based electrode, of the salamis shown in Plate 14.1(e) and (f).

The antimicrobial profiles of slices of mould-ripened salami are similar whether determined by bioluminescent *S. typhimurium* (Plate 14.1(e)) or by *L. monocytogenes* (Plate 14.1(f)). These light profiles mirror the pH profiles obtained from transects of the slices (Fig. 14.1(a) and (b), respectively). The pH around the edge of the slice, and hence at the surface of the mould-ripened salami, is significantly higher than nearer the centre.

Conclusions

By means of bioluminescence, the physiological status of cells of specific bacterial strains may be monitored as individual component species in multixenic microbial systems, including model food systems, biofilms and biofermentors. Use of *lux* genes, under the control of specific promoters, such as those implicated in toxin production – for example, *eta* in *Staphylococcus aureus*

(Sheehan *et al.* 1992) – may allow rapid visualization of conditions and locations in which toxins may be produced. Osmoregulated bioluminescent *E. coli* (Park *et al.* 1989) may facilitate rapid, accurate and simple determination of salt profiles and the a_w in foods.

Studies with photon-counting imaging clearly demonstrate the potential of this technique to obtain near-instantaneous images of dynamic events that previously have been impossible to visualize. Bioluminescence images of inhibitory profiles are evident after only 20 min contact time with the seeded overlay on the food and after 1 min accumulation of light by photon counting. This makes the technique a powerful research tool. The ease of use and the low cost of consumables, as well as the speed with which results are obtained, may be relevant for quality control laboratories in the food industry.

Acknowledgements

We would like to thank MAFF and Campden FDRA for the financial support of Carole Reid, who held a CASE Studentship, and Dr Richard Earnshaw of Campden FDRA for helpful discussions and for supplying the lactic acid bacteria. The Hamamatsu Argus 100 VIM 3 camera was provided through AFRC special equipment support.

References

Ahmad, K.A. & Stewart, G.S.A.B. 1991 The production of bioluminescent lactic acid bacteria suitable for the rapid assessment of starter culture activity in milk. *Journal of Applied Bacteriology* **70**, 113–120.

Blisset, S.J. & Stewart, G.S.A.B. 1989 *In vivo* bioluminescent determination of apparent K_ms for aldehyde in recombinant bacteria expressing *luxA/B*. *Letters in Applied Microbiology* **9**, 149–152.

Dodd, C.E.R. & Waites, W.M. 1989 The use of cryosectioning and histological staining to detect the sites of microbial growth in British fresh sausages. *Journal of Applied Bacteriology* **67**, xvii.

Dodd, C.E.R., Stewart, G.S.A.B. & Waites, W.M. 1990 Biotechnology-based methods for the detection, enumeration and epidemiology of food poisoning and spoilage organisms. *Biotechnology and Genetic Engineering Reviews* **8**, ed. Tombs, M.P., pp. 1–52. Newcastle: Intercept.

Duitschaver, C.L. & Irvine, D.M. 1971 A case study: effect of mould growth on growth of coagulase-positive staphylococci in Cheddar cheese. *Journal of Milk and Food Technology* **34**, 583–585.

Ellison, A., Perry, S.F. & Stewart, G.S.A.B. 1991 Bioluminescence as a real-time monitor of injury and recovery in *Salmonella typhimurium*. *International Journal of Food Microbiology* **12**, 323–332.

Flint, F.O. & Firth B.M. 1988 Improved toluidine blue stain mountant for the microscopy of food products. *Analyst* **113**, 365–366.

Fox, P.A., Dodd, C.E.R., Nicolaides, L. & Waites, W.M. 1989 The use of cryosectioning and histological staining to study microbial growth during fermentation of meat. *Journal of Applied Bacteriology* **67**, xvii.

GRAZIA, L., ROMANO, P., BAGNI, A., ROGGIANI, D. & GUGLIELMI, G. 1986 The role of moulds in the ripening process of salami. *Food Microbiology* **3**, 19–25.

HALPIN-DOHNALEK, M.I. & MARTH, E.H. 1989 *Staphylococcus aureus*: production of extracellular compounds and behaviour in foods – a review. *Journal of Food Protection* **52**, 267–282.

JASSIM, S.A.A., ELLISON, A., DENYER, S.P. & STEWART, G.S.A.B. 1990 *In vivo* bioluminescence: a cellular reporter for research and industry. *Journal of Bioluminescence and Chemiluminescence* **5**, 115–122.

MANIATIS, T., FRITSCH, E.F. & SAMBROOK, J. 1982 *Molecular Cloning: a Laboratory Manual*. New York: Cold Spring Harbor Laboratory Press, p. A1.

MASUKO, M., HOSOI, S. & HAYAKAWA, T. 1991 A novel method for detection and counting of single bacteria in a wide field using an ultra-high-sensitivity TV camera without a microscope. *FEMS Microbiology Letters* **81**, 287–290.

MEIGHEN, E.A. 1991 Molecular biology of bacterial bioluminescence. *Microbial Reviews* **55**, 123–142.

MITCHELL, P. 1991 Detecting the single cell. *Laboratory Equipment Digest* March, 37–38.

NEALSON, K.H. & HASTINGS, J.W. 1979 Bacterial bioluminescence: its control and ecological significance. *Microbial Reviews* **43**, 496–518.

PARK, S.F., STIRLING, D.A., HULTON, C.S.J., BOOTH, I.R., HIGGINS, C.F. & STEWART, G.S.A.B. 1989 A novel, non-invasive promoter probe vector: cloning of the osmoregulated *proU* promoter of *Escherichia coli* K12. *Molecular Microbiology* **3**, 1011–1023.

SCHILLINGER, U. & LÜCKE, F.-K. 1989 Antibacterial activity of *Lactobacillus sake* isolated from meat. *Applied and Environmental Microbiology* **55**, 1901–1906.

SHEEHAN, B.J., FOSTER, T.J., DORMAN, C.J., PARK. S. & STEWART, G.S.A.B. 1992 Osmotic and growth-phase dependent regulation of the *eta* gene of *Staphylococcus aureus*: a role for DNA supercoiling. *Molecular and General Genetics* **232**, 49–57.

STEWART, G.S.A.B. & WILLIAMS, P. 1992 *lux* genes and the applications of bacterial bioluminescence. *Journal of General Microbiology* **138**, 1289–1300.

STEWART, G.S.A.B., DENYER, S.P. & LEWINGTON, J. 1991 Microbiology illuminated: gene engineering and bioluminescence. *Trends in Food Science and Technology* **2**, 7–10.

WALKER, A.J., JASSIM, S.A.A., HOLAH, J.T., DENYER, S.P. & STEWART, G.S.A.B. 1992a Bioluminescent *Listeria monocytogenes* provide a rapid assay for measuring biocide efficacy. *FEMS Microbiology Letters* **91**, 251–256.

WALKER, A.J., HOLAH, J.T., DENYER, S.P. & STEWART, G.S.A.B. 1992b The antibacterial activity of Virkon measured by colony growth and bioluminescence of *lux* recombinant *Listeria monocytogenes*. *Letters in Applied Microbiology* **15**, 80–82.

WICK, R.A. 1989 Photon counting imaging: applications in biomedical research. *BioTechniques* **7**, 262–269.

15: Diarrhoetic Shellfish Poisoning Toxins

A.H.Ritchie
Torry Research Station, PO Box 31, 135 Abbey Road, Aberdeen AB9 8DG, UK

Diarrhoetic shellfish poisoning (DSP), a severe gastrointestinal illness, can be caused by eating shellfish that have been feeding on toxic strains of unicellular marine algae called dinoflagellates. The causes of DSP are phycotoxins that were first identified by Japanese workers in the late 1970s, and 13 of these polyether compounds, comprising three distinct groups, have been described.

The first of the carboxylic toxins to be chemically characterized was dinophysis toxin (DTX-1) (Murata *et al.* 1982), which was isolated from the hepatopancreas of Japanese blue mussels (*Mytilus edulis*) and is a methyl derivative of the parent compound okadaic acid (OA). Subsequently, OA itself was isolated from blue mussels, although it was later shown to be the major toxic component in mussels of European origin (Yasumoto *et al.* 1985; Kumagai *et al.* 1986) and only a relatively minor component in Japanese shellfish. In addition, an unsaturated fatty acid ester of DTX-1, named DTX-3, was isolated from the digestive glands of the scallop (*Patinopecten yessoensis*) (Yasumoto *et al.* 1985). More recently, a new marine toxin, dinophysis toxin-2 (DTX-2), has been isolated and characterized from toxic Irish mussels (Hu *et al.* 1992) (Fig. 15.1). The other two groups of toxins comprise a series of macrolide structures, named pectenotoxins 1–7, and two yessotoxins, characterized by two sulphate ester groups connected to contiguously fused ether rings (Murata *et al.* 1987) (Figs 15.2 & 15.3).

Shellfish Poisoning Toxins

Occurrence

DSP toxins are produced by two genera of dinoflagellates, *Dinophysis* and *Prorocentrum* (Murakami *et al.* 1982; Dahl & Yndestad 1985), which, under

New Techniques in
Food and Beverage Microbiology

0-632-03755-5

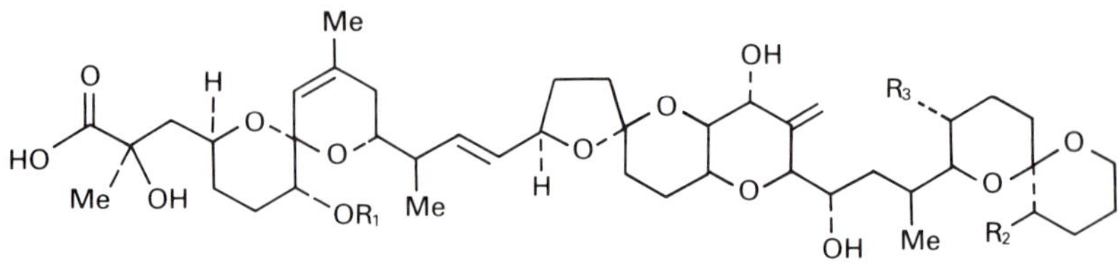

	R_1	R_2	R_3
Okadaic acid (OA)	H	H	CH_3
Dinophysistoxin-1 (DTX-1)	H	CH_3	CH_3
Dinophysistoxin-2 (DTX-2)	H	CH_3	H
Dinophysistoxin-3 (DTX-3)	acyl	CH_3	CH_3

FIG. 15.1.Dinophysis toxins.

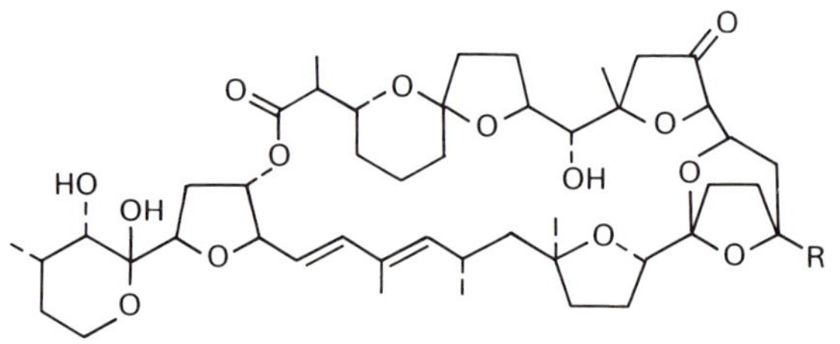

	R
Pectenotoxin-1 (PTX1)	CH_2OH
Pectenotoxin-2 (PTX2)	CH_3
Pectenotoxin-3 (PTX3)	CHO
Pectenotoxin-6 (PTX6)	COOH

FIG. 15.2.Pectenotoxins.

favourable environmental conditions, grow to large numbers to produce algal blooms.

Generally, the predominant species in Western Europe are *Dinophysis* spp., while *Prorocentrum* spp. are found mainly in Japan. In Scandinavia, however, toxic algal blooms simultaneously contain a number of species including *D. acuminata*, *D. acuta*, *D. norvegica* and *P. micans* (Krogh *et al.* 1985). *D. acuminata* is the main component in major algal blooms off the northwestern coast of France (Marcaillou-LeBaut *et al.* 1985), while in Spain in toxic episodes *D. acuminata* and *D. acuta* have been identified (Rodriquez-Vasquez *et al.* 1989).

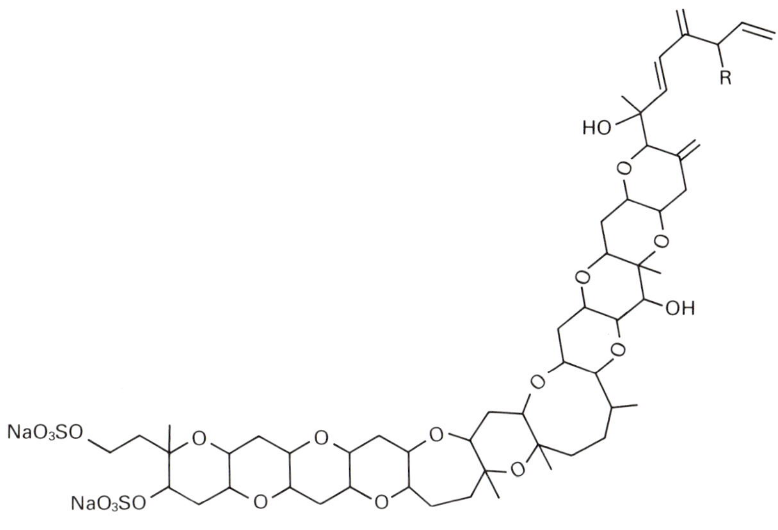

	R
Yessotoxin (YTX)	H
45-Hydroxy yessotoxin (45-OH YTX)	OH

FIG. 15.3.Yessotoxins.

Toxicity

Diarrhoetic shellfish poisoning was first described by Yasumoto *et al.* (1978) in an outbreak of food poisoning associated with eating mussels and scallops. The predominant symptoms were diarrhoea, nausea, vomiting and abdominal pain. The onset of symptoms of the condition, which was never lethal, ranged from 30 min to a few hours after ingestion of the toxic shellfish, with complete recovery within 3 days. The toxins were, however, lethal to mice and the lethal dose by intraperitoneal injection was only 1/16 the oral dose. The minimum dose of OA required to induce toxic symptons in human adults has been estimated at 48 μg (Yasumoto *et al.* 1984). The lethal dose in mice was designated 1 mouse unit (1 MU) and is equivalent to 4 μg OA (Table 15.1). Of the DSP toxins, only OA, DTX-1 and DTX-3 appeared to produce diarrhoea on oral challenge of rats. When tissue containing OA is boiled, very slow decomposition takes place and cooking for nearly 3 h is required to reduce the concentration by 50% (Edebo *et al.* 1988).

In vitro, OA is a powerful inhibitor of protein phosphatases 1 and 2A (Bialojan & Takai 1988; Cohen 1989). These phosphatases are responsible for the phosphorylation of the p-light chains of myosin, which triggers smooth

TABLE 15.1. *Physicochemical and biological characteristics of DSP toxins*

Toxin	Molecular weight	Molecular formula	MP (°C)	$[\alpha]_D^{20}$	Toxicity (μg/kg IP)	Reference
OA	804	$C_{44}H_{68}O_{13}$	156–158	+23°	200	Kumagai *et al.* 1986
DTX-1	818	$C_{45}H_{70}O_{13}$	134	+28°	160	Murata *et al.* 1982
DTX-2	804	$C_{44}H_{68}O_{13}$	128–130	+15.49°	—	Hu *et al.* 1992
DTX-3	—	—	—	—	*c.* 500	Yasumoto *et al.* 1985
PTX-1	874	$C_{47}H_{70}O_{15}$	208–209	+17.1°	250	Yasumoto *et al.* 1985
PTX-2	858	$C_{47}H_{70}O_{14}$	—	+16.2°	230	Yasumoto *et al.* 1985
PTX-3	872	$C_{47}H_{68}O_{15}$	159–160	+2.22°	350	Murata *et al.* 1986
PTX-4	874	$C_{47}H_{70}O_{15}$	—	—	770	Yasumoto & Murata 1990
PTX-6	888	$C_{47}H_{68}O_{16}$	—	—	500	Yasumoto & Murata 1990
YTX	1186	$C_{55}H_{80}O_{21}S_2Na_2$	100	+3.01°	100	Murata *et al.* 1987
45-OH YTX	1202	$C_{55}H_{80}O_{22}S_2Na_2$	—	—	100	Yasumoto & Murata 1990

$[\alpha]_D^{20}$, specific rotation at 20 °C (sodium D line).

muscle contraction. Okadaic acid, therefore, has been used as an extremely useful probe for the study of cellular processes mediated by protein phosphorylation. This phosphorylation activity is believed to be responsible for the oncogenic potential of OA (Cohen *et al.* 1990), and both OA and DTX-1 have potent tumour-promoting activity (Suganuma *et al.* 1988), but little is known about the long-term effects of consuming low levels of these shellfish toxins.

Surveillance

Although illness directly attributable to DSP poisoning has not been notified in the UK, *Dinophysis* blooms have been observed at various coastal locations around the UK. DSP toxin was detected by chance in 1991 in a batch of cockles harvested from the Thames estuary that were being examined for paralytic shellfish toxin (PSP). This episode and impending EC legislation have highlighted not only the lack of current knowledge about the extent of contamination of bivalve molluscs by algal blooms but also the need for regular monitoring and reliable methods for detection and quantification of toxins. In 1992 regular sampling for DSP toxins was initiated.

A new EC Directive (91/492/EEC) (1991), which came into effect in January 1993, states that 'the customary biological testing methods must not give a positive result to the presence of Diarrhetic Shellfish Poison (DSP) in the edible parts of molluscs (the whole body or any part edible separately)'. In order to comply with this Directive, a protocol is being formulated so that a programme for the routine monitoring of shellfish, and also for the presence of potentially toxic algae in seawater, may be carried out.

Toxin Assay

Preparation of 9-anthryldiazomethane derivatives

The hepatopancreas of 12 mussels was excised and finely chopped. A portion of this tissue (1 g) was homogenized with 4 ml of 80% methanol and the mixture was centrifuged at 3000 *g*. The supernatant was removed and 2.5 ml of the suspension extracted three times with petroleum ether. Water (1 ml) was added to the methanol phase and the solution extracted twice with 4-ml volumes of chloroform. The extracts were pooled and made up to 10 ml with chloroform.

A 0.5-ml volume of the chloroform fraction was evaporated to dryness under a stream of nitrogen at room temperature. The sample was spiked with 100 ng deoxycholic acid (DOCA) as internal standard before the sample was dried. 9-Anthryldiazomethane (ADAM) (100 μl 0.1% w/v in methanol) was added to the sample, which was incubated in the dark at 35 °C for 1 h. The reaction mixture was reduced to dryness and transferred to a Mega Bond Elut cartridge (Part No. 1225-6008 Jones Chromatography, Hengoed, Wales, UK) with 3 × 0.25 ml hexane:chloroform (1:1). The cartridge was washed successively with 5 ml hexane:chloroform (1:1) and 5 ml chloroform.

The ADAM–OA conjugate was eluted from the cartridge with 5 ml chloroform:methanol (95:5) and the sample evaporated to dryness under nitrogen at 35 °C. The residue was dissolved in 100 μl methanol and a 10-μl volume injected into the HPLC column.

HPLC analysis was carried out according to the method of Lee *et al.* (1987).

Conclusions

The EC directive on routine monitoring of DSP toxins in shellfish stipulates the use of a bioassay to determine toxicity. Two bioassays are currently in general use in Europe – intraperitoneal injection of mice and oral challenge of rats. In the main, the mouse bioassay is used most frequently because it is more sensitive (4 μg OA) than the rat oral challenge (10 μg OA). However, the mouse test is carried out under different protocols and is subject to false positives because of interference from free fatty acids carried over during the extraction procedure (Table 15.2). To some extent this can be overcome by altering the extraction protocol. Nevertheless, as the basis for control of the fishery in the event of a suspected toxic incident, it is essential to have a reliable consistent alternative assay to confirm positive mouse bioassays. Currently the most suitable alternative is HPLC, which is used in several countries, either routinely on all positive bioassay samples or as a further test if the bioassay cannot be readily interpreted. Methodologies and results by HPLC show considerable variation and these problems are currently being

TABLE 15.2. *Comparison of samples analysed by mouse bioassay and HPLC*

Sample origin	Host	Mouse bioassay	HPLC (μg OA/100 g)
Dutch	Mussels	+ve	335.6
Japanese	Oysters	+ve	178.0
Survey	Mussels	+ve	21.4
Survey	Mussels	+ve	8.9
Survey	Mussels	+ve	13.3
Survey	Mussels	+ve	<0.5
Toxic episode	Queen scallops	+ve	<0.5
Toxic episode	Mussels	+ve	43.0

addressed. One of the underlying problems that prevents quantitative chemical estimation of all DSP toxins is the marked lack of suitable commercial standard reference materials. Furthermore, the HPLC assay is beset by a number of problems relating to sample preparation and storage, extraction of DSP toxins, and derivatization and clean-up procedures. Although DSP toxins are more stable in the dried form, it has been observed that they are extremely difficult to remove from lyophilized tissue. Extraction of the toxins is usually achieved with aqueous methanol, followed by washing with hexane or petroleum ether to remove fat. DSP toxin derivatives are usually prepared with ADAM, but two newer reagents, 1-pyrenyldiazomethane (PDAM) and 1-(bromoacetyl)-pyrene (BAP), have been proposed as suitable alternatives. Accurate values for OA with ADAM reagent can be obtained, provided that an internal standard, usually DOCA, is included in the reaction mixture.

The success of the analysis depends on the subsequent clean-up pro-

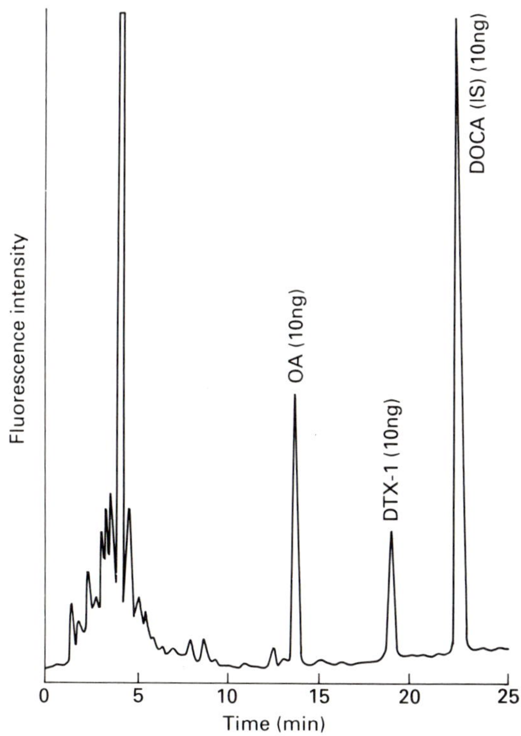

FIG. 15.4.Standards. DTX-1, *Dynophysis* toxin-1; DOCA, Deoxycholic acid; OA, okadaic acid.

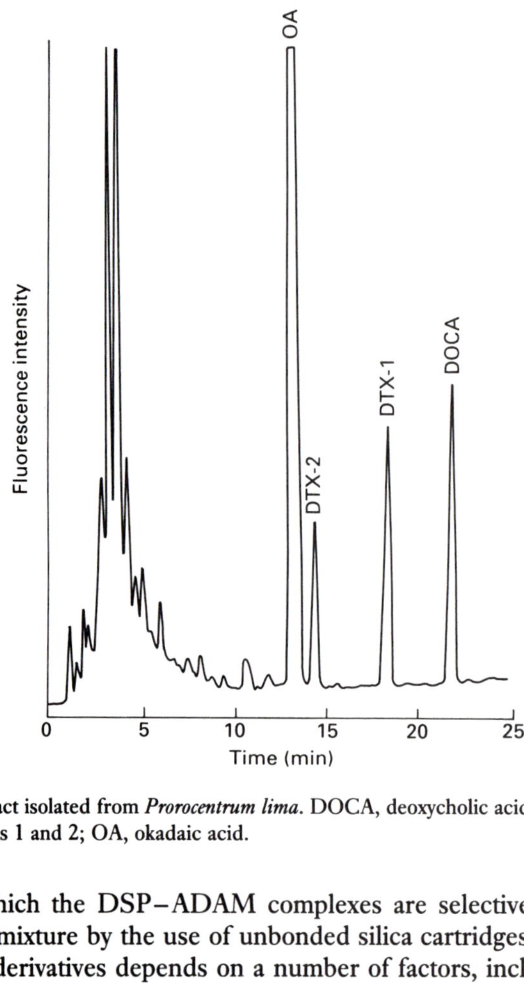

FIG. 15.5. Extract isolated from *Prorocentrum lima*. DOCA, deoxycholic acid; DTX-1, DTX-2, *Dynophysis* toxins 1 and 2; OA, okadaic acid.

cedure in which the DSP–ADAM complexes are selectively removed from the reaction mixture by the use of unbonded silica cartridges. The recovery of the ADAM derivatives depends on a number of factors, including the volume and type of solvents used in the prewash and elution steps, the amount of stationary phase in the cartridge and the speed of elution from the cartridge. Furthermore, ADAM derivatives are broken down in UV light and the manipulation of samples should be carried out in subdued lighting. In spite of all these apparent disadvantages, HPLC can detect less than 0.5 ng of OA–ADAM in a 10-μl injection.

Samples that give a positive reaction by bioassay usually show levels of OA above or near 20 μg/100 g, which is considered the limit for rejection for human consumption. However, when a positive bioassay is accompanied by a low OA value, unidentified peaks, usually eluting between OA and DOCA, are often observed. Unfortunately, the ADAM reagent does not react with

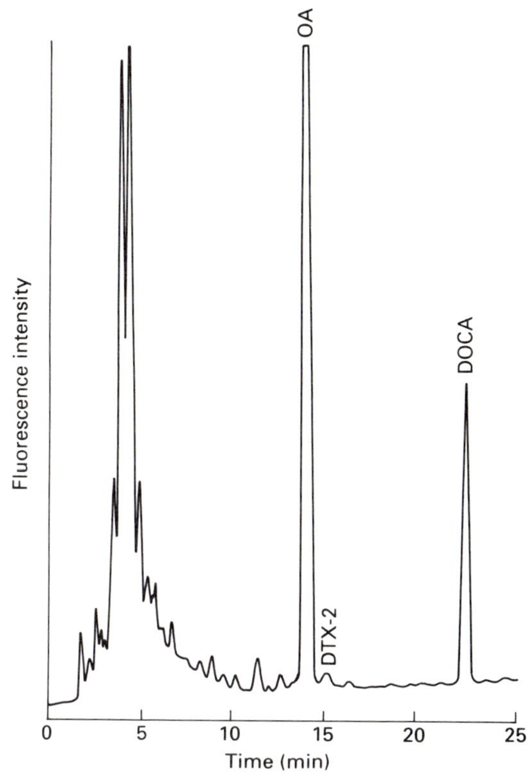

FIG. 15.6. Extract isolated from *P. concavum*. DOCA, deoxycholic acid; DTX-2, *Dynophysis* toxin 2; OA, okadaic acid.

either the PTX or the YTX toxins, but their presence has not been widely observed in shellfish from European waters. Nevertheless, in the case of scallops (Japan) and mussels (Norway), it has been found that YTX toxins account for a considerable part of the lethality in the mouse bioassay.

The algae *Prorocentrum lima* and *P. concavum*, cultivated *in vitro*, were freeze-dried and their aqueous methanol extracts, along with OA, DTX-1 and DOCA standards, derivatized with ADAM and analysed by HPLC (Figs 15.4–15.6).

In the case of *P. concavum*, OA was the main toxin. The DSP derivatives from *P. lima*, however, showed two other peaks apart from OA. The peak eluting at 19–20 min, when compared with standards, was identified as DTX-1 on the basis of retention time. The other major toxin eluting immediately after OA was confirmed to be DTX-2 (E. Nixon, personal communication).

The problem of finding suitable standards for the DTX toxins remains, but it is expected that *P. lima* will provide a suitable source material for some reference toxins.

The mouse bioassay, when used for testing large numbers of samples, is expensive to perform and can give rise to false positive results, because of the presence of non-DSP toxic material in the sample. In addition, it lacks the specificity of the HPLC method, which is quantitative, reliable and fast.

Acknowledgements

Freeze-dried samples of *Prorocentrum lima* and *P. concavum* were supplied by Dr J. Leftley, Dunstaffnage Marine Laboratory, Oban.

References

BIALOJAN, C. & TAKAI, A. 1988 Inhibitory effect of a marine sponge toxin, okadaic acid on protein phosphatases. *Biochemical Journal* **256**, 283–290.

COHEN, P. 1989 The structure and regulation of protein phosphatases. *Annual Reviews in Biochemistry* **58**, 453–508.

COHEN, P., HOLMES, C.F.B. & TSUKITANI, Y. 1990 Okadaic acid: a new probe for the study of cellular regulation. *Trends in Biological Sciences* **15**, 98–102.

DAHL, E. & YNDESTAD, M. 1985 Diarrhetic shellfish poisoning (DSP) in Norway in the autumn of 1984 related to the occurrence of *Dinophysis* spp. *Toxic Dinoflagellates*, eds Anderson, D.M., White, A.W. & Baden, D.G., pp. 495–500. New York: Elsevier.

EC DIRECTIVE (91/492/EEC) 1991 Laying down the health conditions for the production and the placing on the market of live bivalve molluscs. *Official Journal of the European Communities No. L* **268**, 1–14.

EDEBO, L., LANGE, S., LI, X.P. & ALLENMARK, S. 1988 Toxic mussels and okadaic acid induce rapid hypersecretion in the rat small intestine. *APMIS* **96**, 1029–1035.

HU, T., DOYLE, J., JACKSON, D., MARR, J., NIXON, E., PLEASANCE, S., QUILLIAM, M.A., WALTER, J.A. & WRIGHT, J.L.C. 1992 Isolation of a new diarrhetic shellfish poison from Irish mussels. *Journal – Chemical Society Chemical Communications* **1**, 39–41.

KROGH, P., EDLER, L., GRANELI, E. & NYMAN, U. 1985 Outbreak of diarrhetic shellfish poisoning on the west coast of Sweden. *Toxic Dinoflagellates*, eds Anderson, D.M., White, A.W. & Baden, D.G., pp. 501–503. New York: Elsevier.

KUMAGAI, M., YANAGI, T., MURATA, M., YASUMOTO, T., KAT, M., LASSUS, P. & RODRIGUEZ-VAZQUEZ, J.A. 1986 Okadai acid as the causative toxin of diarrhetic shellfish poisoning in Europe. *Agricultural and Biological Chemistry* **50**, 2853–2857.

LEE, J.S., YANAGI, T., KENMA, R. & YASUMOTO, T. 1987 Fluorimetric determination of diarrhetic shellfish toxins by high performance liquid chromatography. *Agricultural and Biological Chemistry* **51** (3), 877–881.

MARCAILLOU-LEBAUT, C., LUCAS, D. & LE DEAN, L. 1985 *Dinophysis acuminata* toxin: status of toxicity bioassays in France. *Toxic Dinoflagellates*, eds Anderson, D.M., White, A.W. & Baden, D.G., pp. 485–488. New York: Elsevier.

MURAKAMI, Y., OSHIMA, Y. & YASUMOTO, T. 1982 Identification of okadaic acid as a toxic compound of marine dinoflagellates *Prorocentrum lima*. *Bulletin – Japanese Society of Scientific Fisheries* **48**, 69–72.

MURATA, M., SHIMATANI, M., SUGITANI, H., OSHIMA, Y. & YASUMOTO, T. 1982 Isolation and structural elucidation of the causative toxin of the diarrhetic shellfish poisoning. *Bulletin – Japanese Society of Scientific Fisheries* **48**, 549–552.

MURATA, M., SANO, M., IWASHITA, T., NAOKI, H. & YASUMOTO, T. 1986 The structure of pectenotoxin-3, a new constituent of diarrhetic shellfish toxins. *Agricultural and Biological Chemistry* **50**, 2693–2695.

MURATA, M., KAMAGAI, M., LEE, J.S. & YASUMOTO, T. 1987 Isolation and structure of yessotoxin, a novel polyether compound implicated in diarrhetic shellfish poisoning. *Tetrahedron* **28**, 5869–5872.

RODRIGUEZ-VAZQUEZ, J.A., MARTINEZ, A.G., OSHIMA, Y., SUGINO, K., LEE, J.S. & YASUMOTO, T. 1989 Analysis of toxins in mussels from the Atlantic coast of Spain. *Mycotoxins and Phycotoxins '88*, eds Natori, S., Hashimoto, K. & Ueno, Y., pp. 367–374. New York: Elsevier.

SUGANUMA, M., FUJIKI, H., SUGURI, H., YOSHIZAWA, S., HIROTA, M., NAKAYASU, M., DJIKA, M., WAKAMATSU, K.; YAMADA, K. & SUGIMURA, T. 1988 Okadaic acid: an additional nonphorbol-12-tetradecanoate-13-acetate-type tumor promoter. *Proceedings of the National Academy of Sciences of the United States of America* **85**, 1768–1771.

YASUMOTO, T. & MURATA, M. 1990 Polyether toxins involved in seafood poisoning. *Marine Toxins*, eds Hall, S. & Stritchartz, G., pp. 120–132. Washington DC: American Chemical Society.

YASUMOTO, T., OSHIMA, Y. & YAMAGUCHI, M. 1978 Occurrence of a new type of shellfish poisoning in the Tokoku district. *Bulletin – Japanese Society of Scientific Fisheries* **44**, 1249–1255.

YASUMOTO, T., MURATA, M., OSHIMA, Y., MATSUMOTO, G.K. & CLARDY, J. 1984 Diarrhetic shellfish poisoning. *Seafood Toxins*, ed. Ragelis, E.P., pp. 207–214. Washington DC: American Chemical Society.

YASUMOTO, T., MURATA, M., OSHIMA, Y., SANO, M., MATSUMOTO, G.K. & CLARDY, J. 1985 Diarrhetic shellfish toxins. *Tetrahedron* **41**, 1019–1025.

16: Tissue Culture Assay for Paralytic Shellfish Poisoning Toxins

SUSAN GALLACHER,[1] KAREN EVANS,[2] M.BLACKBURN,[2] M.WALDOCK[2] AND T.H.BIRKBECK[3]
[1]*Scottish Office Agriculture and Fisheries Department, Marine Laboratory, Victoria Road, Aberdeen AB9 8DB, UK;*
[2]*Ministry of Agriculture, Fisheries and Food, Fisheries Laboratory, Remembrance Avenue, Burnham-on-Crouch, Essex CM0 8HA, UK;*
and [3]*Department of Microbiology, University of Glasgow, Glasgow G12 8QQ*

Paralytic shellfish poisoning (PSP) arises from ingestion of shellfish that have been exposed to toxic dinoflagellates. Poisoning is attributed to a family of 18 or more compounds chemically related to saxitoxin (STX) (Hall *et al.* 1990) which block sodium channels in nerves to differing degrees; the parent compound saxitoxin is the most potent. PSP toxins can accumulate in shellfish and may cause considerable health problems. Outbreaks of PSP reduce consumer confidence in shellfish products, with severe financial losses to the industry, and, because of misinformation, this often extends to the finfish industry (Shumway 1990).

In many countries, shellfish toxicity monitoring is carried out with the mouse bioassay (AOAC 1990) to detect PSP toxins. Apart from the undesirability of toxicity testing in live animals, this method has several disadvantages, notably subjectivity, possible underestimation of toxin by up to 60% at the detection limit (World Health Organization 1984) and its dependency on particular strains of mice. Several alternative methods, such as the use of animal axons (Keynes *et al.* 1967; Moore *et al.* 1967), fluorimetry (Nunez *et al.* 1967), housefly bioassay (Ross *et al.* 1985) and immunoassays (Carlson *et al.* 1984; Cembella *et al.* 1990), have been developed, but they all suffer from some disadvantages that make them unsuitable for large-scale screening programmes.

A high-performance liquid chromatography (HPLC) method developed by Sullivan & Wekell (1984) has a sensitivity range of 10–30 μg STX/100 g flesh, and the method has been extensively evaluated by Fileman (1988) and

New Techniques in
Food and Beverage Microbiology

0-632-03755-5

Waldock *et al.* (1991). They reported several disadvantages of the assay in that some compounds with differing toxicities coeluted, the apparatus was difficult to maintain and only 20 samples could be processed per day. Also, the correlation between the mouse bioassay and the HPLC results for lobster and scallops was low.

We have previously described a tissue culture assay to measure sodium channel-blocking (SCB) toxins produced by bacteria (Gallacher & Birkbeck 1992). The method can detect 3–160 ng/ml saxitoxin and is readily applicable to the detection of PSP toxins in bivalve tissues. The assay is based on the work of Catterall & Nirenberg (1973) on the SCB activity of tetrodotoxin. Kogure *et al.* (1988) developed an assay with mouse neuroblastoma cells that was based on the ability of tetrodotoxin to protect cells treated with an otherwise lethal combination of veratridine and ouabain; veratridine causes opening of sodium channels in nerve cell membranes resulting in an uncontrolled influx of sodium ions and, eventually, cell death. The cell can exchange intracellular Na^+ ions for extracellular K^+ ions against the concentration gradient for these ions by membrane Na^+/K^+ ATPase activity, but this enzyme can be blocked with ouabain. Thus, addition of a sodium channel-blocking toxin, such as tetrodotoxin or saxitoxin, counters the effects of veratridine and ouabain and allows cell survival.

Tissue Culture Assay for Sodium Channel-blocking Toxins

This is shown schematically in Fig. 16.1. Mouse neuroblastoma cells, C-1300 (ATCC CCL131; Flow Laboratories) are routinely cultured at 37 °C in RPMI 1640 medium (Gibco), containing 25 mmol/l HEPES, penicillin (200 iu/ml), streptomycin (200 μg/ml) and 5% fetal calf serum (Flow Laboratories). For routine use, cells are cultured in 175-cm^2 disposable tissue culture flasks (Gibco-Nunc). Sufficient cells can be obtained from one 175-cm^2 flask to seed four 96-well microtitre plates. When confluent monolayers of cells have formed, the cells are removed from the surface of the flask by incubation for 5–10 min in 5 ml trypsin/EDTA solution (0.05% trypsin, 0.02% EDTA in modified Puck's saline; Gibco). The trypsinized cell suspension is centrifuged at 600 *g* for 10 min to collect the cells, which are then resuspended in fresh RPMI medium and adjusted to 1×10^6 cells/ml after determining the cell concentration with a haemocytometer. Microtitre plates (tissue culture grade with flat-bottomed wells, Gibco-Nunc) are seeded with 200 μl of cell suspension and incubated for 24 h at 37 °C in a sealed moist container in an atmosphere of air. The medium is then removed and replaced with 50 μl 0.5 mmol/l ouabain, 50 μl 0.1 mmol/l veratridine and 100 μl of the test sample. All samples should be tested in quadruplicate with controls of: (i) two sets of four wells containing cells with 200 μl RPMI 1640 medium; and (ii) two sets of four wells containing ouabain/veratridine solutions plus 100 μl

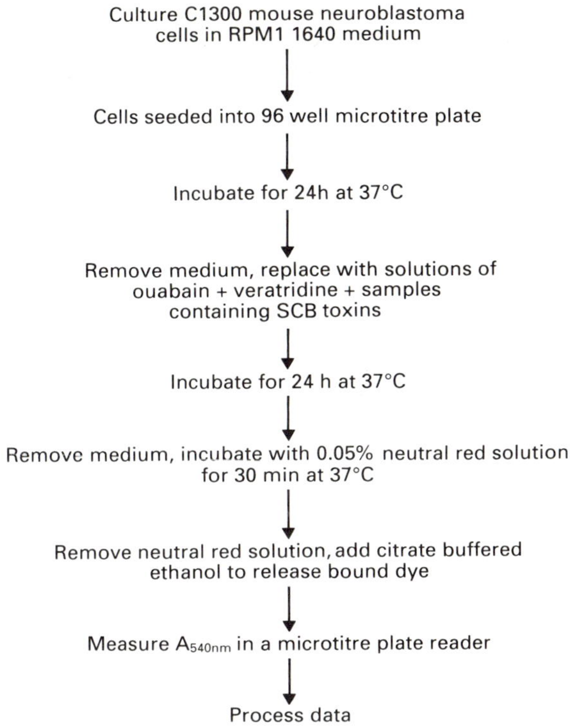

FIG. 16.1. Stages in the tissue culture assay to quantify PSP toxin.

of RPMI medium. In initial experiments, or when fresh batches of fetal calf serum, ouabain or veratridine are used, the optimum concentration of ouabain and veratridine must be determined with 0–1.0 mmol/l ouabain and 0–0.15 mmol/l veratridine in a chequerboard titration (Gallacher & Birkbeck 1992). Typical results are shown in Fig. 16.2. Veratridine alone was toxic to cells at concentrations greater than 0.2 mmol/l and ouabain alone caused >10% cell death above a concentration of 0.5 mmol/l. Combinations that killed 60–70% of cells within 24 h contained 0.05–0.1 mmol/l veratridine and 0.3–0.5 mmol/l ouabain. For routine use, 0.1 mmol/l veratridine and 0.5 mmol/l ouabain normally gave optimum results.

After further incubation for 24 h at 37 °C, the medium is replaced with 200 μl 0.05% w/v neutral red solution in Hank's balanced salt solution (without HEPES). After 30 min at 37 °C, excess dye solution is removed and the cells are rinsed once with Hank's balanced salt solution. Citrate-buffered ethanol (200 μl 50 mmol/l citrate/acetic acid buffer, pH 4.6, 50% ethanol) is added to each well to extract sequestered dye from the cells, and after 1 h at room temperature the absorbance is determined at 540 nm in a Multiskan

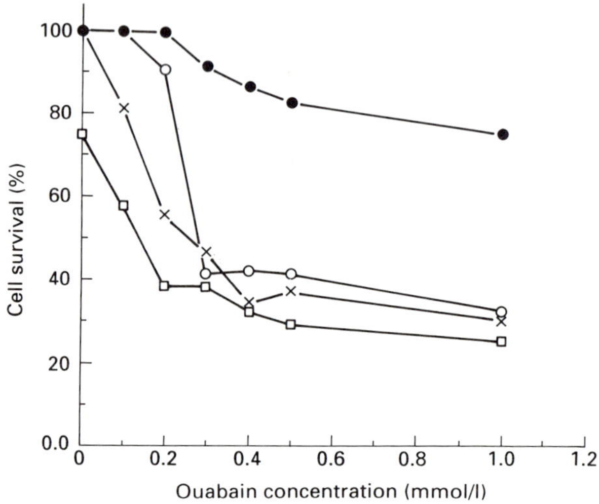

FIG. 16.2. Effect of ouabain and veratridine on mouse neuroblastoma cell viability. Veratridine concentrations: ●, 0 mmol/l; ○, 0.05 mmol/l; ×, 0.1 mmol/l; □, 0.15 mmol/l.

(Titertek) microtitre plate reader. Processing of data is facilitated if the plate reader is interfaced to a microcomputer so that results can be analysed by means of spreadsheet software.

The percentage protection of cells by saxitoxin or test samples can be calculated from the relationship

$$D = \frac{A - B}{C - B} \times 100$$

where D is the percentage protection, A the absorbance of the cells + ouabain + veratridine + test sample, B is the absorbance of cells + ouabain + veratridine and C is the absorbance of cells alone.

Saxitoxin has been tested over the final concentration range 0–100 nmol/l and protected 50% of cells from the lethal action of ouabain and veratridine at a concentration of 25 nmol/l.

Analysis of shellfish for PSP toxins

Samples should initially be processed according to standard procedures (AOAC 1990), diluted in RPMI medium (1/10 to 1/20480) and applied to the tissue culture assay. As noted below, inhibitors of PSP toxins may be present in a small number of mussel extracts, but, when these are destroyed by heating, a conventional dose–response curve is obtained (Fig. 16.3). End-point deter-

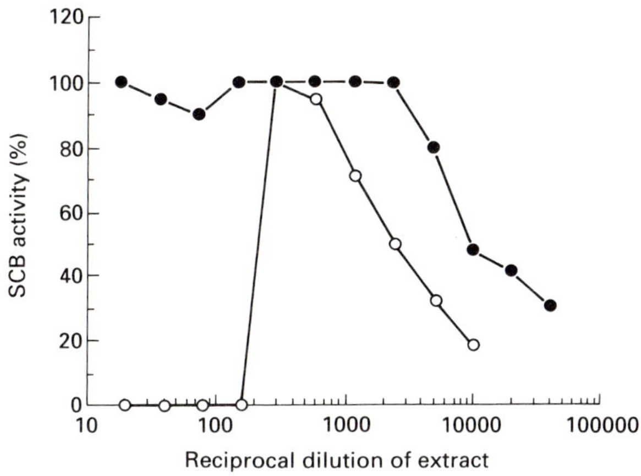

FIG. 16.3. Sodium channel-blocking (SCB) activity of T64 mussel extract, unheated and heated at 100 °C for 60 min. ●, heated; ○, unheated.

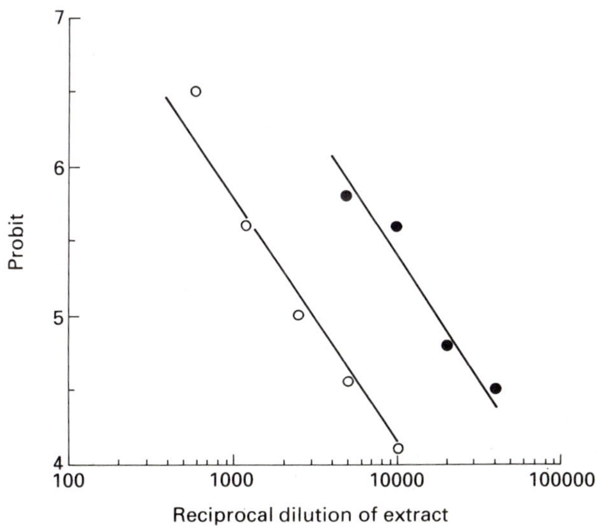

FIG. 16.4. Probit plot of sodium channel-blocking activity of T64 mussel extract. ●, heated; ○, unheated.

mination is best done by probit analysis (Finney 1962; Wardlaw 1985) to alter the normal sigmoidal dose–response curve to a linear response and ascertain the sample dilution that protects 50% of cells from the toxicity of ouabain and veratridine. This is the dilution that contains approximately 25 nmol/l

TABLE 16.1. *The correlations between the mouse bioassay, HPLC and the tissue culture assay for analysis of toxic mussel extracts*

Assay	Correlation coefficient
Mouse bioassay v. HPLC	0.86
Mouse bioassay v. tissue culture	
Unheated	0.94
Heated	0.88
HPLC v. tissue culture	0.93

saxitoxin-equivalent. In Fig. 16.4 the relevant data from Fig. 16.3 have been converted to probits. Since a probit value of 5 corresponds to 50% protection of cells from ouabain/veratridine-induced death, this dilution contains the equivalent of 25 nmol/l saxitoxin, and extract T64 contains 25 000 μg saxitoxin/100 g mussel flesh. Comparative assays of mussel extracts by the mouse bioassay, HPLC and the tissue culture assay showed a higher correlation between the mouse bioassay and the tissue culture assay with 'unheated' samples – that is, not heated more than in the standard AOAC procedures (Table 16.1).

It is important to test a range of dilutions of the extracts, as heat-labile inhibitors of PSP toxins may be present in some bivalve extracts. An example is shown in Fig. 16.3, where sodium channel-blocking activity was not detected in *Mytilus edulis* extract T64, processed by the standard AOAC procedure, at dilutions up to 1/160 unless the sample was further heated. The nature of the heat-labile inhibitor of PSP toxins is not known, but Barber *et al.* (1988) partially characterized an acidic protein of 145 kDa from the shore crabs *Hemigrapsus oregonensis* and *H. nuclus* which was induced by exposure to saxitoxin and conferred resistance to the crabs. Such a protein may also be induced in *Mytilus edulis*. Whilst heating at 100 °C for 60 min destroyed the inhibitor (Fig. 16.3), the sodium channel-blocking activity increased on heating, probably by conversion of the less toxic sulphamate derivatives of saxitoxin to the more toxic carbamates (Hall & Reichardt 1984). A similar rise in PSP toxicity was reported by Nagashima *et al.* (1991). As well as altering the end-point in the assay, the slope of the dose–response curve may change, a further indication that the spectrum of PSP toxins may alter on heating.

It is clear that the tissue culture assay can readily be used to assay PSP toxins in mussel extracts. Further development of the assay is possible to reduce considerably the time required to complete the assay. A similar assay has recently been described by Jellet *et al.* (1992).

References

AOAC 1990 Paralytic shellfish poison. *Association of Official Analytical Chemists: Official Methods of Analysis*, 15th edn, section 959.08, pp. 881–882. Arlington: AOAC.

BARBER, K.G., KITTS, D.D., TOWNSLEY, P.M. & SMITH, D.S. 1988 Appearance and partial purification of a high molecular weight protein in crabs exposed to saxitoxin. *Toxicon* **26**, 1027–1034.

CARLSON, R.E., LEVER, M.L., LEE, B.W. & GUIRE, P.E. 1984 Development of immunoassays for paralytic shellfish poisoning. *Seafood Toxins*, ed. Ragelis, E.P., pp. 181–183. Washington DC: American Chemical Society.

CATTERALL, W.A. & NIRENBERG, M. 1973 Sodium uptake associated with activation of action potential ionophores of cultured neuroblastoma and muscle cells. *Proceedings of the National Academy of Sciences, USA* **70**, 3759–3763.

CEMBELLA, A., PARENT, Y., JONES, D. & LAMOUREUX, G. 1990 Specificity and cross-reactivity of an absorption–inhibition enzyme-linked immunoassay for the detection of paralytic shellfish toxins. *Toxic Marine Phytoplankton*, eds Granelli, G., Sundstrom, B., Edler, L. & Anderson, D.M., pp. 339–344. Amsterdam: Elsevier.

FILEMAN, T.W. 1988 A comparison of HPLC and mouse bioassay techniques for the determination of paralytic shellfish poisoning toxins. *International Council for the Exploration of the Sea, Shellfish Committee Report* CM 1988/E: **10**, 9. Copenhagen: ICES.

FINNEY, D.J. 1962 *Probit Analysis*, 2nd edn. Cambridge: Cambridge University Press.

GALLACHER, S. & BIRKBECK, T.H. 1992 A tissue culture assay for direct detection of sodium channel blocking toxins in bacterial culture supernates. *FEMS Microbiology Letters* **92**, 101–108.

HALL, S. & REICHARDT, P.B. 1984 Cryptic paralytic shellfish toxins. *Seafood Toxins*, ed. Ragelis, E.P., p. 113–123. Washington DC: American Chemical Society.

HALL, S., STRICHARTZ, G., MOEZYDLOWSKI, E., RAVINDRAN, A. & REICHARDT, P.B. 1990 The saxitoxins: sources, chemistry and pharmacology. *Marine Toxins: Origin, Structure and Molecular Pharmacology*, eds Hall, S. & Strichartz, G., pp. 29–65. Washington DC: American Chemical Society.

JELLET, J.F., STEWART, J.E., DOREY, M.L., WATSON-WRIGHT, W. & LAWRENCE, J.F. 1992 Paralytic shellfish poison (saxitoxin family) bioassays: automated endpoint determination and standardization of the in vitro tissue culture bioassay, and comparison with the standard mouse bioassay. *Toxicon* **30**, 1143–1156.

KEYNES, R.D., RITCHIE, J.M. & ROJAS, E. 1967 The binding of tetrodotoxin to nerve membranes. *Journal of Physiology* **213**, 235–254.

KOGURE, K., TAMPLIN, M.L., SIMIDU, U. & COLWELL, R.R. 1988 A tissue culture assay for tetrodotoxin, saxitoxin and related toxins. *Toxicon* **26**, 191–197.

MOORE, J.W., NARAHASHI, T. & SHAW, T.I. 1967 An upper limit to the number of sodium channels in nerve membranes. *Journal of Physiology* **188**, 99–105.

NAGASHIMA, Y., NOGUCHI, T., TANAKA, M. & HASHIMOTO, K. 1991 Thermal degradation of paralytic shellfish poison. *Journal of Food Science* **56**, 1572–1575.

NUNEZ, M.T., FISCHER, S. & JAIMOVICH, E. 1967 A fluorimetric method to determine tetrodotoxin. *Analytical Biochemistry* **72**, 320–325.

ROSS, M., SIGER, A. & ABBOTT, C.C. 1985 The house fly: an acceptable subject for paralytic shellfish toxin bioassay. *Toxic Dinoflagellates*, eds Anderson, D.M., White, A.W. & Baden, D.G. Amsterdam: Elsevier.

SHUMWAY, S.E. 1990 A review of the effects of algal blooms on shellfish and aquaculture. *Journal of the World Aquaculture Society* **21**, 65–104.

SULLIVAN, J.J. & WEKELL, M.M. 1984 Determination of paralytic shellfish poisoning toxins by high pressure liquid chromatography. *Seafood Toxins*, ed. Ragelis, E.P., pp. 197–205. Washington DC: American Chemical Society.

WALDOCK, M.J., EVANS, K.M., LAW, R.J. & FILEMAN, T.W. 1991 An assessment of the suitability of HPLC monitoring techniques for monitoring PSP and DSP on the east coast of England. *Proceedings of a Symposium on Marine Biotoxins*, ed. Freny, J.M., pp. 137–143. Paris: Centre Nationale d'Etudes Vétérinaires et Alimentaires.

WARDLAW, A.C. 1985 *Practical Statistics for Experimental Biologists*. Chichester: J. Wiley.

WORLD HEALTH ORGANIZATION 1984 Aquatic (marine and freshwater) biotoxins. *Environmental Health Criteria* 37. Geneva: World Health Organization.

17: Multilocus Enzyme Electrophoresis

P.G.MURPHY
Bacteriology Department and Northern Ireland Public Health Laboratory, Belfast City Hospital, Lisburn Road, Belfast BT9 7AD, UK

Multilocus enzyme electrophoresis (MEE) has been used for many years to study the population genetics and systematics of eukaryotes. Its application to the study of genetic diversity and population genetics of bacteria is recent (Selander *et al.* 1986) but is rapidly becoming popular with bacterial taxonomists and epidemiologists. A background to the history of MEE and its application to eukaryotes is given by Ferguson (1980), while Selander *et al.* (1986) have provided a seminal account of the application of MEE to bacteria. Many organisms of importance in food microbiology, including *Listeria* spp., *Campylobacter* spp., *Salmonella* spp., *Escherichia coli*, *Shigella* spp. and *Staphylococcus aureus*, have been analysed by MEE. The technique has proved particularly useful for intraspecies clonal analysis for epidemiological purposes, except for *Salmonella* spp., which appear to be clonally homogeneous. The technique has been developed further with *Listeria* spp. and *Campylobacter* spp., for the specific epidemiological investigation of links in the food chain between animal, food and human isolates.

Principle of MEE

The base sequence of structural genes determines the amino acid sequence of the protein for which they code, and any change of base sequence in the structural gene leads to corresponding changes in the amino acid sequence of the protein. Such changes in amino acid sequence are reflected in changes of electrophoretic behaviour, which can easily be measured by simple electrophoretic means designed to detect the overall electrostatic charge and size of the protein or enzyme. Bacterial strains can, therefore, be characterized by the relative electrophoretic mobilities of a range of cellular enzymes or isoenzymes. Mobility variants of an enzyme, termed *electromorphs*, can directly

New Techniques in
Food and Beverage Microbiology

0-632-03755-5

be equated with alleles at the corresponding structural gene locus. In MEE a profile of electromorphs, the *electromorph type* (ET), is derived over the range of enzymes tested. After cluster analysis, the genetic diversity of a locus and the genetic distance between ETs can then be represented as dendrograms or principal coordinate plots. It has been shown that gel electrophoresis can detect 90% of amino acid substitutions in haemoglobin (Ramshaw *et al.* 1979). Some substitutions may, however, be silent, or at least may not affect electrophoretic mobility, and, as a result, the electromorphs may actually be heterogeneous, without this being apparent. Therefore, although electromorph profiles over a range of loci can be equated with multilocus genotypes, and electromorph frequencies with allele frequencies, it should be understood that the alleles recognized may actually be groups of isoalleles.

For pairs of isolates of several species, high correlations have been reported between estimates of the genetic distance by MEE and estimates of divergence in the nucleotide sequence obtained by DNA hybridization (Selander *et al.* 1985). Where discrepancies have been observed between restriction endonuclease digestion of genomic DNA and MEE clonal analysis, it has been observed that the latter is epidemiologically more consistent (Olyhoek *et al.* 1987). It is reasonable to assume, therefore, that a range of enzymes is a representative and random sample of the structural genes of a particular bacterial genome. Such genotypic analysis has two main applications. First, the use of a large number of enzymes and isolates for studies of populations, evolution and taxonomy, and, second, the analysis of a few polymorphic loci to distinguish between possibly related strains for epidemiological purposes.

Method

Preparation of enzyme extracts

Isolates in pure culture are grown under the most appropriate cultural conditions for the species under investigation, to obtain approximately 10^{11} cells. For example, in the case of *Campylobacter*, this requires 20 plates of Mueller–Hinton agar or heart-infusion agar, both containing 5% blood and incubated under microaerophilic conditions at 42 °C for 24–48 h (Patton *et al.* 1991) or 5% blood agar incubated at 42 °C also under microaerophilic conditions (Aeschbacher & Piffaretti 1989). *Listeria* cultures are grown overnight in 100 ml trypticase soy broth (Bibb *et al.* 1990). Though growth conditions do not affect the electrophoretic mobility of enzymes produced, the concentration and, therefore, the intensity of staining may be affected. Cells are harvested by centrifugation of broth cultures or by scraping growth from the agar with the edge of a sterile glass slide into 2 ml of buffer solution. A buffer solution consisting of 10 mmol/l Tris, 1 mmol/l EDTA and 0.5 mmol/l NADP, pH 6.8, is suitable for most organisms.

Various methods can be used to extract enzymes from whole-cell preparations. These range from combinations of freezing and mechanical disruption by various methods, to simple sonication (e.g. two cycles of 30 s) for *Campylobacter* (Aeschbacher & Piffaretti 1989). Sonication is most convenient but, if the equipment is not available, the simpler but more labour-intensive use of pestle and mortar (Nørrung 1992) or vigorous vortexing with glass beads (75–150 μm) for 2–3 min is satisfactory (Bibb *et al.* 1990). The important condition at this and all subsequent stages is that the enzymes are held at low temperatures in ice, or at −70 °C for storage. Whichever method of lysis is used, it must allow enzyme release from the cell, without denaturation, in a wholly active form, so that maximum enzyme activity is preserved and scoring of electromorph mobilities is facilitated.

The enzymes are removed from the lysed cell solution by centrifugation at 15 000–30 000 *g* for up to 20 min. Filter sterilization of the supernatant with a 0.22-μm filter is optional, but it is prudent if the enzymes are to be stored for prolonged periods. Lysates should be stored in small volumes (e.g. 200 μl) to avoid repeated thawing and freezing.

Gel preparation

Starch (Selander *et al.* 1986) and polyacrylamide (McLellan & Ramshaw 1981) are appropriate for enzyme electrophoresis, but the former is preferable because it provides several horizontal slices, such as five 2-mm slices from a single 15-mm gel, which allows use of several types of enzyme detection for each electrophoretic run.

High-grade electrophoresis-quality starch (Connaught Laboratories, catalogue no. 2901-027; Sigma) should be used and made up to approximately 11% in a gel buffer. This can vary, depending on the species and the enzyme to be detected. In the case of *Campylobacter*, three buffer systems are used (Aeschbacher & Piffaretti 1989). Most reports refer to Selander *et al.* (1986) and, if relevant, give additional modifications.

In order to minimize variations in enzyme mobility, it is important to standardize the gel within a study. Therefore, not only the concentration but also the heat source and time of heating should be standardized. Similarly, the aspiration phase, gel solidification and storage should also be standardized. Some minor experimentation in gel preparation may be required to obtain a workable gel of firm consistency for slicing and handling after electrophoresis. There are no published guidelines on gel preparation, but experience suggests that a good general technique is as follows. Starch and diluted buffer are mixed in a 2-litre, thick-walled, heat-resistant, conical flask with a side-arm for a vacuum tube. A ring gas burner without gauze should be used with a total heating time of 5–7 min. During heating, the mixture of starch and buffer must constantly be swirled to obtain a homogeneous gel. The starch begins to increase in viscosity and after 3–4 min the mixture becomes opaque. Swirling and heating must be maintained throughout this stage, until there is gradual decrease in viscosity, and continued for about 1 min until the solution decreases in opacity to form an almost clear solution. This is then degassed by attachment to a vacuum of *c.* 25 psi, until all the small bubbles are removed and only large bubbles remain. It should be emphasized that care is necessary during gel preparation, both during heating and, particularly, during aspiration, when there is a risk of implosion. Heat-resistant gloves, a heavy-duty apron and perspex head guard are recommended.

The molten gel is poured into a horizontal mould (see Fig. 17.1), of which the two ends have been sealed with Teflon tape. It is cooled at room temperature for 2 h, sealed with plastic film and stored at 4 °C overnight for use on the following day. The mould in Fig. 17.1 was constructed by an in-house plastic workshop and is designed to fit tightly the LKB–Pharmacia Multiphor electrophoresis tank. It provides a 15-mm-thick gel, which is loaded as described by Selander *et al.* (1986). This is a simple and effective system which, with fresh enzyme extracts, allows easy detection of enzyme activities. Extracts from up to 20 bacterial strains can be loaded into such a

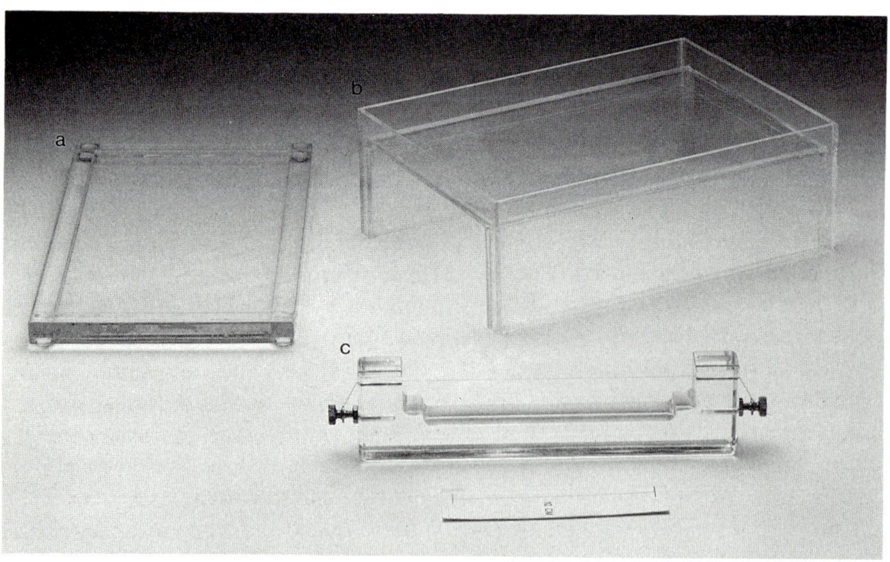

FIG. 17.1. Starch gel mould and slicing apparatus. (a) Slicing board; (b) gel mould; (c) gel slicer.

gel, with control strains near either end and in the middle, in order to reduce scoring variability between gels. Over 15 isolates can be tested for five enzymes in a single gel run. If subsequently it is observed that enzyme staining is weak, a double-soaked filter strip can be 'piggy-backed' on to another strip without disrupting electrical contact in the gel.

Electrophoresis and gel slicing

Constant voltage is maintained during electrophoresis, but this may vary, depending on the enzyme and the buffer system. For example, three buffer systems are used for *Campylobacter* (Aeschbacher & Piffaretti 1989) and two for *Listeria* (Boerlin *et al.* 1991), each at different voltages. Efficient cooling helps to maintain enzyme activity and this can be achieved in several ways. One of the simplest and most efficient methods is to place an aluminium pan of ice on top of the electrophoresis tank cover or on a thin glass plate on the gel mould. Experimentation with variations of methods may be necessary to optimize enzyme detection and the degree of observed polymorphism. For example, gels may be run overnight in a cold room at 4 °C at lower voltages, which may lead to more efficient use of laboratory time.

After electrophoresis, the gel is reduced by trimming the outline of the gel from the mould to give a gel *c.* 150 × 90 mm and 15 mm thick. This is then placed on the gel slicing table (see Fig. 17.1), which has sides raised 2 mm above the inset base. The stainless steel wire, mounted and tensioned on a plastic grip, can then be rested on the raised sides and firmly pulled through the mould to produce a 2-mm-thick gel slice on the bottom of the gel. The top and bottom slices are discarded and the rest of the gel can be sliced to produce five 2-mm slices. Each of these is placed into a plastic box ready for staining.

Gel staining and electromorph mobility scoring

The choice of enzyme depends on the bacterial species under investigation and has been determined empirically, largely by testing many enzymes in various enzyme detection systems. To develop original enzyme systems, useful sources of information are Harris & Hopkinson (1976) and, to a lesser extent, Worthington (1988). However, the seminal paper by Selander *et al.* (1986) is the most useful day-to-day reference. Most materials are easily obtainable from suppliers, such as Sigma, and, apart from the basic electrophoresis equipment, the enzyme staining systems are the only other running expense. Some staining reactions, such as with catalase, can be stopped almost instantly and preserved in water. Others should be stored in an acetic acid : methanol : water (1 : 5 : 5) stopping solution. For enzymes with more diffuse staining reactions, the gel can be held in an agar overlay to score mobilities, but, if resolution is poor, such enzymes are better omitted from the analysis.

For each enzyme system, the bands of stained enzyme in the starch gel each represent a distinctive electromorph. Electrophoretic variants of each enzyme are regarded as alleles of that enzyme and are assigned different ascending allele numbers, based on their increasing anodal migration. The results from enzyme to enzyme for each isolate should be read without knowledge of their identity or position in the gel throughout the MEE procedure. In addition, isolates with different allele scores should be rearranged and rerun on further gels to check whether the differences are reproducible. In one enzyme system, such as, for example, esterase-5, there may be up to four distinctive bands or isoenzymes, but only three may be sufficiently consistent to score as separate electromorphs (ES5-1, ES5-2, ES5-3). Gel-to-gel variation and within-gel variation are minimized by assigning electromorph scores, not only in relation to anodal migration but also with reference to the three internal and three external controls. Migration distances, in millimetres, can be measured from the baseline to the enzyme band centre to compute relative mobility values, but this is generally of little use for determining electromorph scores and, by itself, is poorly reproducible because of gel variation. Enzyme positions are read by eye and a score is ascribed to the centre of the stain. In doing so, some subjectivity may be expected between observers differentiating small differences and those scoring large and more definite differences. Either method will, however, produce the same ET profile analysis. The scores represent relative distances with reference to the control strains. For some electromorphs it may, occasionally, be difficult to ascribe scores because of possible intragel variation. In these instances, extracts should be rerun on another gel in a rearranged order to confirm correct scores.

Electromorphs of an enzyme are equated with alleles at the corresponding structural gene locus and the occasional absence of enzyme activity is attributed to a null allele designated as 0. Such null alleles are rechecked with a higher concentration of enzyme extract, to ensure that the result is due to a true null allele and not simply a weak extract preparation. Each isolate is characterized by its combination of alleles at the various enzyme loci. Distinctive profiles of electromorphs, corresponding to unique multilocus genotypes, are designated as electrophorectic types (ETs).

Data analysis

For each ET group analysed, matrices of allele frequencies at each enzyme locus can be constructed. Genetic diversity at each enzyme locus can be calculated as:

$$h = 1 - x_i^2 [n/n - 1]$$

where x_i is the frequency of the ith allele at the locus, n is the number of isolates or ETs in the population sample and $n/n - 1$ is a correction for bias in small numbers. The mean diversity per locus, H, is the arithmetic average of h over all the loci assayed. Genetic distances can also be calculated as the proportion of enzyme loci at which dissimilar alleles occur (mismatches). This coefficient can be weighted by multiplying the reciprocal of H at the locus in the total population sample being analysed (Selander *et al.* 1985, 1986). In this way, greater weight is given to differences at less variable loci than to those at highly polymorphic loci. A matrix of genetic distances can then constructed for each ET over the other enzyme loci. Cluster analyses may be applied to present this information in a form that is visually more easy to comprehend. A dendrogram can be constructed by the unweighted pair group arithmetic average (UPGMA) clustering method (Sneath & Sokal 1973). Various computer programs are available to help with cluster analysis (Wishart 1978). Genetic distances may also be analysed by principal coordinate analyses (three coordinates) for clarity of visual presentation, according to the method of Gower (1966), and they can be computed by the same programs as are used to produce the dendrograms.

Applications in Food Microbiology

Listeria *spp.*

By serotyping, the vast majority of clinical and food isolates of *Listeria* spp. can only be divided into three or four serotypes, but MEE can demonstrate far greater heterogeneity. While Nørrung (1992) has shown phage typing to be more discriminatory than MEE, MEE has been shown, in terms of clonal analysis, to be largely in agreement with both phage typing and serotyping (Bibb *et al.* 1989; Nørrung 1992). Lawrence & Gilmour (1992) have shown that MEE is more discriminatory than RFLP analysis. Bibb *et al.* (1989, 1990) examined 310 *Listeria* strains, mostly clinical isolates, and produced 56 ETs based on 16 enzyme loci. In their study, three outbreaks had a common source by MEE analysis and in one outbreak a common source was excluded. Pinner *et al.* (1992) studied the *Listeria* isolated from 123 patients with invasive listeriosis, and compared these with strains isolated from food specimens collected from food in their refrigerators. On the basis of the analysis of 15 enzymes, as used by Bibb *et al.* (1989), they found 66 ETs among 226 of 2013 strains from food specimens. Of the refrigerator food isolates, 33% contained at least one with the same ET as the strain from the corresponding patient, and the authors estimated that the probability was extremely low that this was due to chance matching. Boerlin & Piffaretti

(1991) demonstrated 50 ETs in 181 strains from more varied sources, including those from humans, animals, food and the environment. With 21 enzyme loci, they adduced evidence that contamination of meat may originate from the environment in which it is processed, rather than from the animals themselves.

Campylobacter *spp.*

A good reference typing system is urgently required for *Campylobacter* spp. Serotyping systems, which require 60 Penner antisera and 108 Lior antisera, are not widely available. The application of MEE to *Campylobacter* spp. has proved slightly more difficult, because their staining reactions are difficult to interpret.

Aeschbacher & Piffaretti (1989) analysed 125 strains of *C. jejuni* and *C. coli* from human and animal sources over nine polymorphic enzyme loci, which produced 64 ETs in the two principal clusters. After initially screening 38 enzymes, they found that only 10 worked reliably, but polymorphism was high in those enzymes that allowed good discrimination. There were, therefore, a large number of ETs in their collection from sporadic cases of campylobacteriosis and they concluded that disease was caused by a variety of different genotypes, rather than by a small number of distinct ETs or by a particular virulent clone. Isolates from different sources shared common ETs, so confirming the epidemiological link between humans and animals and suggesting that each animal strain is a potential human pathogen. By way of comparison with other typing systems, Patton *et al.* (1991) examined 22 *Campylobacter* isolates from three milkborne and one waterborne outbreak. They screened 40 enzymes and used 11, which yielded nine ETs. They compared four phenotypic tests and six genotypic methods, including phage typing, serotyping, biotyping and DNA ribotyping. Although serotyping was the most discriminating phenotypic method, the most sensitive methods were MEE, ribotyping and restriction endonuclease analysis, when compared with epidemiological evidence as a 'gold standard'.

Salmonella *spp.*

The genus *Salmonella* has a complex taxonomic history, which, in routine clinical laboratories, has been largely overtaken by serotyping as a primary identification. MEE has been used to investigate the genetic relatedness of *Salmonella* isolates and has proved very useful in confirming the findings of DNA–DNA hybridization and, by detecting more subtle genetic differences, has led to new taxonomic proposals (Reeves *et al.* 1989). These workers examined 74 strains, representative of 28 serotypes, by 24 enzyme loci and demonstrated 33 ETs. Each of 26 *S. typhi* strains belonged to the same ET, showing a remarkably stable clone, regardless of their phage types, which were all different, as were their geographic origin and date of isolation (1973–1985). Phage typing is clearly still the most useful typing method for *S. typhi*. Beltram *et al.* (1988) examined 1527 isolates of *Salmonella* other than *S. typhi* and demonstrated 71 ETs. Kapperud *et al.* (1989) applied MEE, along with plasmid analysis, restriction endonuclease analysis, serotyping, phage typing and biotyping, to isolates from an outbreak of *S. typhimurium* associated with chocolate products. Of the 53 strains, 17 were related to the outbreak and were examined with 18 enzymes, and all were monomorphic. Thus, all the isolates had identical ETs. It would seem, therefore, that MEE is of little value in the epidemiological analysis of *Salmonella* spp.

Other organisms

Ochman & Selander (1984) examined 142 isolates of *Escherichia coli* by 12 enzymes and demonstrated 14 ETs. In another study, Ochman *et al.* (1983) examined 1600 human and animal strains of *E. coli* and 123 strains of *Shigella* spp. with 12 enzymes and obtained 279 ETs for *E. coli* and 23 ETs for *Shigella*. These data suggested that the two genera were closely related but that some taxonomic revision should be considered. *Shigella sonnei* was homogeneous but the other strains of *Shigella* were heterogeneous. However, although MEE provided a reasonable degree of typability, no epidemiological significance could be derived from the results.

Many other species have been investigated by MEE but the results are of less relevance in food microbiology. For example, Musser *et al.* (1990) examined 315 isolates of *Staphylococcus aureus* with 20 enzymes, which yielded 49 ETs, but this information was not applied in a food microbiology setting. Nevertheless, the techniques are the same, regardless of the application. Some other techniques, such as polyacrylamide gel electrophoresis and esterase isoenzymes, have usefully been applied to a waterborne outbreak of *Aeromonas hydrophila* infection (Picard & Goullet 1987), but this is a less popular technique than that described with starch.

Conclusions

MEE is emerging as a useful epidemiological tool in the investigation of related isolates, particularly with *Listeria*, *Campylobacter* and *E. coli*. The poor typability seen with salmonellas, and possibly with *S. sonnei*, would appear to make it of little value in these species, at least as a sole typing system, but it is of possible interest as an adjunct when used in combination with other typing systems.

From a practical point of view, the cost of MEE is relatively low in terms of materials and reagents as compared with other genotypic techniques, but analysis of large numbers of strains can be labour-intensive. This is offset to some extent by obtaining up to five gel slices for each run, each of which allows analysis of at least one isoenzyme. It is feasible routinely to run two gels together during each working day, thus allowing approximately 30 isolates to be analysed over at least 10 enzyme loci. MEE is also less hazardous than many other genotypic systems. Nevertheless, in the clinical situation its role currently and in the foreseeable future is in a regional reference laboratory, where the expertise can be maintained by turnover of isolates from various epidemiological investigations.

References

AESCHBACHER, M. & PIFFARETTI, J.C. 1989 Population genetics of human and animal enteric *Campylobacter* strains. *Infection and Immunity* **57**, 1432–1437.

BELTRAM, P., MUSSER, J.M., HELMUTH, R., FARMER, J.J., FRERICHS, W.M., WACHSMUTH, I.K., FERRIS, K., MCWHORTER, A.C., WELLS, J.G., CRAVIOTO, A. & SELANDER, R.K. 1988 Toward a population genetic analysis of *Salmonella*: genetic diversity and relationships among strains of serotypes *S. choleraesuis*, *S. derby*, *S. dublin*, *S. enteritidis*, *S. heidelberg*, *S. infantis*, *S. newport* and *S. typhimurium*. *Proceedings of the National Academy of Sciences USA* **85** 7753–7757.

BIBB, W.F., SCHWARTZ, B., GELLIN, B.G., PLIKAYTIS, B.D. & WEAVER, R.E. 1989 Analysis of *Listeria monocytogenes* by multilocus enzyme electrophoresis and application of the method to epidemiologic investigations. *International Journal of Food Microbiology* **8**, 233–239.

BIBB, W.F., GELLIN, B.G., WEAVER, R., SCHWARTZ, B., PLIKAYTIS, B.D., REEVES, M.W., PINNER, R.W. & BROOME, C.V. 1990 Analysis of clinical and food-borne isolates of *Listeria monocytogenes* in the United States by multilocus enzyme electrophoresis and application of the method to epidemiologic investigations. *Applied and Environmental Microbiology* **56**, 2133–2141.

BOERLIN, P. & PIFFARETTI, J.C. 1991 Typing of human, animal, food and environmental isolates of *Listeria monocytogenes* by multilocus enzyme electrophoresis. *Applied and Environmental Microbiology* **57**, 1624–1629.

BOERLIN, P., ROCOURT, J. & PIFFARETTI, J.C. 1991 Taxonomy of the genus *Listeria* by using multilocus enzyme electrophoresis. *International Journal of Systematic Bacteriology* **41**, 59–64.

FERGUSON, A. 1980 *Biochemical Systematics and Evolution*. Glasgow: Blackie.

GOWER, J.C. 1966 Multivariate analysis and multidimensional geometry. *Statistician* **17**, 13–28.

HARRIS, H. & HOPKINSON, D.A. 1976 *Handbook of Enzyme Electrophoresis in Human Genetics*. Amsterdam: North Holland Publishing Company.

KAPPERUD, G., LASSEN, J., DOMMARSNES, K., KRISTIANSEN, B.E., CAUGANT, D.A., ASK, E. & JAHKOLA, M. 1989 Comparison of epidemiological marker methods for identification of *Salmonella typhimurium* isolates from an outbreak caused by contaminated chocolate. *Journal of Clinical Microbiology* **27**, 2019–2024.

LAWRENCE, L.M. & GILMOUR, A. 1992 Characterisation of *Listeria monocytogenes* from a variety of sources. *Proceedings of the 11th International Symposium on Problems of Listeriosis*, ISOPOL xi, Abstract No. 31.

MCLELLAN, T. & RAMSHAW, J.A.M. 1981 Serial electrophoretic transfers: a technique for the identification of numerous enzymes from single polyacrylamide gels. *Biochemical Genetics* **19**, 647–654.

MUSSER, J.M., SCHLIEVERT, P.M., CHOW, A.W., EWAN, P., KREISWIRTH, B.N., ROSDAHL, V.T., NAIDU, A.S., WITTE, W. & SELANDER, R.K. 1990 A single clone of *Staphylococcus aureus* causes the majority of cases of toxic shock syndrome. *Proceedings of the National Academy of Sciences USA* **87**, 225–229.

NØRRUNG, B. 1992 Characterisation of Danish isolates of *Listeria monocytogenes* by multilocus enzyme electrophoresis. *International Journal of Food Microbiology* **15**, 51–59.

OCHMAN, H. & SELANDER, R.K. 1984 Evidence for clonal population structure in *Escherichia coli*. *Proceedings of the National Academy of Sciences USA* **81**, 198–201.

OCHMAN, H., WHITTAM, T.S., CAUGANT, D.A. & SELANDER, R.K. 1983 Enzyme polymorphism and genetic population structure in *Escherichia coli* and *Shigella*. *Journal of General Microbiology* **129**, 2715–2726.

OLYHOEK, T., CROWE, B.A. & ACHTMAN, M. 1987 Clonal population structure of *Neisseria meningitidis* serogroup A isolated from epidemics and pandemics between 1915 and 1983. *Reviews in Infectious Diseases* **9**, 665–692.

PATTON, C.M., WACHSMUTH, I.K., EVINS, G.M., KIEHLBAUCH, J.A., PLIKAYTIS, B.D., TROUP, N., TOMPKINS, L. & LIOR, H. 1991 Evaluation of 10 methods to distinguish epidemic-associated *Campylobacter* strains. *Journal of Clinical Microbiology* **29**, 680–688.

PICARD, B. & GOULLET, P.H. 1987 Epidemiological complexity of hospital *Aeromonas* infections revealed by electrophoretic typing of esterases. *Epidemiology and Infection* **98**, 5–14.

PINNER, R.W., SCHUCHAT, A., SWAMINATHAN, B., HAYES, P.S., DEAVER, K.A., WEAVER, R.E., PLIKAYTIS, B.D., REEVES, M., BROOME, C.V., WENGER, J.D. & the Listeria Study Group 1992 Role of foods in sporadic listeriosis. *Journal of the American Medical Association* **267**, 2046–2050.

RAMSHAW, J.A.M., COYNE, J.A. & LEWONTIN, R.C. 1979 The sensitivity of gel electrophoresis as a detector of genetic variation. *Genetics* **93**, 1019–1037.

REEVES, M.W., EVINS, G.M., HEIBA, A.A., PLIKAYTIS, B.D. & FARMER, J.J. 1989 Clonal nature of *Salmonella typhi* and its genetic relatedness to other salmonellae as shown by multilocus enzyme electrophoresis, and proposal of *Salmonella bongori* comb. nov. *Journal of Clinical Microbiology* **27**, 313–320.

SELANDER, R.K., MCKINNEY, R.M., WHITTAM, T.S., BIBB, W.F., BRENNER, D.J., NOLTE, F.S. & PATTISON, P.E. 1985 Genetic structure of populations of *Legionella pneumophila*. *Journal of Bacteriology* **163**, 1021–1037.

SELANDER, R.K., CAUGANT, D.A., OCHMAN, H., MUSSER, J.M., GILMOUR, M.N. & WHITTAM, T.S. 1986 Methods of multilocus enzyme electrophoresis for bacterial population genetics and systematics. *Applied and Environmental Microbiology* **51**, 873–884.

SNEATH, P.H.A. & SOKAL, R.R. 1973 *Numerical Taxonomy*. San Francisco: W.H. Freeman & Co.

WISHART, D. 1978 *CLUSTAN User Manual 1, Version 1C, Release 2*, 3rd edn. Edinburgh: University of Edinburgh.

WORTHINGTON, C.C. (ed.) 1988 *The Worthington Enzyme Manual: Enzymes and Related Biochemicals*. Freehold, New Jersey, USA: Worthington Biochemical Corporation.

18: Pyrolysis Mass Spectrometry for the Epidemiological Typing of Foodborne Bacterial Pathogens

PENELOPE R.SISSON,[1] N.F.LIGHTFOOT[1] AND R.FREEMAN[2]
[1]*Regional Public Health Laboratory, Newcastle General Hospital, Westgate Road, Newcastle upon Tyne NE4 6BE, UK;*
and [2]*Department of Microbiology, Medical School, Newcastle upon Tyne NE2 4HH, UK*

Pyrolysis mass spectrometry (PyMS) is a method for the characterization and comparison of large numbers of microorganisms with a precision that equals, and on occasions exceeds, that of conventional typing systems (Gutteridge & Norris 1979; Gould *et al.* 1991). The nature of the method offers both advantages and disadvantages in comparison with conventional techniques. In PyMS, complex material in an inert atmosphere is broken down by heat to produce a series of volatile, low-molecular-weight substances, which are detected and quantified by mass spectrometry. PyMS of bacteria yields bacterial 'fingerprints' that can be compared mathematically for relatedness.

A low-cost, fully automated pyrolysis mass spectrometer, capable of analysing up to 300 samples/day, has been developed (Aries *et al.* 1986). This has allowed the use of PyMS for rapid interspecies and interstrain comparisons of large numbers of isolates pyrolysed as single batches. For our purposes, a batch is defined as a collection of pyrolysis foils, prepared from microorganisms cultured on a standard medium and incubated under identical conditions, pyrolysed on one day as a single machine run. Many different bacterial species, including foodborne pathogens, have been studied (Magee *et al.* 1989; Freeman *et al.* 1990; Gould *et al.* 1991; Orr *et al.* 1991; Sisson *et al.* 1991).

The only specialized equipment required is a pyrolysis mass spectrometer, preferably with a fully automated sampling system. Most recent PyMS studies have used a Horizon Instruments 200X pyrolysis mass spectrometer (Horizon Instruments Ltd, Heathfield, Sussex, UK), supplied with a computer capable

New Techniques in
Food and Beverage Microbiology

0-632-03755-5

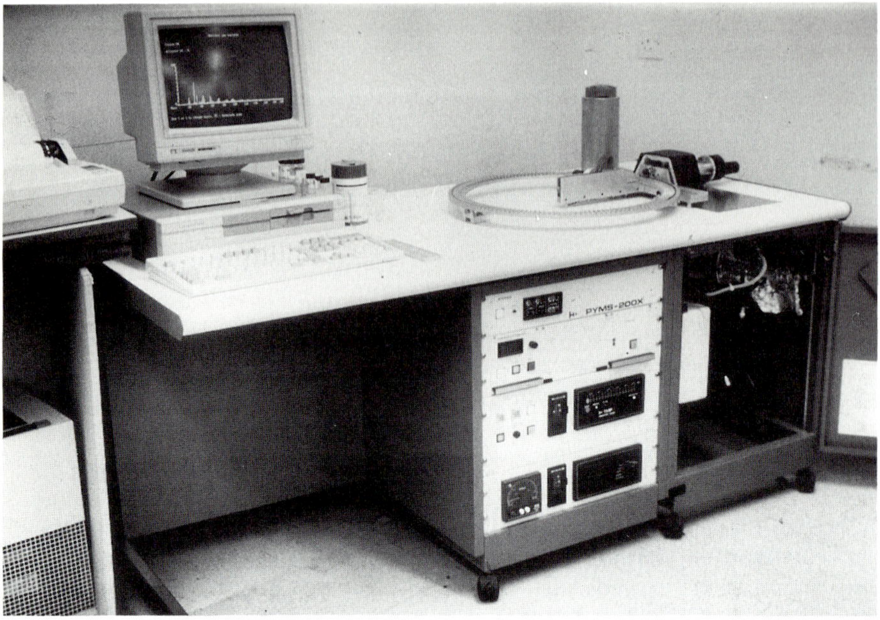

FIG. 18.1. The Horizon 200X pyrolysis mass spectrometer.

of running Genstat 5 software for the data analysis (Fig. 18.1). The method described here assumes the use of this machine.

The materials required are:

1 A collection of similar microorganisms, preferably including reference strains and epidemiologically unrelated isolates of the same species, as well as the isolates from a putative outbreak.

2 Suitable, non-selective media, freshly prepared as a single batch, to allow luxuriant growth of the organisms in the minimum time.

3 Pyrolysis foils of Curie point 530 °C, tubes and O-ring seals (Horizon Instruments Ltd).

4 Liquid N_2 (4 l per pyrolysis run) to cool the condenser plate within the mass spectrometer.

5 A hot-air oven at 80 °C.

Procedure

Microorganisms

Duplicate subcultures of each isolate are prepared and each is pyrolysed separately, so that a measure of machine reproducibility can be obtained.

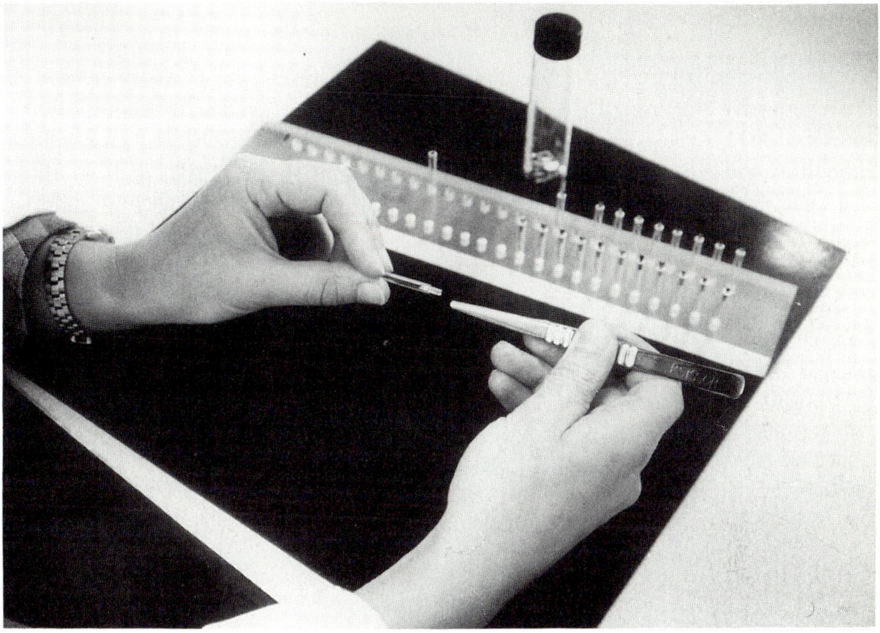

FIG. 18.2.Pyrolysis tubes and pyrolysis foils.

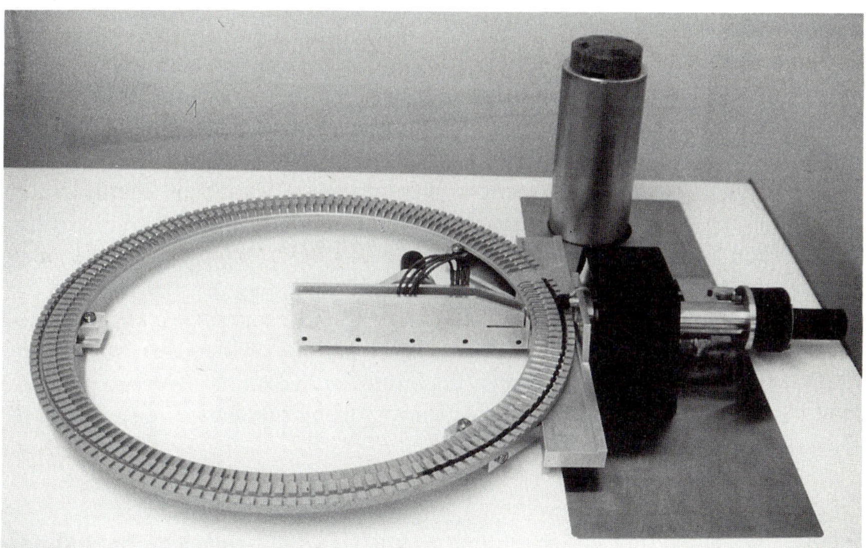

FIG. 18.3.The sampling ring for pyrolysis tubes.

Sample preparation

1 Samples in triplicate are prepared from each subculture of every isolate. A representative portion of the microbial growth is smeared with a sterile wire loop on to a pyrolysis foil, supported in a pyrolysis tube (Fig. 18.2). The inoculum must be sufficient to yield an ion count in the range of 5×10^5–6×10^6 on pyrolysis. It may be necessary to pyrolyse several trial foils to ascertain the appropriate inoculum size. However, generally, a 1-μl loop carries sufficient material for one pyrolysis foil.
2 The loaded foils are dried at 80 °C for 5 min and are placed in a standard position in the pyrolysis tubes with the insertion device provided.
3 An O-ring is fitted to the neck of each tube to provide a seal against the vacuum system of the PyMS machine.
4 The tubes are placed on the PyMS sampling ring (Fig. 18.3) and the microorganisms are pyrolysed in the PyMS group sequence 1, 2, 3 . . . *n*, three times over, rather than in the sequence 1, 1, 1; 2, 2, 2; etc., with the object that the effects of machine drift throughout the pyrolysis run are minimized.

Precautions

1 Interstrain comparisons by PyMS must be made on isolates pyrolysed in a single machine run on one day, because it is a within-batch comparative technique. Since the Horizon Instruments 200X pyrolysis mass spectrometer is capable of analysing 300 samples/day, substantial comparisons can be done at one time.
2 Since PyMS is a highly discriminatory technique, capable of detecting even very small phenotypic differences between microorganisms, it is important to standardize cultural conditions and to minimize phenotypic differences between the isolates to be pyrolysed.
3 The antigenic phase, for example of salmonellas, should be standardized, since the method will discriminate between isolates of a single strain expressing different phase antigens.
4 PyMS comparisons of spore-forming bacteria may be confounded by different rates of sporulation in different cultures. In practice, PyMS comparisons of such organisms have proved to be valid, provided that isolates are cultured on nutritionally rich media for a minimum time to achieve adequate bacterial mass for PyMS sampling, while minimizing the degree of sporulation.

Pyrolysis mass spectrometry

Samples for processing are pyrolysed at 530 °C for 4 s and the products are ionized by collision with a beam of low-energy electrons. The ions are then separated in a quadropole mass spectrometer, which scans the pyrolysate 160 times at intervals of 0.35 s from the start of pyrolysis (Freeman *et al.* 1990). The analysis of each sample takes approximately 2 min. Integrated ion counts, at unit mass intervals from 51 to 200, are recorded on floppy disk, with the pyrolysis sequence number and total ion count for each sample. The duplicate subcultures are pyrolysed sequentially so that isolate 1 is pyrolysed as 1 and 2, isolate 2 as 3 and 4, and so on.

Data analysis

For epidemiological typing, the pyrolysis data are analysed with a Genstat 5 program. The technique and principles involved in data analysis have been previously described (Gutteridge *et al.* 1985; Aries *et al.* 1986).

1 The mass ion counts for each sample are normalized, to account for variations in sample size, and the triplicate spectral data for each subculture are labelled as a separate group.

2 Discriminant analysis examines each mass ion for within-group and between-group variations and ranks them for maximal between-group to within-group ratios.

3 The 150 mass ions are used in principal component (PC) and canonical variate (CV) analyses (Shute *et al.* 1985), which result in a table of Mahalanobis distances.

4 The Mahalanobis distances can be used in an unweighted pair group method with averages (UPGMA) analysis (Sokal & Michener 1958), to create similarity dendrograms (Sisson *et al.* 1991, 1992).

5 The results of PC and CV analyses can be combined and displayed on ordination diagrams, as in Fig. 18.4.

6 The relationships between the pyrolysis groups may be inferred from the relative proximity of their data points on the similarity dendrograms or ordination diagrams. PyMS-derived data points from closely related microorganisms will appear as clusters, while those from markedly dissimilar isolates will appear as ‘outliers’ well separated from each other.

7 If both subcultures from a single isolate have been differentiated as outliers on PyMS analysis, that isolate is considered distinct from the remainder. Its data are edited from the dataset, which is reanalysed further to examine the relationships between the remaining isolates.

8 The end-point of the analysis is the detection of a number of isolates for which the differences between duplicate subcultures of the same isolate are at least as great as those between different isolates. Such isolates are indistinguishable by PyMS and are considered to be a putative strain.

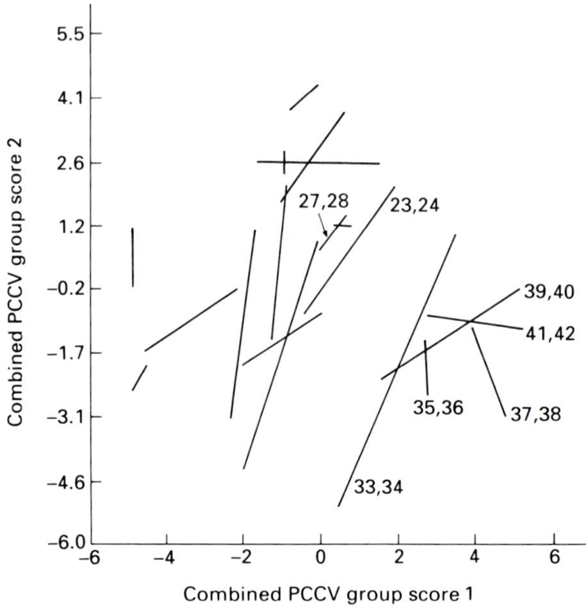

FIG. 18.4.Ordination diagram of the first two principal component canonical variates (PCCV1 & 2) for 19 isolates of *Salmonella enteritidis* examined by PyMS. The means of triplicate analysis of subcultures of the same isolate have been joined together. Numbers represent duplicate subcultures of strains listed in Table 18.1.

Epidemiological Typing of Foodborne Pathogens by PyMS

Outbreaks of infection by a variety of foodborne pathogens have been studied by PyMS, including *Listeria monocytogenes* (Freeman *et al.* 1991), *Bacillus cereus* (Sisson *et al.* 1993) and *Clostridium perfringens* (Sisson *et al.* 1993). The following example illustrates how PyMS may be used to investigate such an outbreak.

Eleven clinical isolates of *Salmonella enteritidis* PT8, epidemiologically related to a food-poisoning outbreak, and one from a suspect food were examined for relatedness by PyMS. Three epidemiologically unrelated isolates each of PT4, 8 and 24 from cases of salmonellosis over the same period were included for comparison (Table 18.1).

Sequential PyMS analyses differentiated two strains (one each of PT8 and PT24) as outliers (data not shown). Two potential clusters were identified within the remainder (Fig. 18.4). One encompassed all the epidemiologically linked strains of *S. enteritidis* PT8, including the isolate from the food source, together with two apparently sporadic isolates of the same phage type. The second, smaller cluster included the three strains of *S. enteritidis* PT4 and the two of PT24.

TABLE 18.1. *Source, phage type and pyrolysis sequence numbers of 21 strains of* S. enteritidis *compared by PyMS*

PyMS nos	Designation	Phage type
1, 2	Outbreak	8
3, 4	Outbreak	8
5, 6	Outbreak	8
7, 8	Outbreak	8
9, 10	Outbreak	8
11, 12	Outbreak	8
13, 14	Outbreak	8
15, 16	Outbreak	8
17, 18	Outbreak	8
19, 20	Outbreak	8
21, 22	Outbreak	8
23, 24	Sporadic	8
25, 26	Sporadic	8
27, 28	Sporadic	8
29, 30	Outbreak	8
31, 32	Sporadic	24
33, 34	Sporadic	24
35, 36	Sporadic	4
37, 38	Sporadic	24
39, 40	Sporadic	4
41, 42	Sporadic	4

In this instance, PyMS was as discriminatory as phage typing. It rapidly confirmed the presence of a local, epidemic strain of *S. enteritidis* and eliminated from further investigations similar but unrelated isolates of this common pathogen, including one strain of the same phage type as the epidemic strain.

Conclusions

The PyMS evidence of identity/non-identity has always been in close agreement with epidemiological data and with available conventional typing systems. The technique is rapid and simple to perform. Preparation and processing of pyrolysis foils is standard, irrespective of the bacteria to be pyrolysed. The consumable costs are less than £1/isolate and the results are available on the day of pyrolysis. However, unlike conventional typing techniques, PyMS interstrain comparisons do not assign permanent type designations, as they apply strictly within each batch and are relative assessments of relatedness.

The speed, simplicity and objectivity of PyMS make it a very attractive technique for initial epidemiological surveillance, especially during the evolution of an outbreak of food-poisoning.

References

ARIES, R.E., GUTTERIDGE, C.S. & OTTLEY, T.W. 1986 Evaluation of a low-cost, automated pyrolysis mass spectrometer. *Journal of Analytical and Applied Pyrolysis* **9**, 81–98.

FREEMAN, R., GOODFELLOW, M., GOULD, F.K., HUDSON, S.J. & LIGHTFOOT, N.F. 1990 Pyrolysis mass spectrometry (Py-MS) for the rapid epidemiological typing of clinically significant bacterial pathogens. *Journal of Medical Microbiology* **32**, 283–286.

FREEMAN, R., SISSON, P.R., LIGHTFOOT, N.F. & MCLAUCHLIN, J. 1991 Analysis of epidemic and sporadic strains of *Listeria monocytogenes* by pyrolysis mass spectrometry. *Letters in Applied Microbiology* **12**, 133–136.

GOULD, F.K., FREEMAN, R., SISSON, P.R., COOKSON, B.D. & LIGHTFOOT, N.F. 1991 Inter-strain comparison by pyrolysis mass spectrometry in the investigation of *Staphylococcus aureus* nosocomial infection. *Journal of Hospital Infection* **19**, 41–48.

GUTTERIDGE, C.S. & NORRIS, J.R. 1979 The application of pyrolysis techniques to the identification of micro-organisms. *Journal of Applied Bacteriology* **47**, 5–43.

GUTTERIDGE, C.S., VALLIS, L. & MACFIE, H.J.H. 1985 Numerical methods in the classification of micro-organisms by pyrolysis mass-spectrometry. *Computer-assisted Bacterial Systematics*, eds Goodfellow, M., Jones, D. & Priest, F.G., pp. 369–401. London: Academic Press.

MAGEE, J.T., HINDMARSH, J.M., BURNETT, I.A. & PEASE, A. 1989 Epidemiological typing of *Streptococcus pyogenes* by pyrolysis mass spectrometry. *Journal of Medical Microbiology* **30**, 273–278.

ORR, K., GOULD, F.K., SISSON, P.R., LIGHTFOOT, N.F., FREEMAN, R. & BURDESS, D. 1991 Rapid inter-strain comparison by pyrolysis mass spectrometry in nosocomial infection with *Xanthomonas maltophilia. Journal of Hospital Infection* **17**, 187–195.

SHUTE, L.A., BERKELEY, R.C.W., NORRIS, J.R. & GUTTERIDGE, C.S. 1985 Pyrolysis mass spectrometry in bacterial systematics. *Chemical Methods in Bacterial Systematics*, eds Goodfellow, M. & Minnikin, D.E., pp. 95–114. London: Academic Press.

SISSON, P.R., FREEMAN, R., GOULD, F.K. & LIGHTFOOT, N.F. 1991 Strain differentiation of nosocomial isolates of *Pseudomonas aeruginosa* by pyrolysis mass spectrometry. *Journal of Hospital Infection* **19**, 137–140.

SISSON, P.R., FREEMAN, R., MAGEE, J.G. & LIGHTFOOT, N.F. 1992 Rapid differentiation of *Mycobacterium xenopi* from mycobacteria of the *Mycobacterium avium–intracellulare* complex by pyrolysis mass spectrometry. *Journal of Clinical Pathology* **45**, 355–357.

SISSON, P.R., KRAMER, J.M., BRETT, M.M., FREEMAN, R., GILBERT, R.J. & LIGHTFOOT, N.F. 1993 Application of pyrolysis mass spectrometry to the investigation of outbreaks of food poisoning and non-gastrointestinal infection associated with *Bacillus* species and *Clostridium perfringens. International Journal of Food Microbiology* **17**, 57–66.

SOKAL, R.R. & MICHENER, C.D. 1958 A statistical method for evaluating systematic relationships. *University of Kansas Science Bulletin* **38**, 1409–1438.

19: Restriction Fragment Length Polymorphism Analysis for Epidemiological Typing of *Listeria monocytogenes*

ANNE M.RIDLEY AND N.A.SAUNDERS*
Laboratory of Microbiological Reagents, Central Public Health Laboratory, 61 Colindale Avenue, London NW9 5HT, UK

The detection of sequence differences in genomic DNA by analysis of restriction fragment length polymorphisms (RFLP) is important for the epidemiological typing of bacterial strains. Bacterial genomes are sufficiently simple for individual bands to be visualized in ethidium bromide-stained gels after endonuclease treatment. However, most enzymes generate a large number of fragments, which yield complex patterns. For this reason, probes able to hybridize with specific restriction fragments that result from cleavage of a small fraction of the genome are often employed. The use of such probes has two main advantages. First, hypervariable regions of the genome may be made the probe target, which results in a high discrimatory power. Second, the complexity – that is, number of bands – of the restriction fragment pattern may be adjusted by the use of a longer or shorter probe sequence.

RFLP analysis of *Listeria monocytogenes* strains has been shown to be of value for epidemiological typing (Saunders *et al.* 1989; Desmarchelier *et al.* 1990; Ridley *et al.* 1992) and for strain identification to the species level (Vogt *et al.* 1990; Graves *et al.* 1991).

Epidemiological Typing of *L. monocytogenes*

Epidemiological typing of *L. monocytogenes* strains is critical to an understanding of listeriosis and, in particular, may answer remaining questions about the sources and routes of infection by this organism. Apart from its use to help confirm relationships between cases and the sources of infection, the

* To whom correspondence should be addressed.

New Techniques in
Food and Beverage Microbiology

0-632-03755-5

typing data may also be helpful in monitoring trends in the occurrence of pathogenic *Listeria* strains (McLauchlin *et al.* 1989). The methods frequently used for subtyping *L. monocytogenes* isolates – serogrouping with polyclonal antisera, phage typing and isoenzyme analysis – rely on the detection of interstrain differences in phenotype. Serotyping is of limited value in epidemiological investigations because the majority of cases of human listeriosis are caused by only four serovars (4b, 1/2a, 1/2b, 1/2c). Phage typing provides increased discrimination and is reproducible (McLauchlin *et al.* 1986a, b), but current limitations are the low percentage of typable serogroup 1/2 strains and the lack of phages for serovars that rarely cause disease. However, the typability of serovar 4b strains is high (85%). Isoenzyme analysis (Bibb *et al.* 1989; Piffaretti *et al.* 1989) is the most discriminatory of the above methods but has not been widely used because it is relatively time-consuming. It is of interest that application of this technique to strains from all the major outbreaks of foodborne listeriosis indicates that such isolates have similar multilocus enzyme profiles (Piffaretti *et al.* 1989).

RFLP Typing

The analysis of RFLP is highly discriminatory because it is possible to distinguish between strains that differ by only one nucleotide residue in their DNA sequences, provided that the varying base is within an appropriate restriction site. In addition, DNA rearrangements, insertions and deletions can also alter the observed restriction fragment patterns.

The discrimination of RFLP analysis depends not only upon the degree of similarity between the base sequences of the strains to be compared but also on the specific probe used. This is because base sequence drift in different parts of the genome, during strain divergence, varies considerably from the highly conserved rRNA genes to parts of the genome that code for gene products with highly variable amino acid sequences. RFLP also arises from rearrangements of the genome and these events also vary in frequency at different genetic loci. Some repeated sequences, for example, are recombination hot spots (Lin *et al.* 1984) and may prove useful target sequences for use as probes.

It is sometimes useful that the degree of pattern variation observed is a crude measure of overall DNA sequence divergence between strains.

Identification of L. monocytogenes *strains*

L. monocytogenes strains are presumptively identified by biochemical methods, including haemolysis on blood agar, Gram stain, positive CAMP reaction, aesculin hydrolysis and motility at room temperature. For serological identification, absorbed hyperimmune rabbit antisera are used in slide agglutination reactions (Seeliger & Hohne 1979).

Bacterial cultivation

DNA can be extracted from *Listeria* cells grown overnight at 37 °C on Marmite agar (Oxoid tryptone, 10 g/l; Marmite, 5 g/l; Oxoid acid-hydrolysed casein, 10 g/l; sodium glycerophosphate, 5 g/l; 70% sodium lactate, 5 ml/l; glycerol, 5 ml/l; magnesium and iron solution ($MgSO_4.7H_2O$, 100 g/l; $FeSO_4.H_2O$, 10 g/l; concentrated sulphuric acid, 1 ml/l), 2 ml/l; New Zealand agar, 12 g/l), but cells grown to stationary phase on any suitable liquid or solid culture medium, including blood agar, nutrient agar, tryptose and brain–heart infusion broth, can be used. Cultures should be fresh, but it is advantageous to avoid cells in exponential growth, because they have a high RNA/DNA ratio. RNA contamination of DNA preparations can be a problem in the DNA isolation procedure described below.

TABLE 19.1. *Protocol for DNA purification*

1 Harvest bacterial cells into a 1.5-ml microcentrifuge tube and pellet by centrifugation at 10 000 g_{max} for 2 min. The pellet size should be between 25 and 50 µl. If only a smaller cell volume is available, the volumes of reagents employed should be proportionately scaled down.
2 Resuspend the cells by vortexing in a solution of lysozyme (4 mg/ml) in sterile distilled water (100 µl) and incubate at room temperature for at least 30 min.
3 Lyse the cells by the addition of 5 mol/l guanidine thiocyanate containing 0.1 mol/l EDTA pH 7.0 (200 µl) with gentle pipetting. The solution should clear rapidly on mixing, but the lysate often remains slightly cloudy and this is not resolved by prolonging the guanidine thiocyanate treatment. The yield of DNA, however is usually satisfactory.
4 Add 7.5 mol/l ammonium acetate (150 µl) and chloroform:isoamyl alcohol (24:1; 450 µl), emulsify by vigorous shaking for 30 s. Separate the phases by centrifugation in a microfuge at 10 000 g_{max} for 2 min and remove the aqueous (top) phase.
5 Precipitate the high-molecular-weight DNA in the aqueous phase, by adding 0.54 volumes of propan-2-ol, and mix by inversion. Remove the precipitated DNA. The DNA, which forms a dense stringy precipitate within a few seconds of mixing, should immediately be separated from the liquid phase with a pipette tip. The DNA should not be collected by centrifugation because:
(a) centrifugation results in the formation of a compacted pellet which is difficult to resuspend;
(b) a cloudy precipitate, probably consisting mainly of RNA and protein, forms in some preparations and this is collected with the DNA if centrifugation is used.
6 Wash the DNA twice in 80% ethanol (1 ml), dry under vacuum and dissolve in TE buffer (10 mmol/l Tris–HCl pH 7.0, 1 mmol/l EDTA). Measure the nucleic acid concentration spectrophotometrically. A 1-mg/ml solution of double-stranded DNA has an absorbance of 20 OD units at 260 nm.

DNA preparation

The preparation of high-molecular-weight DNA, suitable for restriction endonuclease digestion, can be achieved with a modification of the method of Pitcher *et al.* (1989), which is rapid and requires a minimum of technical effort. *Listeria* cells are lysed by treatment with 5 mol/l guanidine thiocyanate, 0.1 mol/l EDTA after lysozyme (5 mg/l) treatment. Lysozyme treatment for 40–60 min is necessary to ensure that efficient cell lysis is achieved. The quantity and quality of the DNA preparation is affected by the concentration of cells in the suspension. In particular, care is necessary to avoid exceeding the recommended quantity of cells (Table 19.1). Deoxyribonuclease (DNase) is inactive under the conditions of extraction described here, but low residual nuclease activity may be present in DNA preparations from some strains

TABLE 19.2. *Preparation of λ phage DNA*

1 Treat the lysed culture with DNase and RNase (1 μg/ml of each) for 30 min at room temperature.
2 Add sodium chloride to 1 mol/l and incubate on ice for 1 h.
3 Remove the cellular debris by centrifugation at 10 000 g_{max} for 10 min at 4 °C. Make the supernatant up to 10% (w/v) in polyethylene glycol 8000 (Sigma) and incubate on ice for 1 h.
4 Collect the precipitate by centrifugation at 10 000 g_{max} for 10 min at 4 °C, resuspend in 25 ml of SM buffer (0.1 mol/l NaCl, 10 mmol/l $MgSO_4$, 50 mmol/l Tris–HCl pH 7.5, 0.01% gelatin) and extract with an equal volume of chloroform.
5 Sediment bacteriophages in the aqueous phase by centrifugation at 120 000 g_{max} at 4 °C for 2 h.
6 Gently resuspend the pellet in 1 ml of SM buffer and disrupt the phages by the sequential addition of 0.5 mol/l EDTA pH 8.0 (40 μl), proteinase K (100 μg), 20% (w/v) SDS (25 μl). Incubate at 65 °C for 1 h.
7 Purify the phage DNA by extraction against equal volumes of:
(a) phenol solution containing 0.8% hydroxyquinolone, equilibrated with 0.1 mol/l Tris–HCl pH 8.0 buffer;
(b) phenol–chloroform solution (1 : 1 v/v);
(c) chloroform.
8 Precipitate bacteriophage DNA in the final aqueous phase by adding 0.1 volumes of 3 mol/l ammonium acetate, pH 6.0, and 2 volumes ethanol; then collect it by centrifugation (10 000 g_{max}, 5 min) and dry *in vacuo*.
9 Analyse on a 1.0% agarose gel for extent of RNA contamination. Check DNA and protein concentrations at 260 and 280 nm respectively. Residual RNA and protein contamination can be removed as follows: dissolve the DNA pellet in SET buffer (0.15 mol/l NaCl, 15 mmol/l EDTA, 50 mmol/l Tris–HCl pH 8.0, 1% SDS) and DNase-free RNase A (50 μg/ml) and incubate at 37 °C for 15 min. Add proteinase K (100 μg/ml) and continue incubation for 30 min at 65 °C. Add 0.5 volumes of 7.5 mol/l ammonium acetate and extract with an equal volume of chloroform : isoamyl alcohol (24 : 1 v/v). Precipitate DNA in the aqueous phase by adding 0.54 volumes of propan-2-ol.
10 Wash the precipitate twice in 80% (v/v) ethanol, dry *in vacuo* and finally dissolve in TE buffer (10 mmol/l Tris pH 7.0, 1 mmol/l EDTA).

(Andersen & Saunders 1990) and this may cause some degradation of the DNA during long-term storage at 4 °C. DNase activity is insufficient for digestion to become apparent during incubation at 37 °C for 2 h, the conditions used for restriction endonuclease treatment. If desired, DNase activity may be eliminated by incubation of the resuspended DNA at 80 °C for 30 min. In our experience, solutions of *Listeria* DNA, without heat treatment to destroy residual DNase, are stable for several years when frozen at −20 °C in microcentrifuge tubes. Repeated freezing and thawing of the DNA also result in its fragmentation and should be avoided. If necessary, contaminating RNA can be removed from the DNA by treatment with RNase and proteinase K (Table 19.2).

Restriction endonuclease digestion

Preliminary studies have revealed that many of the commonly employed restriction endonucleases fail to cleave the DNA isolated from some or all *Listeria* strains, and thus the choice of endonuclease is limited. The failure of some enzymes to digest DNA from *L. monocytogenes* may be due to methylation of nucleotides at the relevant restriction sites. A further consideration in the choice of restriction endonuclease is the size range of restriction fragments generated. *NciI* produces restriction fragments primarily in the size range 500–10 000 bp. Molecules of this size are well resolved in approximately 1% agarose gels, which have sufficient strength for easy handling during the subsequent processing steps, and they allow efficient blotting of the DNA on to a solid support.

NciI, which cuts target DNA at the sequence 5′-CC$^{G}/_{C}$GG-3′, was chosen after comparison of the results obtained with several different enzymes. The DNA of various *Listeria* strains was digested with the enzymes and the restriction fragment patterns obtained after application of the λ32/λ35 probe were compared. The restriction fragments generated by *NciI* form relatively uncomplicated patterns of bands, which are, nevertheless, sufficient to provide discrimination between strains. In practice, the number of restriction fragments that make up the patterns is limited by the ability of the system to resolve bands and the requirement that the pattern is amenable to comparison.

Listeria DNA (5 μg) is digested in 50-μl reaction volumes as follows: 5 μl of the 10× reaction buffer (supplied by the manufacturer) is added to DNA in 45 μl of distilled water. Alternatively, if buffer is not supplied, one of the restriction endonuclease buffers described by Sambrook *et al.* (1989) may be used. An excess (1.2–2 units/μg DNA) of *NciI* is added and the mixture is incubated at 37 °C for 2 h and then kept on ice. A volume (5 μl) is withdrawn and subjected to agarose gel electrophoresis for 1 h at 5 V/cm (see below), to check that the digestion is complete. A sample of the undigested DNA should be run in an adjacent track as a control. The gel is stained with ethidium bromide (0.5 μg/ml in water) and examined under UV transillumination. The DNA fragments should migrate faster than the undigested DNA and it should be possible to distinguish a few high-molecular-size bands, and noticeable areas of high and low intensity in the pattern at lower size. These features are not usually present if the DNA is only partially digested. If no digestion is apparent, the activity of the enzyme should be checked against another DNA sample (e.g. commercial λ phage or plasmid DNA) and, if the activity of the enzyme is shown to be weak or absent, alternative sources should be explored. The possibility that the DNA is contaminated with a general inhibitor of endonuclease activity, such as chemicals used in the extraction process, can be investigated by attempting digestion with another enzyme. Contaminants can often be removed by reprecipitation of the DNA from 66% ethanol containing 0.3 mol/l sodium acetate (pH 6.0). If neither the enzyme nor the DNA is the cause of the problem, it is likely that the relevant restriction sites are methylated. Incomplete digestions may often be taken to completion by the addition of a second volume of enzyme and incubation for a further 2 h. Problems with endonuclease digestion are usually due to chemical contamination of the DNA substrate or to low enzyme activity. When the digestion is complete 0.5 volumes (25 ml) of a solution containing 10% Ficoll and 0.05% bromophenol blue are added. The restriction fragments can then be stored at −20 °C or below until required.

Agarose gel electrophoresis

Many different types of apparatus are available for running horizontal agarose slab gels submerged beneath the surface of the electrophoresis buffer. Good results are obtained with 0.5-cm-wide slots. The use of smaller slots has the advantage that more samples can be processed simultaneously, but the effect of trailing at the edges of the slots is greater and the results are less clear. The agarose concentration used determines the size range of fragments that can be resolved on a gel. Gels with an agarose concentration between 0.8 and 1% are easy to handle and give reasonable resolution of fragments in the size range 0.5–10 kb, even when the gel is run at a high voltage. Many types of agarose are available and there is considerable variation between grades and batches. Some agarose preparations contain impurities that interfere with the binding of DNA to Hybond-N filters in vacuum blotting. Fortunately, different batches of 'Ultrapure' agarose (BRL, Life Technologies Inc., Gaithersburg, MD, USA) have all given satisfactory results. Agarose gels 20 × 20 × 0.5 cm in TBE buffer (89 mmol/l Tris; 89 mol/l borate; 2.5 mmol/l EDTA Na_2, pH 8.3) are subjected to electrophoresis either at 1.5 V/cm for approximately 16 h or at 6 V/cm for approximately 3 h, until the marker has migrated 75% of the length of the gel. The resolution of the high-molecular-weight bands is lower at high voltage but the bands obtained are sharper. Sample DNA (1 μg in 15 μl) is applied to each slot. Restriction fragments of known molecular size should be applied to slots adjacent to each sample. If the probe is a recombinant λ phage, it is convenient to use a digest of commercial λ phage DNA as the standard. The fragments produced by separate digestion with the enzymes *PstI* and *EcoRI* can be mixed (1 ng of each digest in a total volume of 10–15 μl) to produce markers of a useful size range. However, digestion of λ DNA with many other enzymes or combinations of enzymes will yield satisfactory marker fragments.

Southern blotting

Transfer of DNA fragments from agarose gels to a hybridization membrane can be accomplished satisfactorily by capillary (Southern 1975), vacuum or electroblotting. Vacuum blotting is preferred, because it is rapid and gives excellent resolution of closely spaced bands. Many of the different hybridization membranes available are suitable for use in blotting experiments. Hybond-N (Amersham International, Amersham, UK), a nylon-based membrane, is convenient, because it is very strong and has high binding capacity for nucleic acids which can be fixed to it covalently by a short exposure to ultraviolet light. A standard vacuum-blotting procedure is summarized in Table 19.3. The hybridization membrane should be handled carefully to avoid scratches and

TABLE 19.3. *Transfer of DNA to hybridization membrane*

1 Soak the gel twice in 2.5 gel volumes of 0.25 mol/l HCl for 15 min with gentle agitation. This results in the partial depurination of the DNA, which is then susceptible to alkaline hydrolysis in the next stage.
2 Denature the DNA by soaking the gel twice in 2.5 volumes of 0.5 mol/l NaOH, 1.5 mol/l NaCl, for 15 min with gentle agitation.
3 Neutralize twice in 2.5 volumes of Tris–HCl pH 8.0, 1.5 mol/l NaCl, for 15 min with gentle agitation. If necessary, the gel may be left in this buffer for several hours, but gradual diffusion of the DNA will occur, resulting in reduced resolution of the bands.
4 Assemble the gel on to a vacuum blotting apparatus (suitable apparatus can be obtained from several manufacturers, for example the Vacublot (Anderman Co. Ltd, Kingston upon Thames, UK)) and blot the restriction fragments on to Hybond-N with 10× SSC as transfer buffer (1× SSC is 0.15 mol/l NaCl, 15 mmol/l sodium citrate, pH 7.0).
5 Wash the membrane in 2× SSC, air-dry, wrap in ultraviolet light-transparent plastic film and place, DNA side down, on a UV transilluminator. The duration of ultraviolet light exposure required for maximal binding of hybridizable target DNA should be determined for each transilluminator and type of plastic wrap. It is usually 2–5 min.

fingermarks on its surface. Such marks often stain intensely when the blot is developed and may obscure less intense bands on the blot.

Cloning and selection of probes

A full description of the various methods available for cloning bacterial DNA is beyond the scope of this chapter and the standard techniques required are the subject of some very helpful texts (for example, Karn *et al.* 1983; Kaiser & Murray 1985; Sambrook *et al.* 1989). Bacteriophage λ vectors are particularly suitable for the preparation of cloned probes for RFLP analysis, rather than the M13, plasmid and cosmid alternatives. The advantages are as follows:

1 Relatively large DNA fragments can be cloned with great efficiency. The small fragments readily cloned in M13 and plasmid vectors can be used as probes in RFLP analysis, but such probes, if they represent unique sequences, hybridize to few restriction fragments in the optimal size range and therefore few RFLPs are revealed.

2 There is no need to purify the cloned bacterial DNA from the vector. The vector sequences become labelled and will hybridize with restriction fragments of bacteriophage λ used as molecular weight standards and run in electrophoresis tracks adjacent to the restriction fragments derived from strains of interest. When used in this way, these sequences act as controls for the efficiency of the probe labelling, vacuum blotting, hybridization and detection procedures.

3 Purification of phage DNA in high yield and of suitable purity is easily accomplished.

4 Screening of large numbers of plaques is easier than screening bacterial colonies.

We have prepared libraries of *L. monocytogenes* genomic DNA, cleaved with either *Sau3A* (partial) or *EcoRI* and cloned into the *BamHI* site of λ phage vector EMBL3 (Kaiser & Murray 1985) and the *EcoRI* site of λgt.WES.λB (Tiemeier *et al.* 1976), respectively, with equally satisfactory results.

Identification of clones carrying repeated sequences

The use of probes for sequences present within the genome as multiple copies offers the possibility of enhanced discriminatory power. Recombinants carrying such sequences can be readily identified by hybridizing the cloned DNA to labelled total DNA (probe) from the parent organism. The rate of hybridization depends on the concentration of cloned sequence in the total DNA probe. Repeated sequences, therefore, hybridize faster. Thus, if an incubation period insufficient to allow complete reassociation is employed, the signal will be highest for multiple-copy genes.

TABLE 19.4. *Biotin/digoxigenin detection*

1 Immediately after the posthybridization washes at the required stringency, immerse the blot in a 1% (w/v) solution of blocking agent (Boehringer Mannheim GmbH, Mannheim) in buffer A containing Tris–HCl pH 7.5 and NaCl (1 ml buffer/10 cm^2 filter) in a sealed hybridization bottle (Hybaid, Teddington, Middlesex) or bag, and incubate at room temperature for 30 min.
2 Drain off excess blocking buffer. Dilute the conjugate to working strength in buffer A (a 10^{-3} dilution of the BRL streptavidin–alkaline phosphatase conjugate stock or a 5×10^{-3} dilution of anti-(DIG) alkaline phosphatase) and apply to the blot (1 ml diluted conjugate/20 cm^2 filter), with constant agitation, at room temperature for 10 min.
3 Remove excess conjugate from the membrane by washing, with constant agitation, in two changes (15 min each) of buffer A (1.5 ml buffer/cm^2 filter), followed by 10 min in buffer B containing 0.1 mol/l Tris–HCl pH 9.5, 0.1 mol/l NaCl and 50 mmol/l $MgCl_2$.
4 Develop by application of 1 ml dye solution/20 cm^2 filter, containing 330 mg/l nitroblue tetrazolium (NBT) and 150 mg/l 5-bromo-4-chloro-3-indolyl phosphate (BCIP) disodium salt in buffer B. The dye solution is made up immediately before use. Stock solutions of NBT (75 mg/ml in 70% (v/v) dimethylformamide, stored at 4 °C) and BCIP (50 mg/ml in water, stored frozen at −20 °C)).
5 Leave bands to develop to a satisfactory level at room temperature under subdued lighting conditions in a sealed plastic bag. Most colour development occurs within 2 h, but some darkening of the bands, and background, is seen for several days.
6 Wash thoroughly under running tap water, air-dry and store in a plastic bag.

To screen a library of partial *Sau3A* digestion fragments, cloned into the vector λEMBL3, the phages are plated out at high cell density in top agar, to ensure that the plaques obtained are small and of uniform size (Kaiser & Murray 1985). *Escherichia coli* NM539 is used to ensure that only the recombinant phages form plaques. Hybond-N nylon filters are laid on the plates and allowed to absorb moisture and phage particles from the agar. The filter is peeled off and floated on denaturing solution for 1 min and transferred to neutralizing solution for 1 min before immersion in 2 × SSC (1 × SSC is 0.15 mmol/l NaCl, 15 mmol/l sodium citrate, pH 7.0) for up to 5 min (Table 19.2). The plate is stored at 4 °C until the required plaques have been identified. After drying, the DNA attached to the filter is fixed by exposure to UV. The filter is prehybridized at 68 °C for 1 h and the probe (0.1 μg) is added and allowed to hybridize with the filter at that temperature for 20 min. The probe consists of chromosomal DNA from the *L. monocytogenes* strain used to prepare the gene library, labelled by primer extension (Table 19.4) with digoxigenin. After the 20-min exposure, the filter is processed through to colour development by the procedure summarized in Table 19.4. The filters are examined for spots of high colour intensity, corresponding to plaques on the lawn of NM539 cells. It is assumed that, after such a short hybridization period, only phage bearing repeated sequences will produce strong signals. The corresponding plaques are picked and stored in a microfuge tube containing SM (0.1 mol/l NaCl, 0.01 mol/l $MgSO_4.7H_2O$, 50 mmol/l Tris (pH 7.5), 2% gelatin (5 ml)) and a drop of chloroform. DNA from such phages is then isolated and labelled for use as probes (Table 19.2).

Labelling of probe and hybridization

In our studies of *L. monocytogenes*, we have used phages λ32/λ35 selected from a library of *EcoRI* fragments inserted into λgtWES.λB and two clones, λAR 3.1 and λAR 5.2, carrying repeated sequences, which were selected from a library of partial *Sau3A* fragments in the vector λEMBL3.

Recombinant DNA is extracted from the bacteriophage λ particles by the pelleting method (Table 19.2), which is modified from Sambrook *et al.* (1989) and is recommended because high yields of good-quality DNA are rapidly obtained. λ phage DNA is labelled with biotin by primer extension at 37 °C for 2 h, although longer incubation periods can be used.

For biotinylation the reaction mixture (100 μl) contains: 1 μg DNA (heat-denatured at 95 °C for 10 min in TE buffer), 50 μg BSA, 100 μg salmon sperm primers, 25 μmol/l dATP, 25 μmol/l dCTP, 25 μmol/l dGTP, 5 μmol/l dTTP, 20 μmol/l biotin-11-dUTP, 50 mmol/l Tris-HCl (pH 8.0), 2 mmol/l dithiothreitol, 5 mmol/l $MgCl_2$, 40 mmol/l KCl, 10 units Klenow fragment.

The digoxigenin labelling mixture, in a final volume of 50 μl, contains: 1 μg heat-denatured DNA, 2 μl hexanucleotide mix (Boehringer Mannheim, Mannheim, Germany), 2 μl digoxigenin(DIG)-dUTP (containing 1 mmol/l each of dATP, dCTP, dGTP; 0.65 mmol/l dTTP and 0.35 mmol/l digoxigenin dUTP; pH 6.5), 4 units Klenow fragment. Prolonged incubation, for up to 20 h, can increase the yield of digoxigenin-labelled product. Labelled probe is used directly or stored frozen at −40 °C until required.

Blots (15 × 18 cm) are interleaved with nylon mesh, rolled up, placed into hybridization bottles and incubated in a hybridization oven (Hybaid Ltd, Teddington, Middlesex, UK) at 68 °C for 1 h in hybridization buffer (20 ml) containing 5 × SSC; blocking agent 1% (w/v); *N*-lauroylsarcosine 0.1% (w/v); SDS 0.02% (w/v). If a hybridization oven is not available, good results can be obtained with heat-sealable polythene bags, which may be incubated immersed in a water bath. After prehybridization, the bottle (or bag) is opened and the buffer is thoroughly drained off, but care must be taken to prevent the filter from drying out. Labelled probe DNA (100–200 ng) is denatured by heating in a boiling water bath for 10 min, cooled on ice, mixed with hybridization buffer (1 ml) and added to the bottle (or bag). If bags are used, bubbles should be carefully removed. Hybridization is carried out at 68 °C for 16 h. Blots are washed under non-stringent conditions at 20 °C with 150 ml of 2 × SSC, 0.1% SDS (10 min), followed by two washes in 150 ml of 0.2 × SSC, 0.1% SDS (2 × 10 min). Non-stringent conditions are used so that the homologous sequences, from divergent strains, will hybridize with the probe. In practice, with the probes used to date, this does not result in staining of large numbers of background bands.

Probe detection

The biotin- or digoxigenin-labelled probe/target duplexes are detected with the appropriate antibody–enzyme conjugates. For biotin detection, a monoclonal antibiotin polyalkaline phosphatase conjugate (D.Samuel and P.Vincent, personal communication) is employed, whilst digoxigenin label is detected with an antidigoxigenin alkaline phosphatase conjugate (Boehringer Mannheim) (Table 19.4). Successful blocking of non-specific binding sites can be achieved with the blocking agent supplied by Boehringer Mannheim (1%) or a mixture of casein (2%, Sigma, Poole) and BSA fraction V (2%, Sigma). After colour development, blots can be kept stored in the dark for several years without fading.

Interpretation of Results

The blot is assessed by reference to the λ phage restriction fragment markers, which are a particularly useful indicator of quality. The marker bands should be present in their correct relative positions, corresponding to each of the phage DNA fragments. The intensity of the bands should also be proportional to the size of the DNA fragments. If the expected pattern of bands is present, all the steps in the restriction fragment analysis, from probe labelling and gel electrophoresis through to colour development, have clearly been successful. If the anticipated pattern of marker bands is achieved, but without signal from the test DNA, it may be concluded that the DNA samples applied to the gel were defective or that sequences complementary to the probe were not present within them. Preliminary inspection of the patterns should also show whether partial or non-specific digestion of the test DNA has occurred. Partial digests are normally characterized by the presence of multiple high-molecular-weight bands with the low-molecular-size fragments present at reduced intensity. However, in our experience, occasional incomplete digests of *L. monocytogenes* strains produce patterns difficult to distinguish from those of complete digests (see Fig. 19.3).

Patterns resulting from the analysis of partially degraded DNA samples may be recognized by the reduced intensity of the larger fragments, together with an increase in intensity of the background corresponding to fragments of low molecular size.

It is relatively easy to compare, by eye, the restriction fragment patterns revealed with probes, because they are designed to consist of a small number of well-spaced bands. Comparison is facilitated by drawing a representation of the observed bands against a molecular weight scale derived from the λ phage standards. A solid line should be drawn for strong bands and a dotted line for weak bands – those that are less than approximately 10% of the intensity of the most intense band. In this way, the restriction fragment patterns of strains analysed on different gels can be conveniently and rapidly compared. Alternatively, comparison between large numbers of RFLP patterns can be effected with computer programs from the PHYLIP 3.4 package (J.Felsenstein, University of Washington). The MIX and DOLLOP programs are each designed to estimate phylogenetic evolutionary relationships from discrete character data with two states (0 and 1) by means of various parsimony methods. Although the restriction fragment data obtained for typing purposes are generally unsuitable for the determination of phylogenies, these programs can nevertheless be used to show homology between patterns and, in addition, they may provide limited insights into the evolutionary relationships between strains. It is convenient to section blots, according to the positioning of the λ standard bands, into 25 compartments. The presence (1) or absence (0) of a band(s) in each compartment is then recorded to produce a 25-character string that consists of the values 0 and 1. The data for test strains, together with a string for each of the designated patterns, are then subjected to analysis with either of the two programs. A consensus tree is next prepared with the CONSENSE program and drawn with DRAWTREE. The branching of the resulting tree indicates which of the designated patterns are most similar to that derived from the test strain. These relationships may then checked by eye. The assumptions of the Wagner (MIX) or Dollo (DOLLOP) parsimony methods are probably the most appropriate to assess the relationship between restriction fragment patterns.

In some instances a subtle difference between the patterns of two strains may not be apparent, when the strains are analysed on different gels. In such cases, the strains should be subjected to repeated restriction fragment analysis in adjacent tracks of the same gel or, alternatively, minor differences may be ignored and the strains classed as being indistinguishable.

The reproducibility of RFLP patterns obtained from a typing system should be checked to ensure that indistinguishable patterns are obtained (Saunders *et al.* 1990) for: (i) different DNA preparations from the same strain; (ii) different batches of probe DNA; and (iii) different batches of restriction endonuclease.

The stability of the patterns obtained from individual *Listeria* strains can be assessed by comparison of the restriction fragments obtained from: (i) *Listeria* strains subcultivated at intervals over an extended period of time; and (ii) epidemiologically associated strains – for example, isolates from outbreaks of listeriosis, sporadic incidents such as cases of mother-to-baby transmission and strains isolated from a single patient at different stages of illness.

Discriminatory ability of RFLP typing can be assessed by calculation of Simpson's index as described by Hunter & Gaston (1988). The probability (*D*) that two unrelated strains have indistinguishable restriction fragment patterns, is calculated from the formula:

$$D = 1 - \frac{1}{N(N-1)}\sum_{j=1}^{s} x_j(x_j - 1)$$

where N is the size of the population, s is the number of types and x_j is the number of population members falling into the jth type.

Examples of RFLP Typing

The restriction fragment patterns of 16 strains of *L. monocytogenes* are shown in Fig. 19.1. The DNA was digested with the restriction endonuclease *NciI* and the probe used was a mixture of DNA (λ32/λ35) isolated from two

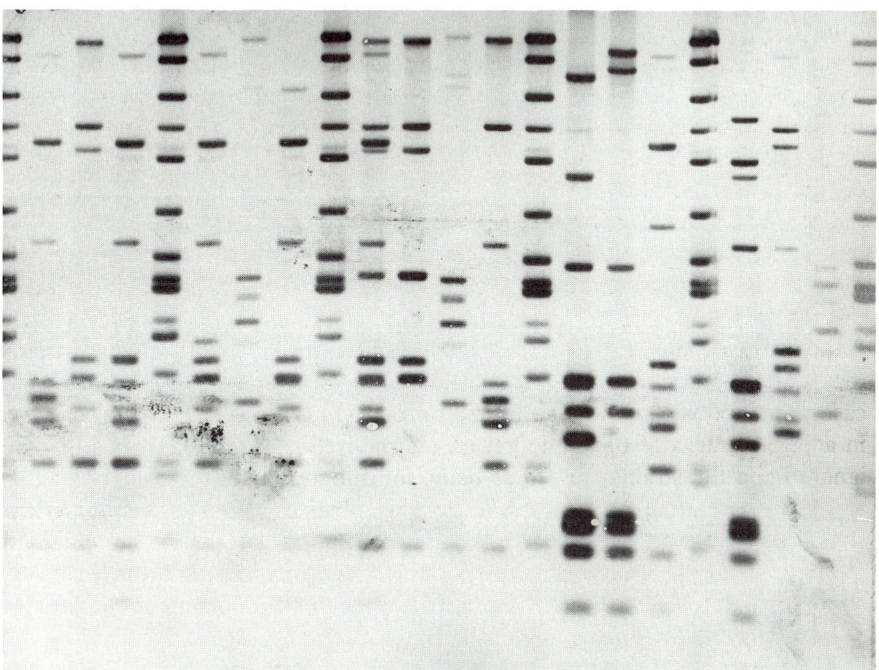

FIG. 19.1. A selection of RFLP patterns from *L. monocytogenes*. Analysis of 16 *L. monocytogenes* strains by RFLP typing with a digoxigenilated probe consisting of DNA from λNS32 and λNS35 (300 ng). The agarose gel (1%) was electrophoresed at 1.5 V/cm for 16 h. Tracks 1, 5, 9, 14, 18 and 22 are of a mixture of *EcoRI*- and *PstI*-digested λ phage DNA (1 ng of each digest). The remaining tracks show *NciI*-digested DNA (1 μg) from *L. monocytogenes* strains.

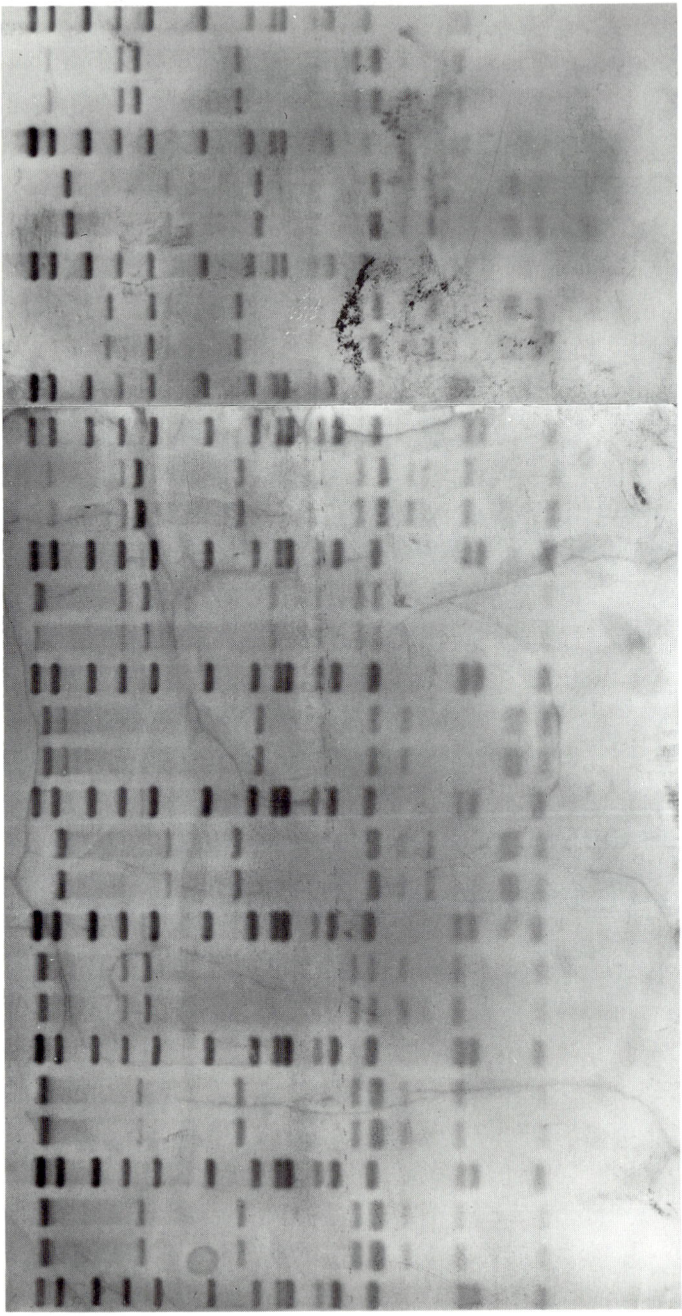

FIG. 19.2. Blot demonstrating reproducibility of the *L. monocytogenes* typing system. Analysis of 20 *L. monocytogenes* serogroup 1/2 strains by RFLP typing with a digoxigenin-labelled probe consisting of DNA from λNS32 and λNS35 (300 ng). The agarose gel (1%) was electrophoresed at 1.5 V/cm for 16 h. Tracks 1, 4, 7, 10, 13, 16, 19, 22, 23, 26, 29 and 32 show a mixture of *EcoRI*- and *PstI*-digested λ phage DNA (1 ng of each digest). The remaining tracks show *NciI*-digested DNA (1 μg) from *L. monocytogenes* serogroup 1/2 strains.

clones selected from a library of *L. monocytogenes* genomic fragments in the vector λgtWES.λB (Saunders *et al.* 1989) and labelled with digoxigenin. Fifteen different restriction fragment patterns can be seen (tracks 2, 3, 4, 6, 7, 8, 10, 11, 13, 15, 16, 17, 19, 20 and 21). The pattern obtained from the strain represented in track 7 is indistinguishable from that of the strain in track 12. Studies on this probe-based typing method (Saunders *et al.* 1989) have shown that its discrimination is serogroup-dependent, with strains of serovar 4b predominantly of a single RFLP type, while the serogroup 1/2 strains are distributed among many types.

The results of analysis of *L. monocytogenes* isolates used to establish the reproducibility of RFLP typing using this probe are demonstrated in Fig.

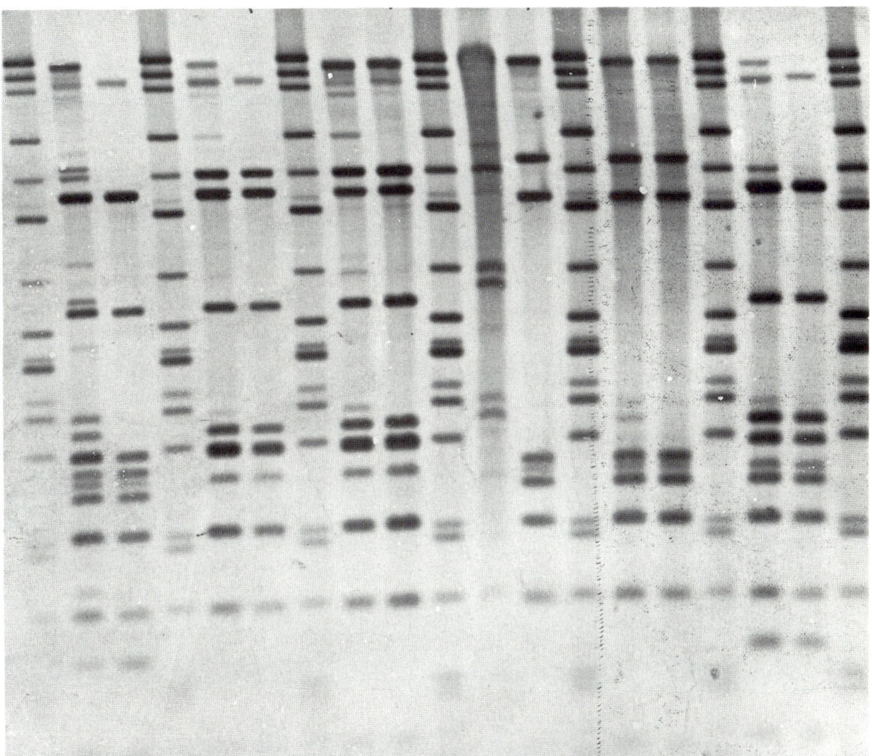

FIG. 19.3.Blot illustrating the problems encountered with typing of incompletely digested *L. monocytogenes* DNA. Analysis of seven strains of *L. monocytogenes* isolated from environmental sources in Belfast, digested with *NciI*, showing unusual RFLP patterns with the λ32/λ35 probes (tracks 2, 5, 8, 11, 14 and 17). When new DNA preparations of the same strains were prepared and digested with a fresh batch of *NciI* enzyme, the RFLP patterns shown in tracks 3, 6, 9, 12, 15 and 18 were obtained. Tracks 1, 4, 7, 10, 13, 16 and 19 show λ markers cut with *EcoRI/PstI*.

19.2. Tracks 20 and 21 illustrate the indistinguishable patterns of two isolates from a single patient and tracks 24, 25, 27 and 28 show pairs of strains from patients and associated foods. Analyses of DNAs obtained from cells before and after multiple passage are shown in tracks 30 and 31. The remaining tracks are the patterns obtained from strains with a suspected epidemiological association. Tracks 2, 3, 5 and 6 represent strains isolated from patients living in the Carlisle area in 1980. Tracks 8, 9, 11, 12, 14, 15, 17 and 18 represent cases of suspected cross-infection. In each case, related strains were indistinguishable by RFLP typing, thus confirming the linkage between them.

A potential problem of RFLP typing is illustrated in Fig. 19.3. The results of analysis of a set of *L. monocytogenes* strains from milk are demonstrated. Subsequent analysis has shown that some of these patterns (tracks 2, 5, 8, 11, 14 and 17) resulted from incomplete digestions of their DNA by *NciI*. The

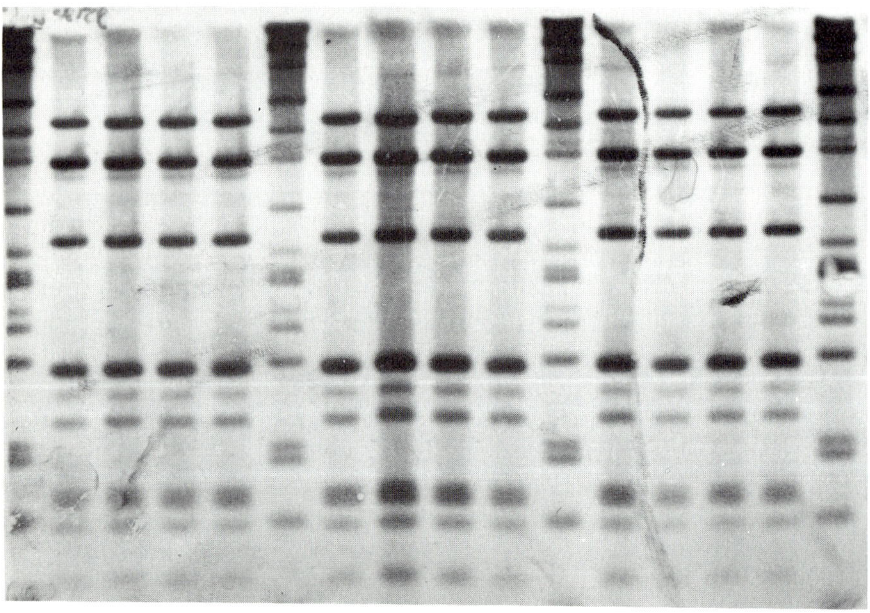

(a)

FIG. 19.4.Comparison of three different probes for RFLP typing of the same group serovar 4b *L. monocytogenes*. The extracted DNA was digested with *NciI* and the blots were probed with three different biotin-labelled probes. The digested DNA preparations were run in the equivalent tracks on each blot. Tracks 2, 3, 4, 5, 7, 8, 9, 10, 12, 13, 14 and 15 are of the serovar 4b strains; tracks 1, 6, 11 and 16 show λ markers cut with *EcoRI/PstI*. Blot (a) shows the RFLP patterns with the randomly screened probe λ32/λ35. Blot (b) shows the same digests probed with the repeated-sequence probe A 3-1. Blot (c) shows the same digests probed with a different repeated-sequence probe, A 5-2.

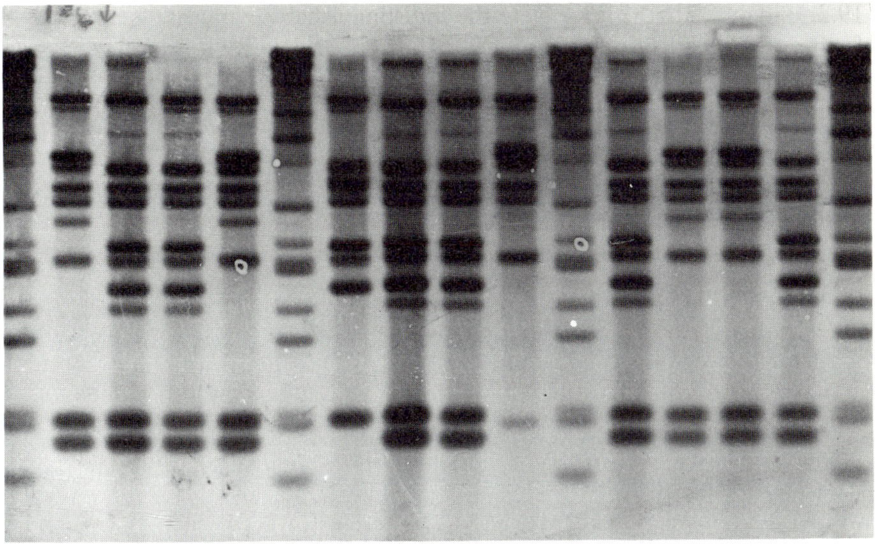

(b)

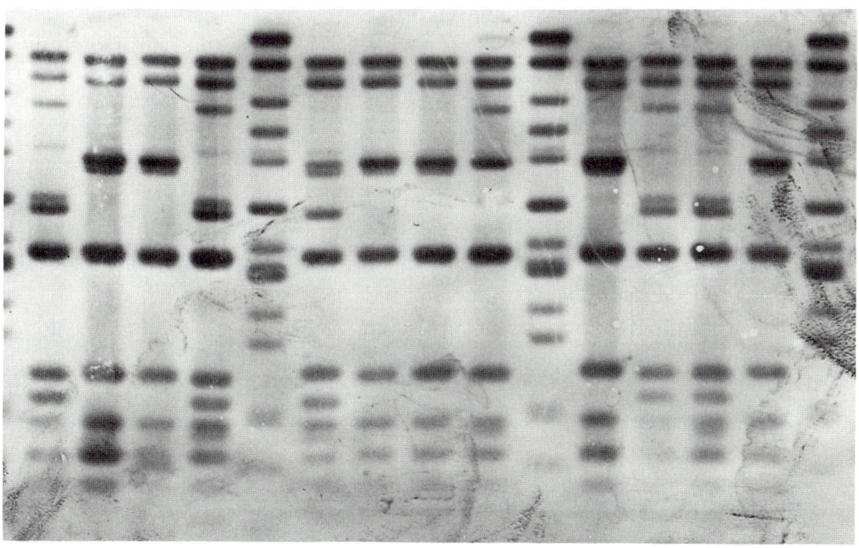

(c)

FIG. 19.4.*Continued*

incomplete nature of the digest was clearly evident in one instance (track 11) but difficult to identify in others (tracks 2 and 17). The first track of each of the seven pairs shows the incompletely digested DNA and the second shows the true RFLP pattern. Only in track 11 is the plethora of high-molecular-weight bands evident, which typically result from partially digested DNA. The problem of incomplete digestion can be minimized by overdigestion and ensuring that traces of contaminating solvents are removed from the DNA before digestion.

Analysis of *L. monocytogenes* strains with three different biotin-labelled probes is illustrated in Fig. 19.4. Blot (a) shows 12 *L. monocytogenes* serovar 4b strains examined with the randomly selected λ32/λ35 probe. Blots (b) and (c) represent the same 12 strains examined respectively with the repeated-sequence probes A 3-1 and A 5-2. The results demonstrate that the repeated-sequence probes provide better discrimination between 4b strains than the probe (λ32/λ35) used for routine typing of serogroup 1/2 isolates.

Conclusion

Comparison of *L. monocytogenes* isolates by the analysis of RFLP with probes, to detect restriction endonuclease fragments derived from a limited portion of the genome, is a powerful epidemiological typing method for closely related strains, which can also be used for species identification. The main disadvantages of RFLP analysis for typing *L. monocytogenes* are that it is a multistep procedure, which must be adhered to closely and is therefore relatively time-consuming, and that the discrimination between strains of some serovars is low. For these reasons it is not usually appropriate for large numbers of bacterial isolates, though it gives reliable and reproducible results. Despite these drawbacks, the major advantage of the technique is that the results are based on reliable genotypic characteristics rather than on the phenotype. This is sufficient to ensure that it will continue to be of value in various applications. Furthermore, the introduction of repeated-sequence probes should lead to improved discrimination between strains.

References

ANDERSEN, J.K. & SAUNDERS, N.A. 1990 Epidemiological typing of *Yersinia enterocolitica* by analysis of restriction fragment length polymorphisms with a cloned ribosomal RNA gene. *Journal of Medical Microbiology* **32**, 179–187.

BIBB, W.F., SCHWARTZ, B., GELLIN, B.G., PLIKAYTIS, B.D. & WEAVER, R.E. 1989 Analysis of *Listeria monocytogenes* by multilocus enzyme electrophoresis and application of the method to epidemiologic investigations. *International Journal of Food Microbiology* **8**, 233–239.

DESMARCHELIER, P.M., LEW, A.E., SAUNDERS, N.A. & SZABO, E.A. 1990 The use of a DNA probe for the typing of *Listeria monocytogenes*. *Australian Microbiologist* **11**, 249.

GRAVES, L.M., SWAMINATHAN, B., REEVES, M.W. & WENGER, J. 1991 Ribosomal DNA fingerprinting of *Listeria monocytogenes* using a digoxigenin-labelled DNA probe. *European Journal of Epidemiology* **7**, 77–82.

HUNTER, P.R. & GASTON, M.A. 1988 Numerical index of the discriminatory ability of typing systems: an application of Simpson's index of diversity. *Journal of Clinical Microbiology* **26**, 2465–2466.

KAISER, K. & MURRAY, N.E. 1985 The use of phage lambda vectors in the construction of representative genomic DNA libraries. *DNA Cloning*, ed. Glover, D.M., ch. 1. Oxford: IRL Press.

KARN, J., BRENNER, S. & BARNETT, L. 1983 New bacteriophage lambda vectors with positive selection for cloned inserts. *Methods in Enzymology*, eds Wu, R., Grossman, L. & Moldave, K., ch. 3. London: Academic Press.

LIN, R.J., CAPAGE, M. & HILL, C.W. 1984 A repetitive DNA sequence, *rhs*, responsible for duplications within the *Escherichia coli* K-12 chromosome. *Journal of Molecular Biology* **177**, 1–18.

MCLAUCHLIN, J., AUDURIER, A. & TAYLOR, A.G. 1986a Aspects of the epidemiology of human *Listeria monocytogenes* infection in Britain 1967–1984: the use of serotyping and phage-typing. *Journal of Medical Microbiology* **22**, 367–377.

MCLAUCHLIN, J., AUDURIER, A. & TAYLOR, A.G. 1986b The evaluation of a phage-typing system for *Listeria monocytogenes* for use in epidemiological studies. *Journal of Medical Microbiology* **22**, 357–365.

MCLAUCHLIN, J., CROFTS, N. & CAMPBELL, D.M. 1989 A possible outbreak of listeriosis caused by an unusual strain of *Listeria monocytogenes. Journal of Infection* **18**, 179–187.

PIFFARETTI, J.-C., KRESSBUCH, H., AESCHBACHER, M., BILLE, J., BANNERMAN, E., MUSSER, J.M., SELANDER, R.K. & ROCOURT, J. 1989 Genetic characterization of clones of the bacterium *Listeria monocytogenes* causing epidemic disease. *Proceedings of the National Academy of Sciences (USA)* **86**, 3818–3822.

PITCHER, D.G., SAUNDERS, N.A. & OWEN, R.J. 1989 Rapid extraction of bacterial genomic DNA with guanidium thiocyanate. *Letters in Applied Microbiology* **8**, 151–156.

RIDLEY, A.M., SAUNDERS, N.A. MCLAUCHLIN, J. & TAYLOR, A.G. 1992 Evaluation of an RFLP typing method for *Listeria monocytogenes. Listeria 1992. The Eleventh International Symposium on Problems of Listeriosis* 215–216. Copenhagen: Eigtved's Pakhus.

SAMBROOK, J., FRITSCH, E.F. & MANIATIS, T. 1989 *Molecular Cloning: a Laboratory Manual.* New York: Cold Spring Harbor Laboratory Press.

SAUNDERS, N.A., RIDLEY, A.M. & TAYLOR, A.G. 1989 Typing of *L. monocytogenes* for epidemiological studies using DNA probes *Acta Microbiologica Hungarica* **36**, 205–209.

SAUNDERS, N.A., HARRISON, T.G., HATHTHOTUWA, A., KACHWALLA, N. & TAYLOR, A.G. 1990 A method for typing strains of *Legionella pneumophila* serogroup 1 by analysis of restriction fragment length polymorphisms. *Journal of Medical Microbiology* **31**, 45–55.

SEELIGER, H.P.R. & HOHNE, K. 1979 Serotyping of *Listeria monocytogenes* and related species. *Methods in Microbiology*, eds Bergan, T. & Norris, J.R., ch. 31. New York: Academic Press

SOUTHERN, E.M. 1975 Detection of specific sequences among DNA fragments separated by gel electrophoresis. *Journal of Molecular Biology* **98**, 503–517.

TIEMEIER, D., ENQUIST, L. & LEDER, P. 1976 Improved derivative of a phage EK2 vector for cloning recombinant DNA. *Nature* **263**, 526–527.

VOGT, R.L., DONNELLY, C., GELLIN, B., BIBB, W. & SWAMINATHAN, B. 1990 Linking environmental and human strains of *Listeria monocytogenes* with isoenzyme and ribosomal RNA typing. *European Journal of Epidemiology* **6**, 229–230.

20: Pulsed-field Gel Electrophoresis for the Analysis of Bacterial Populations

FIONA M.THOMSON-CARTER, P.E.CARTER
AND T.H.PENNINGTON
Department of Medical Microbiology, University of Aberdeen Medical School, Foresterhill, Aberdeen AB9 2ZD, UK

The separation of DNA fragments by gel electrophoresis is an increasingly common requirement in modern microbiology. In conventional gel electrophoresis, DNA molecules are made to migrate through a solid matrix by application of a static electric field. This technique is, however, limited to resolution of fragments of less than 50 kilobase pairs. In efforts to separate larger DNA fragments, the application of two alternating (or pulsed) electric fields proved to be successful (Schwartz *et al.* 1983). The original concept of pulsed-field gel electrophoresis (PFGE) has undergone dramatic development and DNA fragments up to 10 megabase pairs can be separated. In microbiology, use of PFGE and restriction endonucleases, which cleave bacterial DNA infrequently, has made possible the generation of restriction patterns of large DNA fragments that represent the complete bacterial genome. Several methods have been devised based on the principle of separation of DNA fragments by application of alternating electric fields, including orthogonal-field-alternation gel electrophoresis (OFAGE), field-inversion gel electrophoresis (FIGE), transversely alternating-field electrophoresis (TAFE), and contour-clamped homogeneous electrical field (CHEF) (Cantor *et al.* 1988b). Further information on the theory of pulsed-field separation and the above techniques has been published (Cantor *et al.* 1988a, b; Mathew *et al.* 1988a, b, c). This chapter will consider only CHEF. This was first described by Chu *et al.* (1986) and represents a significant improvement in the resolution of DNA by a pulsed field. Because it uses a hexagonal array of fixed electrodes, a homogeneous electrical field is created, which enhances the resolution of a mixture of DNA fragments.

The methods described below have been used principally for the analysis

New Techniques in
Food and Beverage Microbiology

0-632-03755-5

of *Staphylococcus* spp. but, with minor modifications, they can be applied to a wider variety of bacterial genera.

Pulsed-field Gel Electrophoresis

Bacterial cultures

Single colonies picked from an overnight agar culture are inoculated into 10 ml nutrient broth. The culture is incubated with shaking overnight at 37 °C and 2 ml of an overnight culture are added to 8 ml fresh nutrient broth and incubated with shaking at 37 °C to an OD_{650} of 0.8–1.0. This takes 2–3 h and represents an approximate cell density of 1.5×10^8 cfu/ml. Penicillin G is added to the culture to a final concentration of 2 μg/ml. After incubation for a further 1 h at 37 °C, cultures are harvested by centrifugation at 10 000 *g* for 15 min, resuspended in 2 ml EET (0.1 mol/l EDTA, 0.01 mol/l EGTA, 0.01 mol/l Tris–HCl, pH 8.0) and washed twice at 10 000 *g* for 15 min. The cell pellets are finally resuspended in 0.8 ml EET.

It is essential that cultures are grown to late exponential phase to maximize their DNA content before harvesting. The above cell suspension (10 ml) at a density of 1.5×10^8 cfu/ml provides a maximum of 20 inserts (200 μl) with 1–2 μg DNA per insert. Ideally, insert portions containing 0.1–0.2 μg DNA should be loaded per gel track.

The growth procedure described above should be suitable for most bacteria but adjustments may be necessary for particular growth requirements. Bacteria can also be harvested directly with moistened swabs from agar plates and then resuspended and washed with an appropriate buffer. This procedure is especially useful for pathogenic bacteria, to avoid large volumes of liquid culture. However, DNA yields from cultures grown on solid media may be lower than those grown in liquid media, because of nutrient limitation. In order to obtain a sufficient amount of DNA for subsequent pulsed-field analysis, liquid cultures should be in mid to late exponential growth before antibiotic treatment of harvesting. Certain bacteria, especially Gram-positive organisms, may require preincubation in penicillin to aid subsequent lysis.

Insert preparation

One part cell suspension should be mixed with 4 parts of 1% (w/v) FMC InCert agarose solution (Flowgen, Sittingbourne, Kent, UK) prepared in sterile distilled water, melted at 65 °C and cooled to 37 °C. The mixtures are immediately dispensed into a mould (BioRad, Hemel Hempstead, Herts., UK) with 200-μl wells and allowed to set for 15 min at 4 °C.

Cells suspended in the agarose inserts are lysed by incubation at 37 °C for at least 16 h in 1 ml EET containing 1 mg lysostaphin (Sigma, Poole, Dorset, UK), 1 mg lysozyme and 2 μl RNase (final concentration 20 μg/ml). The agarose inserts are extensively washed with TE (0.001 mol/l EDTA, 0.01 mol/l Tris–HCl, pH 8.0) at 25 °C and then incubated at 50 °C for at least 16 h in 1 ml ESP (960 μl EET containing 40 μl 25% (w/v) sodium dodecyl sulphate and 1 mg proteinase K (Boehringer Mannheim, Lewes, East Sussex, UK)). Proteinase K is inactivated by washing the inserts twice for 1 h at 25 °C in TE containing 0.001 mol/l phenyl methyl sulphonyl fluoride (Sigma), followed by three washes of at least 1 h each at 25 °C in TE alone. Agarose inserts containing DNA are stored at 4 °C in TE. Immediately before restriction endonuclease digestion, the inserts are washed with fresh TE.

The agarose used to prepare inserts should be of sufficiently high quality to prevent inhibition of restriction endonuclease activity. Lysis of bacteria *in situ* can be achieved with a number of reagents or enzymes. Gram-negative bacteria can usually be lysed with lysozyme and detergents such as SDS, Sarkosyl and Brij. Gram-positive bacteria may, however, require pretreatment with enzymes, such as lysostaphin and lysozyme for staphylococci or mutanolysin for streptococci, before incubation with SDS. Incubation with lytic enzymes should be sufficiently long to ensure penetration of the enzymes into the inserts. The requisite times will vary according to the bacteria used, but 4 h should be regarded as a minimum. Inserts may be incubated in ESP solution for 16–48 h, to ensure cell lysis and digestion of cellular proteins. After phenyl methyl sulphonyl fluoride treatment and TE washes, agarose inserts may be stored in TE at 4 °C for up to 3 months without significant deterioration of DNA quality.

More rapid methods of bacterial DNA preparation for PFGE, which omit one or several of the steps described above, have been proposed. Though these methods may save time, the quality of the banding patterns obtained may be compromised.

Restriction endonuclease digestion

The choice of restriction endonuclease, for digestion of genomic DNA, depends primarily on the guanine and cystosine (G + C) content. When this is low, as in staphylococci (36%) or campylobacters (33%), endonuclease recognition sequences rich in G + C will be comparatively rare. Hence, the use of 'rare-cutting' endonucleases, which recognise base pair sequences of 6 or longer, such as *Sma* I (recognition sequence: GGGCCC) or *Nae* I (recognition sequence: GCCCCG), will result in the requisite small number of large fragments for separation by pulsed-field electrophoresis. Many highly purified restriction endonucleases suitable for genome analysis are now available. To assess the suitability of an endonuclease for use in pulsed-field analysis, several factors should be considered, including the recognition sequence of the endonuclease, the predicted size and number of fragments that will be generated and the size and G + C content of the genome. Partial digestion of genomic DNA may be prevented by preincubation of the insert with the reducing agent dithiothreitol (DTT) and bovine serum albumin, before addition of the endonuclease. Manufacturers usually provide specific instructions for each enzyme.

A portion of the insert (*c.* 1/4–1/3) is used for each digestion. The use of small portions of inserts minimizes the amount of enzyme which must be added but provides sufficient material to load three or four gel tracks. Preincubation of the insert portions and subsequent incubation with restriction endonucleases are most easily performed in sterile microfuge tubes. Where preincubations are recommended by the manufacturer, sufficient preincubation solution is added to immerse the insert portion completely. After 4 h incubation at 37 °C, the preincubation mix is pipetted off and the insert portion equilibrated in the appropriate restriction endonuclease buffer. Some manufacturers recommend preequilibration of inserts with enzyme buffer for 30–60 min on ice, before adding restriction endonuclease. Enzyme (20 units/μg) should be added to the DNA to be digested. The final reaction volume should be twice the volume of the insert portion. Digestion of DNA is achieved by incubation overnight at 37 °C or any other specified temperature. It is important that the length of incubation is adequate, to ensure complete enzyme penetration of the agarose insert. Shorter incubation times may lead to incomplete DNA digestion. The reaction can be terminated by addition of 25 μl 0.05 mol/l EDTA, pH 8.0. The insert portion containing digested genomic DNA may be loaded immediately on to a pulsed-field gel or stored in TE at 4 °C for future analysis. We have found that *Sma* I is most suitable for cleavage of staphylococcal and campylobacter genomic DNA.

Gel electrophoresis

Pouring a pulsed-field gel

Agarose gels for PFGE are made up in the same way as gels for conventional DNA electrophoresis. Good-quality agarose (e.g. Sigma Type II Medium EEO) should be used. The agarose is melted in half-strength TBE (0.089 mol/l borate, 0.089 mol/l Tris, 0.002 mol/l EDTA, pH 8.3), the buffer recommended for PFGE, and poured into the casting stand provided by the manufacturer. The agarose concentration may be varied according to the size of DNA fragments to be separated (see Table 20.1 for parameters). An agarose concentration of 1% (w/v) is generally suitable for separation of bacterial DNA fragments.

TABLE 20.1. *Pulsed-field gel electrophoresis run parameters*

	DNA (Mb)			
	<0.1	0.1–2	2–4	>4
% Agarose	1.5	1–1.5	0.8–1	0.5–0.8
TBE (Buffer)	0.5×	0.5×	0.5×	0.5×
Temperature	14 °C	14 °C	14 °C	14 °C
Voltage (V)	200	150–200	125–175	50–75
Pulse parameters	1–10 s	50–100 s	120–180 s	10–60 min
Run times	3–12 h	18–24 h	24–36 h	3–6 days

Loading a pulsed-field gel

In order to obtain good banding patterns, great care is necessary while the gel is being loaded. Inserts should be placed on a clean surface, such as Parafilm, and cut to the appropriate size with a glass coverslip that has previously been dipped in alcohol and flamed. Metal implements, such as razor blades and scalpels, should not be used to manipulate inserts, because their constituent

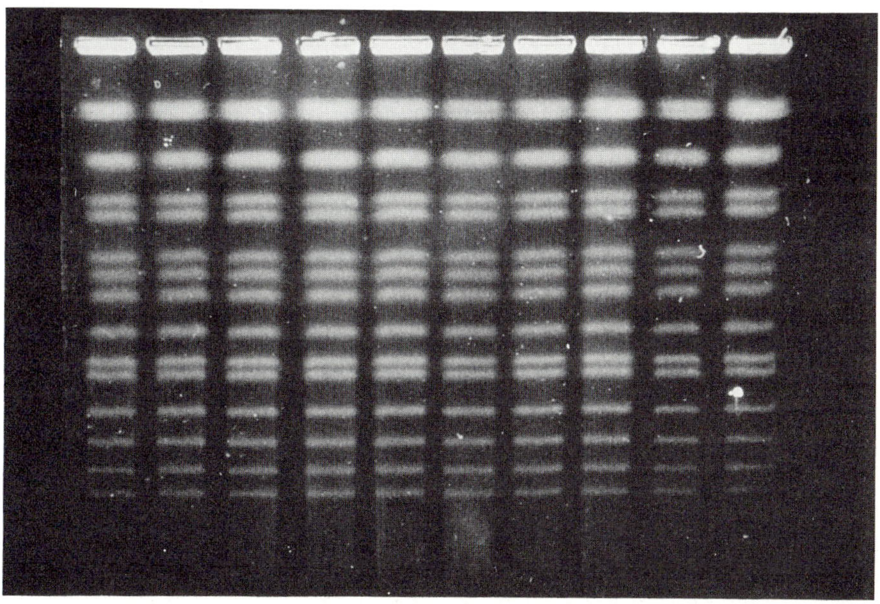

FIG. 20.1. PFGE of yeast chromosomes. Running conditions as described in Table 20.2. Band sizes range from 200 kb to 2.2 Mb.

divalent ions may damage the native DNA structure. An insert slice of approximately 20 μl (or 1/10 of the original 200 μl insert) is usually loaded per well. Insert slices can be loaded by gently easing them from their cover-slips into the wells with an alcohol-dipped, flame-sterilized bent-glass Pasteur pipette or similar implement. It is most important that insert slices should be positioned at the front and the bottom of the well; otherwise the samples will not electrophorese satisfactorily. When all samples have been loaded in a similar manner, the wells are filled with agarose solution to fix the insert slices in place. A 0.8–1% (w/v) agarose solution of low gelling temperature in half-strength TBE is used. The agarose used to fill the wells should be cooled to approximately 40 °C before use, to prevent well distortion. After 15 min at room temperature, the agarose will have hardened and the gel is ready for electrophoresis. Depending on the electrophoresis system used, up to 30 samples may be run on each gel.

DNA standards, such as λ concatemers or yeast chromosomes (Fig. 20.1), are available commercially as preformed inserts and should be handled and loaded as described for sample inserts.

Running a pulsed-field gel

The selection of the conditions, including temperature, agarose concentration, voltage, etc., under which any pulsed-field gel is run will vary according to the

TABLE 20.2. *Examples for selection of pulsed-field gel electrophoresis running conditions*

DNA	DNA size range (kb)	Agarose concentration (%)	Pulse time (s)	Run time (h)	Voltage (V)
High + low DNA standards	0.2–23	1.5	1–4[a]	5	200
λ ladder	50–1000	1.0	50–90[b]	24	200
Saccharomyces cerevisiae	200–2000	1.0	60[c] 90	15 9	200
Candida albicans	2000–5000	0.8	120[d] 180	24 12	150
Schizosaccharomyces pombe	3000–9000	0.6	3600[e]	160	50

[a] Ramped pulse from 1 to 4 s over 5 h.
[b] Ramped pulse from 50 to 90 s over 24 h.
[c] 60-s pulse time for 15 h followed by a 90-s pulse time for 9 h.
[d] 120-s pulse time for 24 h followed by a 180-s pulse time for 12 h.
[e] 60-min single pulse time for 6.6 days.

needs of the individual experiment. Guidelines for the selection of conditions for a range of applications are given in Tables 20.1 and 20.2. Variation of any of these factors will affect the electrophoresis run of a gel. Optimal running conditions for individual needs must be determined by trial and error within the described guidelines.

Satisfactory separation of bacterial genomic DNA fragments can usually be achieved under the following conditions. The buffer of choice for PFGE is half-strength TBE, precooled to 10–14 °C before the gel is placed in the apparatus (Fig. 20.2). It is important that the temperature is maintained within this temperature range throughout the electrophoresis run, by circulating the buffer through a cooling system. The buffer type and run temperature are the only factors that do not vary across a wide range of pulsed-field applications (Tables 20.1 and 20.2). Broadly speaking, as the sizes of the fragments to be separated increase, the run parameters vary as follows: (i) the agarose concentration decreases; (ii) the voltage applied decreases; (iii) the pulse time increases; and (iv) the total run time increases. The most critical factor for optimal resolution is the pulse time, that is, the time for which the alternating electrical fields A and B (Fig. 20.3) are applied. There are three options for pulse time selection: (i) the pulse time can remain fixed for the duration of the electrophoresis run; (ii) the run can be divided into a number of segments and a different pulse time chosen for each segment; or (iii) an initial and a final pulse time can be set before a run commences; during the run, the pulse time increases linearly from the initial preset value to the final preset value, a process termed 'ramping'. The electrophoresis equipment can be programmed for one of the options described, and the run parameters chosen are auto-

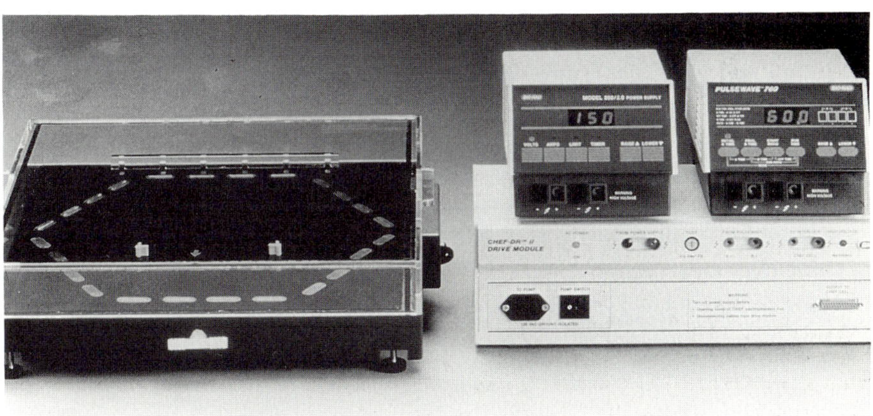

FIG. 20.2. The CHEF-DR II electrophoresis system. The apparatus consists of a gel tank with electrodes, drive module, Pulsewave 760 switcher and power pack.

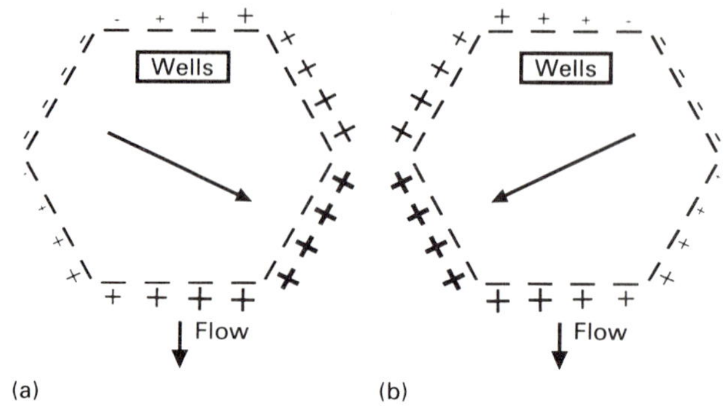

FIG. 20.3.Voltage clamping of the CHEF-DR II system. (a) on – relative electrode potentials and field vector when A channel of the Pulsewave 760 switcher is on. (b) on – relative electrode potentials and field vector when B channel is on. The average vector during the run is in the direction of the arrow marked 'Flow'.

matically applied to the gel. Specific examples of these conditions are shown in Tables 20.1 and 20.2.

We have been particularly interested in pulsed-field analysis of staphylococci and campylobacters and routinely use pulse times ramped linearly to increase from 12 to 17 s for staphylococci and, similarly, 20 to 25 s for campylobacter DNA fragments, over a run time of 22–24 h. These conditions are sufficient to resolve DNA fragments between 700 and 20 kb in size. Ramping of pulse times results in improved band definition, compared with that observed with a single preset pulse time. The effects of variation of run parameters on DNA migration in PFGE are discussed elsewhere (Cantor *et al.* 1988a; Mathew *et al.* 1988a, b, c).

A problem we have encountered after repeated PFGE runs with the CHEF-DR II apparatus (BioRad) is the thinning and eventual breakage of the platinum electrodes. Although damage to one or two electrodes may not adversely affect an electrophoresis run, they should be replaced as soon as possible. New electrodes are available from the manufacturers, but a length of 0.3 mm diameter platinum wire held in place using silicone cement is an effective substitute.

Staining and photography of gels

After electrophoresis, the gel can be removed from the apparatus and stained, in a manner similar to conventional agarose gels, with ethidium bromide solution at a final concentration of 20 µg/ml in half-strength TBE. Stained

DNA fragments can be visualized with UV illumination and photographed with a Polaroid camera or video camera system.

Analysis of PFGE restriction profiles

The banding patterns of restriction fragments obtained by PFGE are suitable for analysis with the computer programs available for smaller restriction fragments.

Blotting of pulsed-field gels

Pulsed-field gels may be blotted according to protocols similar to those for conventional agarose gels. The critical step is the pretreatment required to 'nick' and denature the large DNA fragments, to enable them to move out of the gel matrix during the blotting procedure. This may be accomplished by acid (HCl) treatment or UV irradiation. We have found that the UV method gives more consistent results. After normal staining with ethidium bromide, the gel is irradiated with a UV dose equivalent to $60\,mJ/cm^2$ at 300 nm in a UV cross-linker or similar UV box. Individual UV boxes would have to be calibrated to prevent overexposure (Smith *et al.* 1988). The gel is then soaked in 0.4 mol/l NaOH, 1.5 mol/l NaCl, for 15–30 min. A conventional Southern blotting arrangement of paper towels, filter paper, membrane, gel and wick is assembled. The gel is inverted so that the membrane is placed on the lower side of the gel. We usually use Hybond-N (Amersham Labs, Amersham, Bucks., UK), but nitrocellulose may also be used. Blotting proceeds overnight (approximately 16 h) by capillarity, after addition of 0.4 mol/l NaOH, 1.5 mol/l NaCl, as the transfer buffer. When transfer is complete, DNA is fixed on to the membrane by conventional UV cross-linking procedures. These filters may then be probed according to any of the numerous protocols available.

The Uses of PFGE

The ability of pulsed-field gel electrophoresis to separate large DNA fragments provides the opportunity to study complete microbial genomes. As a consequence, this powerful technique has many applications in modern microbiology. Individual researchers can adapt the methods described according to individual requirements. Below we describe several situations in which we have successfully applied PFGE, but this list is by no means comprehensive.

PFGE as a typing method

PFGE can be used as a typing method, either alone or in combination with conventional techniques. In the pulsed-field analysis of staphylococci, we have established that different staphylococcal species have characteristic restriction profiles by which isolates may be identified, although intraspecies variation in banding patterns occurs (Fig. 20.4) (Thomson-Carter & Pennington 1991). PFGE may have limited use as a typing method for genera with less well-defined species, in that sufficient differences may not exist to form the basis of a typing scheme or, conversely, intraspecies profile variation is so great that species-specific banding patterns cannot be discerned. In these circumstances PFGE may only be useful as a supplementary typing method. In our studies

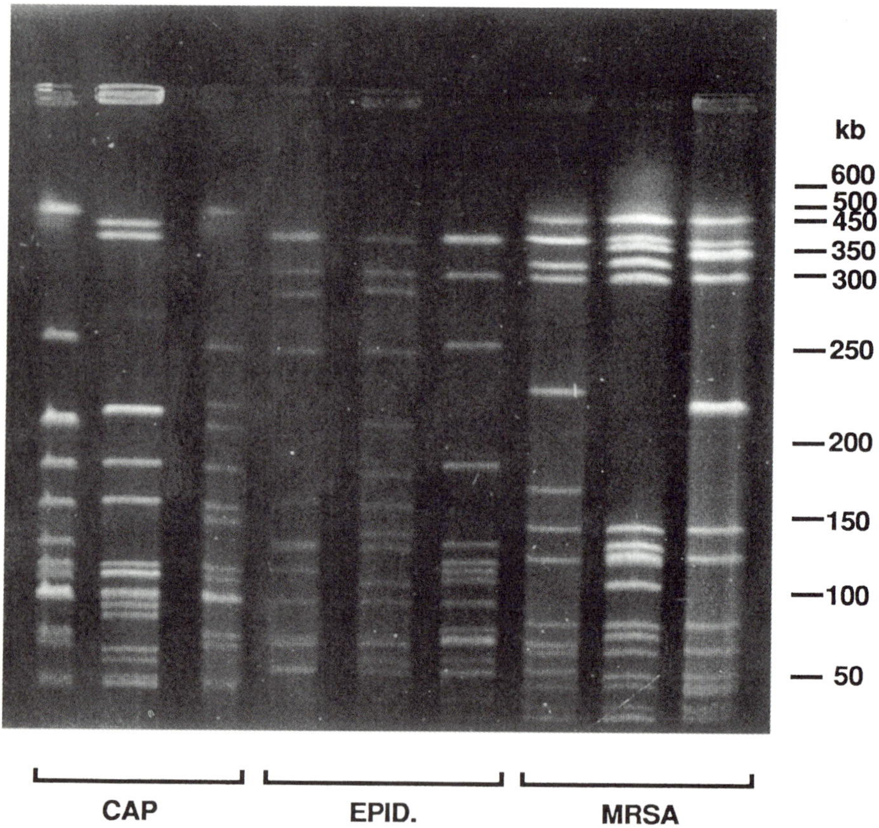

FIG. 20.4.PFGE of staphylococci. Restriction profiles obtained after *Sma* I digestion of staphylococcal genomic DNA of three different species: CAP, *Staph. capitis*; EPID, *Staph. epidermidis*; MRSA, *Staph. aureus* (methicillin-resistant isolates).

with staphylococci isolated from normal human microflora, diagnostic clinical material and members of enclosed communities, PFGE has proved to be capable of greater discrimination and characterization than conventional diagnostic methods.

Staphylococcus hominis *endocarditis*

Staphylococci from six successive blood cultures from a patient with native valve endocarditis were initially characterized as *Staph. intermedius* by API-Staph strips (Thomson-Carter *et al.* 1993a). Genomic DNA analysis by PFGE demonstrated that these isolates did not share a restriction profile with the *Staph. intermedius* type strain. Apparently only one infective clone was present, because all the PFGE profiles of the isolates were identical. Tube coagulase tests confirmed that the type strain was coagulase-positive but that the blood culture isolates were coagulase-negative. Subsequent API tests did not satisfactorily identify the isolates, because discrimination between *Staph. epidermidis*, *Staph. lugdunensis* and *Staph. hominis* was low. Further PFGE analysis, including the type strains of these species, showed that the PFGE profile of the blood culture isolates shared significant similarities with that obtained for the *Staph. hominis* type strain. PFGE, therefore, speciated the clinical isolates unequivocally, in contrast to conventional biochemical methods.

Atypical staphylococci from an enclosed community

A group of staphylococcal isolates collected from members of an enclosed community were selected for PFGE studies, because of atypical results in biochemical identification tests (Thomson-Carter *et al.* 1993a). Initially speciated as *Staph. capitis*, these isolates were correctly identified as *Staph. capitis* ssp. *ureolyticus* by PFGE. Comparison of the restriction profiles of the atypical isolates with that obtained for the *Staph. capitis* ssp. *ureolyticus* type strain revealed overall similarity of banding pattern, thus confirming the superior discrimination of PFGE.

PFGE as an epidemiological tool

The ability of PFGE to provide species-specific profiles for staphylococci has been invaluable in studies of their epidemiology. If it is assumed that isolates with identical profiles are either clonal or closely related, the occurrence and movement of clones within bacterial populations can be monitored. In hospitals, staphylococci resistant to one or more antibiotics pose particular problems in infection control. One of our recent studies demonstrated that an outbreak of methicillin-resistant *Staph. aureus* infection in an orthopaedic ward was caused by one infective clone, characterized by identical PFGE profiles obtained for seven different isolates (Thomson-Carter *et al.* 1993b). Similarly, staphylococcal clones in normal human microflora can be identified and their transfer among members of an enclosed community monitored with PFGE (Bright *et al.* 1993). The application of PFGE in epidemiological studies is preferable to the use of conventional techniques, because it provides genetic data on the isolates studied rather than being only descriptive.

PFGE analyses of other bacterial genera

Methods similar to those described above have been used successfully in the PFGE analysis of many other bacteria, including *Pseudomonas* spp. (Grothues & Tummler 1991), *Escherichia coli* (Arbeit *et al.* 1990) and *Shigella* (Soldati & Piffaretti 1991). Other microorganisms, such as fungi and protozoa, have also been studied by PFGE (Smith *et al.* 1988).

PFGE as a preparative technique

In addition to its typing and epidemiological applications, PFGE can also be used for more specific studies, such as localization of particular genes and the generation of physical maps of bacterial genomes. Pulsed-field gels may be blotted, as described above, and the location of a gene on a particular genomic fragment identified. Similar studies have allowed mapping of individual genes on bacterial chromosomes, to create complete genome maps (Chang & Taylor 1990; Smith & Condemine 1990).

Conclusion

The scope of PFGE and its advantages over more conventional techniques make it an attractive option for the microbiologist. Its advantages are that all isolates are typable, since, under the correct conditions, all isolates provide a PFGE restriction profile. The method offers superior discrimination when compared with phenotypic methods, such that identification of subspecific groups is possible. Using PFGE, the epidemiology of infection can be accurately defined. The results obtained are susceptible to population genetic analysis, and allow quantification of data. The application of PFGE facilitates better characterization of the individual bacterium and the population in which it exists.

Acknowledgements

We would like to thank Tracey Richmond of BioRad UK for her advice in the preparation of this manuscript. Part of this work was funded by a grant

(K/MRS/50/C1263) awarded by the Biomedical Research Committee of the Scottish Home and Health Department.

References

ARBEIT, R.D., ARTHUR, M., DUNN, R., KIM, C., SELANDER, R.K. & GOLDSTEIN, R. 1990 Resolution of recent evolutionary divergence among *Escherichia coli* from related lineages: the application of pulsed-field electrophoresis to molecular epidemiology. *Journal of Infectious Diseases* **161**, 230–235.

BRIGHT, S., THOMSON-CARTER, F.M. & PENNINGTON, T.H. 1992 Geographical distribution of *S. epidermidis* clones identified by pulsed-field gel electrophoresis. *The Staphylococci: Proceedings of the VIIth International Symposium on Staphylococci and Staphylococcal Infections*, ed. Molby, R., In press.

CANTOR, C.R., GAAL, A. & SMITH, C.L. 1988a High resolution separation and accurate size determination in pulsed-field gel electrophoresis of DNA. 3. Effect of electrical field shape. *Biochemistry* **27**, 9216–9221.

CANTOR, C.R., SMITH, C.L. & MATHEW, M.K. 1988b Pulsed-field gel electrophoresis of very large DNA molecules. *Annual Review of Biophysical Chemistry* **17**, 287–304.

CHANG, N. & TAYLOR, D.E. 1990 Use of pulsed-field gel electrophoresis to size genomes of *Campylobacter* species and to construct a *SalI* map of *C. jejuni* UA580. *Journal of Bacteriology* **172**, 5211–5217.

CHU, G., VOLLRATH, D. & DAVIS, R. 1986 Separation of large DNA molecules by contour clamped homogeneous electric fields. *Science* **234**, 1582–1585.

GROTHUES, D. & TUMMLER, B. 1991 New approaches in genome analysis by pulsed-field gel electrophoresis: application to the analysis of *Pseudomonas* species. *Molecular Microbiology* **5**, 2763–2776.

MATHEW, M.K., SMITH, C.L. & CANTOR, C.R. 1988a High resolution separation and accurate size determination in pulsed-field gel electrophoresis of DNA. 1. Size standards and the effect of agarose and temperature. *Biochemistry* **27**, 9204–9210.

MATHEW, M.K., SMITH, C.L. & CANTOR, C.R. 1988b High resolution separation and accurate size determination in pulsed-field gel electrophoresis of DNA. 2. Effect of pulse time and electric field strength and implications for models of the separation process. *Biochemistry* **27**, 9210–9216.

MATHEW, M.K., HUI, C.-F., SMITH, C.L. & CANTOR, C.R. 1988c High resolution separation and accurate size determination in pulsed-field gel electrophoresis of DNA. 4. Influence of DNA topology. *Biochemistry* **27**, 9222–9226.

SCHWARTZ, D.C., SAFFRAN, W., WELSH, J., HAAS, R., GOLDENBERG, M. & CANTOR, C.R. 1983 New techniques for purifying large DNAs and studying their properties and packaging. *Cold Spring Harbor Symposium on Quantitative Biology* **47**, 189–195.

SMITH, C.L. & CONDEMINE, G. 1990 New approaches for physical mapping of small genomes. *Journal of Bacteriology* **172**, 1167–1172.

SMITH, C.L., KLCO, S.R. & CANTOR, C.R. 1988 Pulsed-field gel electrophoresis and the technology of large DNA molecules. *Genome Analysis: a Practical Approach*, ed. Davies, K.E., pp. 41–72. Oxford: IRL Press.

SOLDATI, L. & PIFFERETTI, J.C. 1991 Molecular typing of *Shigella* strains using pulsed-field gel electrophoresis and genome hybridisation with insertion sequences. *Research in Microbiology* **142**, 489–498.

THOMSON-CARTER, F.M. & PENNINGTON, T.H. 1991 Chromosomal DNA analyses of staphylococci. *The Staphylococci: Proceedings of the VIth International Symposium on Staphylococci and Staphylococcal Infections*, *Zentralblatt fur Bakteriologie*, *Supplement 21*, eds Jeljaszewicz, J. & Ciborowski, P., pp. 81–82. Stuttgart: Gustav Fischer Verlag.

THOMSON-CARTER, F.M., PURCELL, T.M., CARTER, P.E. & PENNINGTON, T.H. 1993a Genomic DNA analyses of staphylococcal populations. *The Staphylococci: Proceedings of the VIIth International Symposium on Staphylococci and Staphylococcal Infections*, ed. Molby, R., In press.

THOMSON-CARTER, F.M., CARTER, P.E., PURCELL, T.M. & PENNINGTON, T.H. 1993b Application of genotypic techniques in analyses of staphylococcal infections. *Molecular Pathogenesis of Surgical Infections*, eds Wadstrom, T., Ljungh, A., Holder, I. & Eliasson, I., In press.

21: Ribotyping and Arbitrary-primer PCR Fingerprinting of Campylobacters

R.J.OWEN[1] AND J.HERNANDEZ[2]
[1]*National Collection of Type Cultures, Central Public Health Laboratory, London NW9 5HT, UK;*
and [2]*Departamento de Biotecnologia, Laboratorio de Microbiologia, Universidad Politecnica de Valencia, Camino de Vera 14, 46022 Valencia, Spain*

Bacteria of the genus *Campylobacter* are of worldwide significance in human and animal disease. In particular, *C. jejuni* is recognized as one of the commonest causes of acute bacterial diarrhoea in man, with about 3000 cases per month reported in the UK (Skirrow 1987; Healing *et al.* 1992). While milk and environmental sources, such as water, have been implicated in some large outbreaks, the majority of *Campylobacter* infections in developed countries are apparently sporadic, but they occur constantly (Cover & Blaser 1989; Griffiths & Park 1990). Campylobacters are widespread in nature, particularly in birds. Sources of infection remain largely undetermined, but contaminated foods, notably poultry and raw milk, are widely regarded as important vehicles of infection (Franco 1988; Butzler & Oosterom 1991; Park *et al.* 1991).

Accurate methods of strain identification are essential to determine the routes and vehicles of transmission of campylobacters, and to assess and monitor the effectiveness of intervention procedures. Conventional typing methods, such as biotyping and serotyping, have been applied to campylobacters for over a decade, but are now being challenged by new molecular methods that avoid dependence on expressed and possibly variable phenotypic traits, such as somatic and flagella antigens. This chapter will describe two molecular methods that provide novel but precise and stable strain markers applicable to pure cultures of campylobacters, namely ribosomal RNA (rRNA) gene restriction patterns (ribopatterns), and a method that depends on polymerase chain reaction (PCR)-generated fingerprints, derived from random primer sequences.

New Techniques in
Food and Beverage Microbiology

0-632-03755-5

TABLE 21.1. *Campylobacters and their clinical importance*

Established pathogens	
C. jejuni subsp. *jejuni*	Acute enteritis
C. coli	Acute enteritis
C. fetus subsp. *fetus*	Systemic diseases in compromised hosts
Potential pathogens	
C. upsaliensis	Enteritis and bacteraemia
C. hyointestinalis	Proctitis and diarrhoea
Pathogens of uncertain significance	
C. jejuni subsp. *doylei*	Enteritis
C. lari	Enteritis
C. concisus	Periodontal disease
C. curvus	Blood and oral lesions
C. rectus	Periodontal disease
Commensal	
C. sputorum biovar *sputorum*	Gingival crevices and faeces
Not found in man	
C. helveticus	
C. mucosalis	
C. fetus subsp. *venerealis*	
C. sputorum biovar *bubulus*	
C. sputorum biovar *fecalis*	

Campylobacter Identification and Typing

The genus *Campylobacter* has undergone considerable revision over the past five years, mainly as a result of phylogenetic analysis of 16S rRNA sequences (Thompson *et al.* 1988; Vandamme *et al.* 1991). Twelve species are currently recognized (Table 21.1). *C. jejuni* accounts for about 90% of all human infections, though several other species are to a lesser extent associated with disease in man. Conventional physiological, biochemical and morphological characteristics for identification of these species of *Campylobacter* are described elsewhere (Penner 1988; Skirrow 1990).

Over the past 20 years, a range of methods have been applied to the epidemiological typing of strains of *Campylobacter*, in particular those of *C. jejuni* (Table 21.2). Most of these methods have proved to be of limited utility, but biotyping, serotyping and, more recently, phage typing have been used, often in combination, in specialized reference centres. Recently,

TABLE 21.2. *Main phenotyping methods for the subspecific identification and typing of campylobacters of pathogenic or potentially pathogenic importance*

C. jejuni and *C. coli*
Serotyping
  Heat-stable LPS antigens (Penner & Hennessy 1980)
  Thermostable flagella antigens (Lior *et al.* 1982)
Bacteriophage typing (Grajewski *et al.* 1985; Salama *et al.* 1990)
Biotyping (Skirrow & Benjamin 1981; Lior 1984; Hebert *et al.* 1985)
Resistotyping (Bolton *et al.* 1984)
Auxotyping (carbon-substrate assimilation) (Tenover *et al.* 1985)
Electrophoresis of total proteins (Owen *et al.* 1988a)
Multilocus enzyme electrophoresis (Patton *et al.* 1991)

C. fetus
Biotyping and resistotyping (Bolton *et al.* 1984; Skirrow 1990)

C. lari
Biotyping (Lior 1984)
Electrophoresis of total proteins (Owen *et al.* 1988b)

C. upsaliensis
Serotyping – thermostable antigens (Lior & Woodward 1991)
Electrophoresis of total proteins (Owen *et al.* 1984)

molecular genotyping methods have gained acceptance as a means of discriminating isolates from food, clinical and environmental sources, because they avoid dependence on phenotypic schemes. As fewer than 50% of campylobacters carry plasmids (Tenover *et al.* 1984), plasmid profiling has not been developed to any extent as a typing method, although its potential in *C. upsaliensis* has been demonstrated (Owen & Hernandez 1990).

Ribotyping of *Campylobacter*

Culture of bacteria

Pure cultures of pathogenic or potentially pathogenic *Campylobacter* from food, clinical and environmental sources are cultivated at 37 °C for 48 h on 5% (v/v) defibrinated horse or sheep blood agar under microaerophilic conditions in an anaerobic jar (catalyst removed), evacuated to a pressure of 560 mmHg (*c.* 74.7 kPa) and filled with 10% CO_2, 5% O_2 and 85% N_2. Alternatively, cultures are incubated at 37 °C under microaerobic conditions (5% O_2, 5% CO_2, 2% H_2, 88% N_2) in a variable atmosphere incubator (Don Whitley Scientific Ltd, Shipley, Yorks.). Strains can be preserved at −70 °C on glass beads on nutrient broth no. 2 (Oxoid CM67), containing 10% (v/v) glycerol, and reference strains lyophilized in 5% (w/v) inositol serum.

Chromosomal DNA extraction

Relatively pure chromosomal DNA is necessary for digest analysis and blot hybridization. The following two simple and rapid methods are suitable for extracting and purifying DNA from cells of *Campylobacter.*

Guanidium isothiocyanate (GES) method

This method, originally described by Pitcher *et al.* (1988), depends on the proteolytic and denaturing properties of guanidium isothiocyanate (GCN).

The reagents are as follows:

1 TE buffer (10 mmol/l Tris, 1 mmol/l ethylenediaminetetra-acetic acid (EDTA)), adjusted to pH 8.0 with HCl. Prepare 1 mmol/l Tris and 0.5 mmol/l EDTA as stock solutions. Adjust the pH of the Tris to 8.0, dilute and add the EDTA.

2 GES reagent (prepare in fume hood). Add 60 g (5 mol/l) guanidium thiocyanate, 3.7 g (0.1 mol/l) EDTA to ~20 ml water. Heat in 65 °C water bath with mixing until dissolved (use fume hood and mix on heated stir plate). Add 1.7 ml of 30% sarcosyl (cool), make up to 100 ml and mix well. Filter through a 0.45-μm filter.

3 Ammonium acetate (7.5 mol/l).

4 Isopropanol.

5 Chloroform:2-pentanol (24:1).

6 70% ethanol.

Store all reagents at 4 °C, except the GES reagent, which should be stored at room temperature.

The following method is used:

1 A pellet the size of a rice grain is required (heavy growth is needed, since DNA yield is growth-dependent). Resuspend the pellet in 100 μl 1× TE buffer in a McCartney bottle. It is essential to resuspend thoroughly.

2 Washing is unnecessary; add 0.5 ml GES reagent (use gloves when handling GES) to the cells and resuspend thoroughly (on vortex).

3 Shake gently until lysed.

4 Add 0.25 ml cold 7.5 mmol/l ammonium acetate and 0.5 ml chloroform reagent, vortex until phases mix and leave on ice for 10 min. Transfer to 1.5-ml Eppendorf tube.

5 Spin at fast speed for 10 min (microfuge, 13 000 *g*).

6 Transfer supernatant carefully to a fresh tube, add 0.54 vols cold isopropanol, mix well (usually 0.8 ml supernatant plus 0.43 ml isopropanol). Place on multimix rollers to mix until DNA has precipitated.

7 Remove fluid with a pipette; avoiding sucking up the DNA pellet. DNA sticks to side of tube, so pull to the side while pipetting.
8 Wash pellet three times in 70% ethanol.
9 Dry under vacuum 5–10 min.
10 Redissolve in 100 μl of sterile 1× TE buffer overnight at 4 °C to give, ideally, 0.5–1.0 μg/μl (A_{260} = 0.5 for a 1/100 dilution of sample). Next day, centrifuge briefly and read absorbance at 260, 280 and 230 nm.
11 Calculate the absorbance ratios 230/260 and 280/260. Both should be approximately 0.5. If the ratios are not satisfactory (i.e. >0.6), add one-third volume of 7.5 mmol/l ammonium acetate to sample and 1 volume chloroform reagent, vortex and leave on ice for 10 min. Reprecipitate with 1 vol 2-propanol and wash twice in 70% ethanol. Redissolve in 1× TE buffer. Repeat absorbancy check. Store at 4 °C if using immediately or −20 °C if longer storage is necessary. Sample will keep for 2–3 weeks at 4 °C. Freeze in appropriate aliquots to be used without having to remeasure when testing.

Comments. The GES method has been successfully used to isolate DNA from campylobacters in several studies (Owen *et al.* 1990; Hernandez *et al.* 1991a, b; Fayos *et al.* 1992). A problem encountered with the method when applied to some strains is that the DNA becomes completely degraded. Such strains produce DNase, which apparently is not effectively eliminated by guanidium isothiocyanate (GCN). The alternative DNA purification method outlined below is recommended for such problematic strains. Also, DNA prepared with GCN appears to be more susceptible to partial cutting by *Hin*dIII, in which case the alternative cetyltrimethylammonium bromide (CTAB) method must be used. CTAB complexes with both polysaccharides and residual proteins, so facilitating their subsequent removal (Wilson 1987). It should be noted that neither of these methods gives sufficient DNA for conventional DNA–DNA hybridization experiments or DNA base composition estimations (Owen & Pitcher 1985) unless the volumes given in the protocols are scaled up. For the latter type of experiments, it is preferable to use the method of DNA isolation described previously (Owen & Borman 1987).

Cetyltrimethylammonium bromide (CTAB) method

The reagent used is CTAB/NaCl solution (10% CTAB in 0.7 mmol/l NaCl). Dissolve 4.1 g NaCl in 80 ml water and slowly add 10 g CTAB (cetyltrimethylammonium bromide) while heating and stirring. If necessary, heat to 65 °C to dissolve. Adjust final volume to 100 ml.

The method is as follows:
1 Suspend culture in 900 ml 1× TE.
2 Centrifuge 5 min at 12 000 *g*.
3 Remove supernatant and resuspend pellet in 567 μl 1× TE.

4 Add 30 μl 10% sodium dodecyl sulphate (SDS) and 3 μl of 20 mg/ml proteinase K.

5 Mix and incubate for 1 h.

6 Add 100 μl 5 mmol/l NaCl and mix, then add 80 μl CTAB/NaCl and mix. Incubate for 10 min at 65 °C.

7 Add equal volume (0.7 ml) of chloroform/isoamyl alcohol. Mix.

8 Centrifuge for 5 min at 12 000 *g*.

9 Remove supernatant and add equal volume of 25:24:1 phenol:chloroform:isoamyl alcohol. Mix and repeat step 8 above.

10 Collect supernatant, add 0.6 volume isopropanol to precipitate DNA.

11 Centrifuge 5 min to collect DNA and wash once with 0.5 ml 70% ethanol.

12 Dry under vacuum and redissolve DNA in 100 μl 1× TE.

DNA Restriction Endonuclease Analysis (REA)

The aim of this technique is to digest the DNA with a suitable restriction endonuclease to yield a distinctive pattern of fragments after agarose gel electrophoresis. The DNA digest pattern obtained provides a characteristic fingerprint of a bacterial strain.

DNA digestion and electrophoresis

The chromosomal DNA (5 μg) is digested with the appropriate restriction endonuclease (see Table 21.3) – *c.* 1 unit/μg DNA – for 4 h at 37 °C in the buffer recommended by the manufacturers. The digested DNA is electrophoresed at 30 V for 16 h in horizontal 0.8% (w/v) agarose (Gibco-BRL Ltd; ultrapure, electrophoresis grade) gel in 1× TBE buffer containing 89 mmol/l Tris, 89 mmol/l boric acid, 2 mmol/l disodium EDTA (pH 8.3). After electrophoresis, the gels are stained in ethidium bomide (1 μg/ml) and photographed for a permanent record.

Comments. Species studied by REA include *C. jejuni*, *C. upsaliensis*, *C. coli*, *C. lari* and *C. fetus* subsp. *fetus* (Table 21.3). The DNA digest patterns obtained with common restriction endonucleases, such as *Hae*III, *Hind*III, *Bst*EII, *Cfo*I, *Xho*I and *Bgl*II, appear to be highly stable and indicate that they provide a sensitive means of differentiating individual strains. Such patterns are generally complex (>20 bands) and strain differences can be most readily discerned in the higher-molecular-weight region (10–12 kb) of the electrophoretic profile. Another complicating factor with REA is the possible presence of plasmid DNA. Generally, however, the band patterns are treated in the same manner, whether plasmid DNA is present or absent, because naturally occurring *Campylobacter* plasmids appear to be stable strain markers.

TABLE 21.3. *Restriction endonucleases for DNA digest analysis and for ribotyping of* Campylobacter

Species	Restriction endonuclease(s)
C. jejuni	*Hae*III*, *Hind*III*, *Xho*I, *Bgl*II, *Pvu*II*, *Pst*I*
C. coli	*Hae*III*, *Hind*III, *Hha*I
C. upsaliensis	*Hae*III, *Pvu*II
C. fetus	*Hae*III*, *Hha*I, *Bst*EII, *Xho*I
C. mucosalis	*Bgl*II*, *Hind*III*

* Suitable for ribotyping.

Strain comparison and data analysis

Generally, restriction digest patterns of strains are compared visually and, because of their overall complexity, only broad similarities and differences can be detected. Such comparisons are only feasible when samples are run on the same agarose gel. Intergel comparisons are highly subjective and inconsistent. In order to achieve more objective assessments of pattern similarities, densitometric scanning and computer-assisted methods of data recording and numerical analysis have been applied successfully to REA of DNA from several species of *Campylobacter* (Bruce *et al.* 1988; Owen *et al.* 1989; Hernandez *et al.* 1991b). As the methods are complex and highly specialized, they are not appropriate for general identification and typing purposes, but may provide a basis for defining future genomic types.

Ribotyping probes

Nucleic acid probes are used in Southern blot hybridization analyses to highlight specific DNA restriction-site heterogeneities, which in turn can be used to detect strain differences both within and between species. The advantage of this technique is that it greatly reduces the number of bands in the fingerprint and facilitates comparisons between strains. Its disadvantage is that it provides information about specific regions only within the genome and is technically more complicated than REA. The various probes used for this purpose with campylobacters include cloned random chromosomal sequences and rRNA sequences. The rRNA-based probes provide a widely applicable system to investigate the molecular epidemiology of diverse bacteria (Grimont & Grimont 1986), whereas other probes may be more limited in being species-specific or specific only for strains within a particular species.

The rationale for using rRNA genes in DNA fingerprinting has its basis in evolutionary history, because such sequences are known to be highly conserved (Woese 1987). A bacterial rRNA cistron has a total sequence of about 4600 bases comprising several subspecies, which include genes for 23S rRNA (~2500 bases), 16S rRNA (~1500 bases), 5S rRNA (~120 bases) and various intervening and flanking sequences. In eubacteria, the rRNA genes are a small (*c.* 0.1%), but highly conserved, part of the genome and are found in clusters (rRNA operons), within which they are usually linked in the order 16S–23S–5S. Furthermore, the rRNA cistrons are present in multiple copies, which vary from species to species but in *Campylobacter* are present in two to four copies (Romaniuk & Trust 1987).

Many of the rRNA cistron sequences appear to have changed little during bacterial evolution, and so DNA probes specific for these sequences can detect a wide range of bacteria containing similar sequences. Evidence for their conservation is provided by computed homologies between known sequences of rRNA molecules from different species. The 16S rRNA sequences of species in the genus *Campylobacter* are well documented (Thompson *et al.* 1988), and phylogenies based on such data have been constructed. Homologies between 16S rRNA sequences of different *Campylobacter* species are in the range 85–95%, and there is a significant degree of homology of about 80% between these and *Escherichia coli*, and representatives of other Gram-negative prokaryotes, even though they are phylogenetically distant from *Campylobacter* (Thompson *et al.* 1988). The close homologies with *E. coli* have led to the notion of using rRNA genes from *E. coli* as a 'universal probe'. This has the advantage of convenience, since the rRNA is readily available from commercial suppliers.

Biotinylated probe cDNA – universal probe

The biotinylated copy DNA (cDNA) probe is prepared from 1 μg 16 + 23S rRNA from *E. coli* with Moloney mouse leukaemia virus reverse transcriptase (Gibco-BRL Ltd, Poole, UK). It is simultaneously biotinylated by the incorporation of biotin-16-dUTP (Gibco-BRL Ltd) (Pitcher *et al.* 1987). This type of probe has been used successfully for ribotyping campylobacters (Owen *et al.* 1990; Hernandez *et al.* 1991a, b; Fayos *et al.* 1992).

Biotinylated probe cDNA – specific probe

Ribosomal RNA isolated from *C. jejuni* NCTC 11168 has also been used in our laboratory for the preparation of a cDNA probe by reverse transcription, by the methods given above. An evaluation of the two probes was performed by comparing *Hae*III digests of a set of nine strains of *C. jejuni*, some of which were epidemiologically related. Both probes yielded ribopatterns with between five and eight bands, with similar degrees of intensity over the 1–12 kb range. The two probes, despite their different phylogenetic origins, yielded comparable levels of discrimination within *C. jejuni* isolates, and it was concluded that the 'universal' *E. coli* rRNA gene probe could be applied to *Campylobacter* without any apparent loss in discrimination.

Synthetic oligomer probes

The ability to design specific and defined probes provides a significant advance in ribotyping. Romaniuk & Trust (1987) described a 17-nucleotide probe complementary to a unique sequence in the 5′ domain of the 16S ribosomal RNA of campylobacters. Use of this probe provided information about copy number and indicated differences between different species and between identical strains of the same species. Other probes specific for 16S rRNA genes of *Campylobacter* have also been described (Moureau *et al.* 1989; Patton *et al.* 1991). They provide simple one- or two-band profiles that are species-characteristic, but they do not provide sufficient discrimination to separate individual strains.

Southern blot hybridization

After photography, to record the total digest pattern, the restricted DNA in the agarose gel is transferred to Hybond-N membrane (0.45-μm pore size; Amersham International) by vacuum-assisted transfer (Vacu-Gene XL; Pharmacia-LKB), involving depurination by treatment with 0.25 N HCl for 20 min, denaturation in 0.5 mol/l NaOH, 1.5 mol/l NaCl, for 20 min and neutralization in 0.5 mol/l Tris–HCl, 1.5 mol/l NaCl, 1 mmol/l disodium EDTA (pH 7.2), for 1 h. The membranes are washed once in 2× SSC (standard saline citrate – 0.15 mol/l NaCl, 0.015 mol/l trisodium citrate), air-dried and baked at 80 °C for 2 h. Prehybridization (42 °C for 3–4 h) and hybridization (42 °C for 18 h) are carried out according to standard procedures (Pitcher *et al.* 1987). The hybridization reactions are visualized colorimetrically with the BluGene (Gibco-BRL Ltd) non-radioactive detection kit. The pattern of bands containing rRNA gene sequences is designated the ribopattern.

DNA fragment size estimation

DNA band sizes in the Southern blot hybridization patterns are calculated from migration distances, measured direct and converted into molecular size

TABLE 21.4. *DNA band-size windows used for the numerical analysis of* HaeIII *ribopatterns of* C. jejuni *and* C. coli

Band window (locus)	Band size (kb) range	Band window (locus)	Band size (kb) range
1	1.01–1.17	13	3.11–3.38
2	1.18–1.34	14	3.39–3.59
3	1.35–1.49	15	3.60–3.78
4	1.50–1.64	16	3.79–4.01
5	1.65–1.82	17	4.02–4.38
6	1.83–2.01	18	4.39–4.80
7	2.02–2.18	19	4.81–5.18
8	2.19–2.28	20	5.19–5.80
9	2.29–2.43	21	5.81–6.37
10	2.44–2.59	22	6.38–7.50
11	2.60–2.75	23	7.51–8.90
12	2.76–3.10	24	8.91–9.99

with one of several standard algorithms available (Owen & Beck 1987) or by an automated gel reader and analysis system (IBI, New Haven, Conn., USA). Biotinylated λ phage (Gibco-BRL Ltd), digested with *Hin*dIII, is used to provide the size markers. Three lanes (outside left and right, and centre) containing the markers are included on each gel.

Quantification of strain similarities

Coding of data

To compare ribopatterns from different membranes, the bands are coded according to size, to minimize errors when determining similarities by computer-assisted methods of analysis. The blot hybridization patterns are screened for bands within 24 different size ranges, from 1 kb up to 10 kb (Table 21.4), and positive (presence) and negative (absence) results are recorded. Bands of faint intensity, which are probably partial digests, are excluded. Double bands falling within a given range are scored as a single band.

Computations

Computed similarities among strains are estimated by means of the Dice coefficient (negative matches excluded), and clustering of strains is based on the unweighted pair group method to facilitate the generation and the plotting of a dendrogram (Sneath & Sokal 1973). All computations are performed with the DNAGE program, a modification of previously described software for analysis of electrophoretic protein patterns (Costas 1991).

Practical applications

Typing of C. jejuni *and* C. coli

The rRNA gene polymorphisms of about 150 strains of *C. jejuni* and *C. coli* have been determined, and considerable diversity amongst strains is apparent (Fig. 21.1). Our investigations have been based mainly on the use of *Hae*III

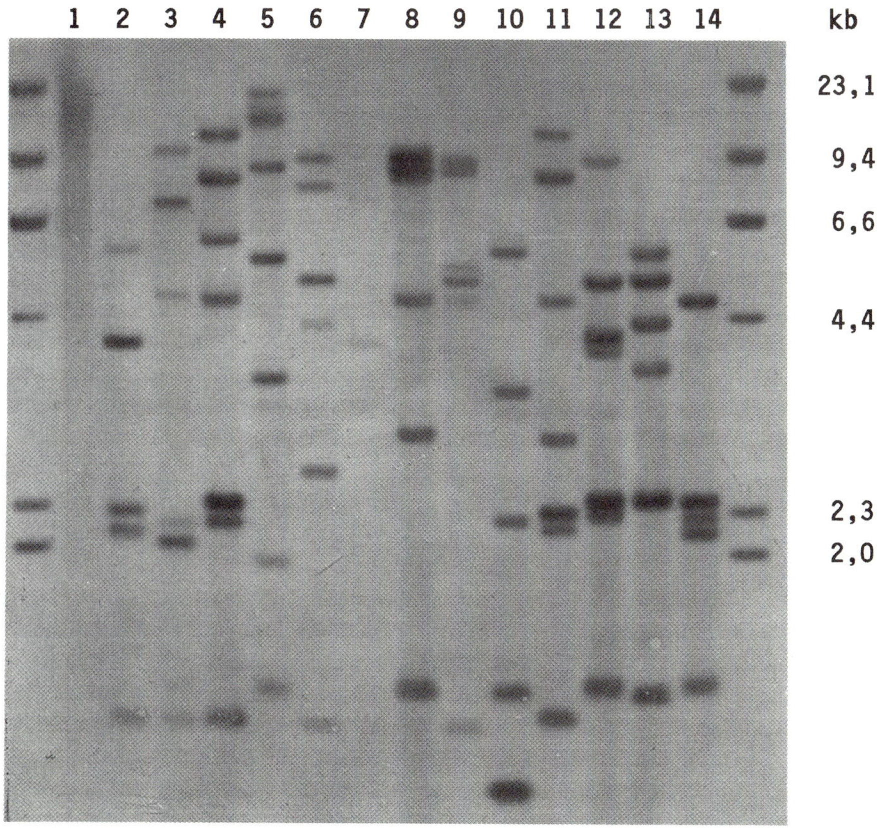

FIG. 21.1. The rRNA gene patterns (ribopatterns) for *Hae*III digests of *Campylobacter* chromosomal DNA probed with cDNA from 16 + 23S rRNA from *Escherichia coli*. Lanes 1–14, DNA from various human faecal isolates of *C. jejuni*. Sizes (kb) indicated are for bacteriophage λ *Hin*dIII DNA digests. Note: DNA in lanes 1 and 7 was degraded.

digests, but Patton *et al.* (1991) concluded that *Pvu*II and *Pst*I digests also provide a sensitive ribotyping system. From comparative studies of these endonucleases, we believe that *Hae*III ribopatterns are the most suitable for data analysis purposes, because they provide a spread of about 21–24 well-defined bands in the 1–10 kb region. Several important observations can be made from present data:

1 Strains of *C. jejuni* and *C. coli* exhibit a diversity of ribopatterns.

2 Sporadic isolates rarely have the same ribopattern, even when they have the same biotype and serotype.

3 Similar ribopatterns amongst epidemiologically related isolates strongly indicate a common source outbreak.

4 Ribopatterns are comparable with bacteriophage typing in the achievable level of discrimination.

5 Ribopatterns provide an accurate and discriminatory method of 'typing' strains derived from single outbreaks, but a general-purpose typing scheme based on ribopatterns has yet to be elaborated for *C. jejuni*.

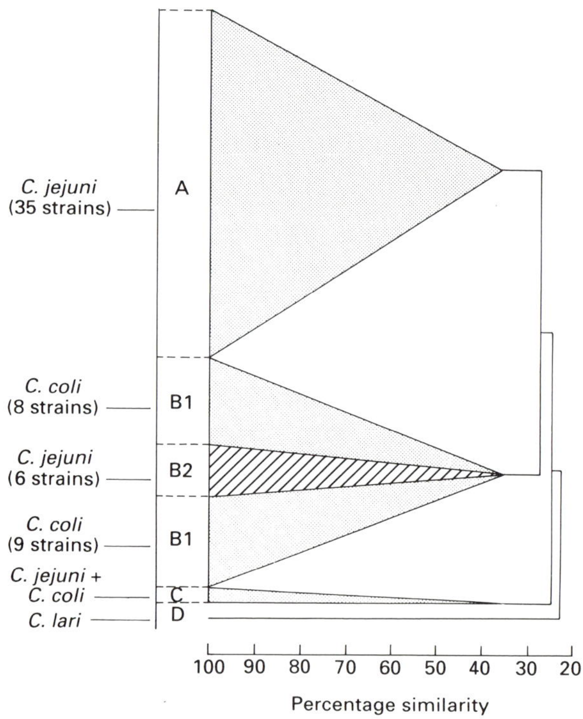

FIG. 21.2.Simplified dendrogram of a numerical analysis based on *Hae*III ribopatterns of *Campylobacter* isolates (vertical axis). The numbers on the horizontal axis indicate the percentage similarities.

Identification of atypical strains

There is an excellent correlation between phenotypically defined species of *Campylobacter*, namely *C. jejuni* (A), *C. coli* (B1 and B2) and *C. lari* (D), and groupings based on numerical taxonomic analysis of *Hae*III ribopatterns (Fig. 21.2). However, a small proportion (<10%) of strains appear to be atypical, either because they are hippurate-positive but are genomically allied to *C. coli* (group B2, Fig. 21.2) or because they have a low relatedness to *C. jejuni*, *C. coli* and *C. lari* (group C, Fig. 21.2). These strains are currently under investigation and may provide the basis for novel genomic groups within the thermophilic campylobacters.

Definition of species-specific markers

Detailed numerical analysis of ribopatterns and construction of databases of band (fragment)-size profiles provide a basis for describing species composite band profiles, reflecting the overall distribution of bands within different species. The composite riboband profiles of *C. jejuni* and *C. coli* are compared in Fig. 21.3. For *C. jejuni*, 50% or more of strains have characteristic bands of 3.0, 5.0 and 9.3 kb, whereas the strains of *C. coli* have bands of 1.5, 2.2, 2.3 and 4.7 kb. A band of 1.3 kb is common to both species. Such fragments provide novel chemotaxonomic markers for these two species. Work is currently in progress to evaluate such markers for other species of *Campylobacter*.

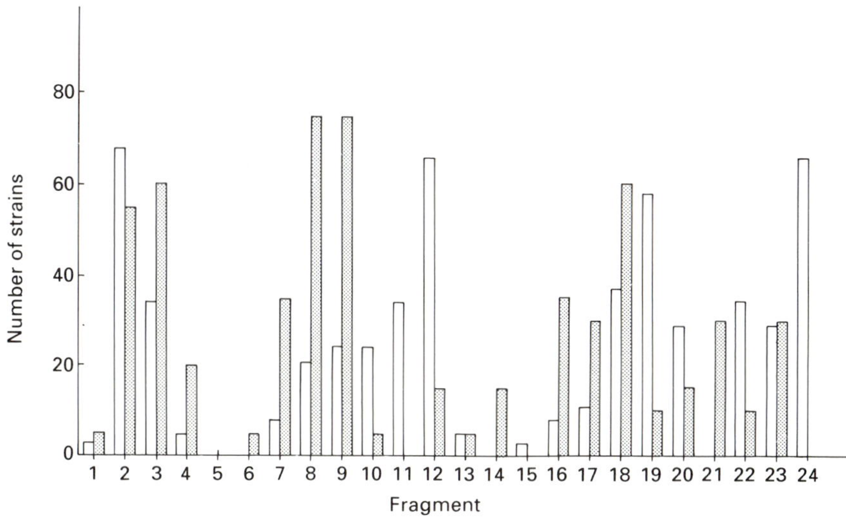

FIG. 21.3. Composite ribopatterns of *C. jejuni* and *C. coli*. The band loci listed relate to the size fragment (kb) bands given in Table 21.4. □, *C. jejuni*; ▩, *C. coli*.

Fingerprinting of *Campylobacter* by RAPD-PCR

Background

The method for *in vitro* amplification of a specific sequence from a portion of DNA, now generally referred to as the polymerase chain reaction (PCR), was first described in 1985 (Saiki *et al.* 1985). It uses a DNA-dependent DNA polymerase to generate additional new copies of a DNA sequence, by elongation of two synthetic oligonucleotides that have been annealed to flank the sequence to be amplified (Saiki *et al.* 1988). In repeated cycles of low- and high-temperature reactions, million-fold amplification of a single copy of a DNA fragment can be accomplished, opening the way for the use of this technique for diagnostic purposes.

DNA amplification by PCR has been successfully applied to the detection of low numbers of bacteria in various specimens, for example, *Listeria monocytogenes* in soft cheeses (Wernars *et al.* 1991). More recently, PCR has been explored as a potentially useful tool for the epidemiological typing of micro-organisms, and a technique has been developed to obtain randomly amplified polymorphic DNA (RAPD) fingerprints (Williams *et al.* 1990). This RAPD analysis has recently been applied to fingerprint several bacteria, including *Campylobacter* (Mazurier *et al.* 1992) and *Listeria* (Mazurier & Wernars 1992). Without the need to extract total DNA, the RAPD technique provides a reproducible strain-specific fingerprint, with whole bacterial cells as the starting material.

Principle of technique

The procedure we have developed is based on the use of the Operon 10-mer primer kit (Operon Technologies Inc., Alameda, CA). A single 10-base oligonucleotide primer is used to amplify genomic DNA. A DNA amplification product is generated for each genomic region that by chance is flanked by a pair of 10-base priming sites in the appropriate orientation and which are within 5000 base pairs of each other. Amplification products are analysed by agarose gel electrophoresis. Genomic DNA from two different bacterial strains often produce different amplification fragment patterns. A particular DNA fragment that is generated for one strain but not for another represents a DNA polymorphism and can be used as a genetic marker.

Methods

Reagents

1 *Taq* DNA polymerase incubation buffer (10× concentrated):

Tris–HCl	100 mmol/l
$MgCl_2$	15 mmol/l
KCl	500 mmol/l
Gelatin	1 mg/ml
pH 8.3 (20 °C)	

2 dNTP solution:

dATP	10 mmol/l
dCTP	10 mmol/l
dGTP	10 mmol/l
dTTP	10 mmol/l

Mix the four solutions and distribute in 20 µl aliquots.

3 Primers. Each Operon 10-mer sample is supplied in a quantity of 15–17 µg/tube. Each primer (Table 21.5 lists base sequences) is resuspended in 500 µl of ultrapure water to give a final concentration of 30 ng/µl, distributed in 20-µl aliquots, and stored at −20 °C.

4 *Taq* DNA polymerase, 5 units/µl.

5 Stop-mix solution:

Bromophenol blue 1%	2.5 ml
Ficoll	2.5 g
EDTA 0.5 mmol/l, pH 8.0	1.0 ml

Make up to 10 ml with Tris–HCl 10 mmol/l.

TABLE 21.5. *Oligoprimers tested on RAPD-PCR of* C. jejuni

Lanes		Primer*	
V12881	NCTC 11168	Code	5′ to 3′
Fig. 21.4(a)			
1	2	OPA-01	CAGGCCCTTC
3	4	OPA-02	TGCCGAGCTG
5	6	OPA-03	AGTCAGCCAC
7	8	OPA-04	AATCGGGCTG
9	10	OPA-05	AGGGGTCTTG
Fig. 21.4(b)			
1	2	OPA-06	GGTCCCTGAC
3	4	OPA-07	GAAACGGGTG
5	6	OPA-08	GTGACGTAGG
7	8	OPA-09	GGGTAACGCC
9	10	OPA-10	GTGATCGCAG

* Operon Technologies Inc., Alameda, CA, USA.

Preparation of bacteria for PCR

The strains are grown on 5% (v/v) defibrinated horse blood agar as described. Chromosomal DNA is extracted and rapidly purified, after lysis of the bacteria with guanidium thiocyanate reagent (Pitcher *et al.* 1988). DNA concentrations are adjusted to 20 ng/µl (A_{260} = 0.4).

PCR protocols

DNA amplification is performed in solution with a total volume of 100 µl containing 100 ng of DNA (5 µl of a 20 ng/µl stock solution), 10 µl of 10× *Taq* polymerase buffer (100 mmol/l Tris–HCl (pH 8.3), 15 mmol/l $MgCl_2$, 500 mmol/l KCl, 0.1% (w/v) gelatin), 60 ng of each primer (2 µl of a 30 ng/µl stock solution), 4 µl of the deoxynucleoside triphosphates (10 mmol/l each in stock solution), 2 units of *Taq* DNA polymerase (0.4 µl of a 5 units/µl stock solution) and 78.6 µl of ultrapure (double-distilled) water. Paraffin oil (200 µl) is added to each mixture to avoid evaporation. For each reaction, the following reagents are added in sequence (on ice):

Ultrapure water	78.6 µl
Taq DNA polymerase buffer (10×)	10 µl
dNTP solution (10 mmol/l each)	4 µl
Primer (30 ng/µl)	2 µl
Taq DNA polymerase (5 units/µl)	0.4 µl
DNA solution (20 ng/µl)	5 µl
Paraffin oil	200 µl

Samples are amplified in an IHB 2024 thermocycler (Pharmacia LKB). After an initial cycle at 94 °C for 5 min, a total of 45 PCR cycles are run under the following conditions: DNA denaturation at 94 °C for 1 min, primer annealing at 36 °C for 1 min, and DNA extension at 72 °C for 2 min. After the final cycle, the reaction is terminated by holding at 72 °C for 10 min. The PCR products are stored in the cycler at 4 °C until collected.

Cycle 1 (1 loop)	5 min 94 °C
Cycle 2 (45 loops)	1 min 94 °C
	1 min 36 °C
	2 min 72 °C
Cycle 3 (1 loop)	10 min 72 °C

Agarose gel electrophoresis

The PCR-amplified samples are analysed by agarose gel electrophoresis on a horizontal 1.2% (w/v) agarose (Gibco-BRL Ltd; ultrapure, electrophoretic

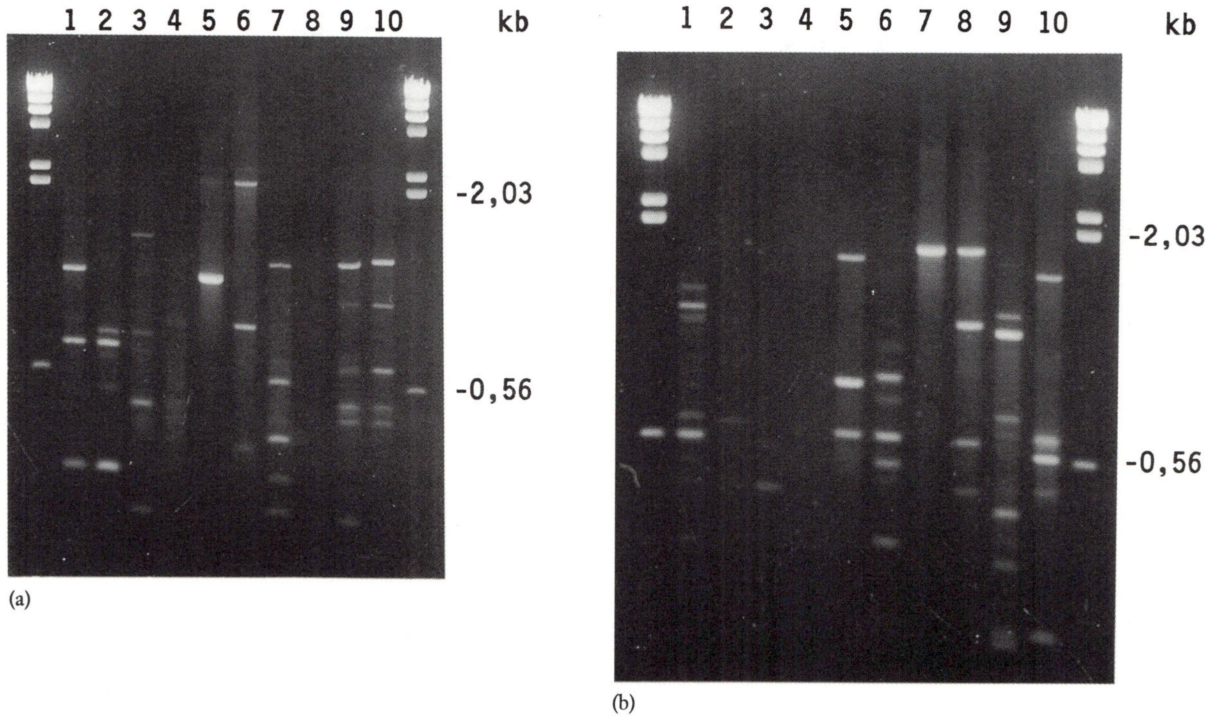

FIG. 21.4. RAPD fingerprints of *C. jejuni* with the different OPA primers. Strains tested were NCTC 11168 and V12881 (faecal isolate, Valencia). See Table 21.5 for the primers used.

grade) gel in 1× TBE buffer with 0.003% (w/v) ethidium bromide incorporated for DNA staining.

Undiluted PCR products (20 µl) are mixed with 4 µl of stop-mix solution, and 20 µl of sample is applied to each well. Bacteriophage λ *Hin*dIII digest fragments are included for molecular weight determination. Gels are run in 1× TBE buffer at 100 V for 2.5 h. The PCR products are visualized and photographed under UV transillumination.

Results and applications

Each of the 10 primers was tested in PCR with template DNA from two reference strains of *C. jejuni*, in order to determine which of the primers was most suitable for fingerprinting purposes. Most of the primers yielded a multiband pattern (Fig. 21.4 (a) and (b)). On the basis of the intensities and distributions of the bands, and the overall pattern reproducibility in repeat PCR, primers OPA-7 and OPA-8 were selected for further use. The PCR fingerprints of the two reference strains were consistent with their ribopattern differences. When OPA-8 was used in PCR of pure DNA from a set of different *C. jejuni* isolates, their patterns were all different, so demonstrating the high discriminatory potential of the technique.

Future Prospects

The substantial body of data now available on the ribopatterns of *C. jejuni* and *C. coli* shows members of these species to be highly diverse in the organization of their *rna* genes and flanking regions. For *C. jejuni*, the data indicate that ribopatterns provide essentially a strain-specific fingerprint. Although such ribopatterns cannot provide the basis of a general molecular typing scheme with, for example, internationally accepted types, they do provide a highly reproducible and accurate means of identifying and defining common source outbreaks and of subsequent strain monitoring after intervention. Ribotyping techniques are technically complex, time-consuming and expensive, and hence unsuitable for large-scale studies of campylobacters in food and the environment. In contrast, the use of PCR fingerprinting, which is significantly faster, provides an important advance in the application of high-resolution identification methods. Initial results on *C. jejuni* indicate that RAPD-PCR fingerprints offer levels of discrimination that are comparable to ribotyping, although further comparative studies are under way to determine the relative advantages of the two techniques.

Acknowledgements

R. Hammond, L. Jeans and N. Dalal are thanked for their assistance in the preparation of this article. We are indebted to the North Atlantic Treaty

Organization (NATO) for a Collaboration Research Grant (No. 890495). We also thank Meeta Desai and Alicia Fayos for their assistance.

References

BOLTON, F.J., HOLT, A.V. & HUTCHINSON, D.N. 1984 Campylobacter biotyping scheme of epidemiological value. *Journal of Clinical Pathology* **37**, 677–681.

BRUCE, D., HOOKEY, J.V. & WAITKINS, S.A. 1988 Numerical classification of campylobacters by DNA-restriction endonuclease analysis. *Zentralblatt für Bacteriologie, Mikrobiologie und Hygiene, Series A* **269**, 284–297.

BUTZLER, J.P. & OOSTEROM, J. 1991 *Campylobacter*: pathogenicity and significance in foods. *International Journal of Food Microbiology* **12**, 1–8.

COSTAS, M. 1991 Microcomputers in the comparative analysis of one-dimensional electrophoretic patterns. *Microcomputers in Biochemistry: a Practical Approach*, ed. Bryce, C.F.A., pp. 189–213. Oxford: IRL Press.

COVER, T.L. & BLASER, M.J. 1989 The pathology of *Campylobacter* infections in humans. *Annual Review of Medicine* **40**, 269–285.

FAYOS, A., OWEN, R.J., DESAI, M. & HERNANDEZ, J. 1992 Ribosomal RNA gene restriction fragment diversity amongst Lior biotypes and Penner serotypes of *Campylobacter jejuni* and *Campylobacter coli*. *FEMS Microbiology Letters* **95**, 87–94.

FRANCO, D.A. 1988 *Campylobacter* species: considerations for controlling a food borne pathogen. *Journal of Food Protection* **51**, 145–153.

GRAJEWSKI, B.A., KUSEK, J.W. & GELFAND, H.M. 1985 Development of bacteriophage typing system for *Campylobacter jejuni* and *Campylobacter coli*. *Journal of Clinical Microbiology* **22**, 13–18.

GRIFFITHS, P.L. & PARK, R.W.A. 1990 Campylobacters associated with human diarrhoeal disease. *Journal of Applied Bacteriology* **69**, 281–301.

GRIMONT, F. & GRIMONT, P.A.D. 1986 Ribosomal ribonucleic acid gene restriction patterns as potential taxonomic tools. *Annales de l'Institut Pasteur Microbiologie* **137B**, 165–175.

HEALING, T.D., GREENWOOD, M.H. & PEARSON, A.D. 1992 Campylobacters and enteritidis. *Reviews in Medical Microbiology* **3**, 159–167.

HEBERT, G.A., HOLLIS, D.G. & WEAVER, R.E. 1985 Biotyping schemes for *Campylobacter jejuni*. *Journal of Clinical Microbiology* **22**, 326–327.

HERNANDEZ, J., FAYOS, A. & OWEN, R.J. 1991a Biotypes and DNA ribopatterns of thermophilic campylobacters from faeces and seawater in Eastern Spain. *Letters in Applied Microbiology* **13**, 207–211.

HERNANDEZ, J., OWEN, R.J., COSTAS, M. & LASTOVICA, A. 1991b DNA–DNA hybridization and analysis of restriction endonuclease and rRNA gene patterns of atypical (catalase-weak/negative) *Campylobacter jejuni* from paediatric blood and faecal cultures. *Journal of Applied Bacteriology* **70**, 71–80.

LIOR, H. 1984 New extended biotyping scheme for *Campylobacter jejuni*, *Campylobacter coli*, and *Campylobacter laridis*. *Journal of Clinical Microbiolgy* **20**, 636–640.

LIOR, H. & WOODWARD, D.L. 1991 A serotyping scheme for *Campylobacter upsaliensis*. *Microbiology and Ecology in Health and Disease* **4**, S89.

LIOR, H., WOODWARD, D.L., EDGAR, J.A., LAROCHE, L.J. & GILL, P. 1982 Serotyping of *Campylobacter jejuni* by slide agglutination based on heat-labile antigenic factors. *Journal of Clinical Microbiology* **15**, 761–768.

MAZURIER, S. & WERNARS, K. 1992 Typing of *Listeria* strains by random amplification of polymorphic DNA. *Research in Microbiology* **143**, 499–505.

MAZURIER, S., VAN DE GIESSEN, A., HEUVELMAN, C.J. & WERNARS, K. 1992 RAPD analysis of

Campylobacter isolates: DNA fingerprinting without the need to purify DNA. *Letters in Applied Microbiology* **14**, 260–262.

MOUREAU, P., DERCLAYE, T., GREGOIRE, D., JANSSEN, M. & CORNELIS, G.R. 1989 *Campylobacter* species identification based on polymorphism of DNA encoding rRNA. *Journal of Clinical Microbiology* **27**, 1514–1517.

OWEN, R.J. & BECK, A. 1987 Evaluation of three procedures using a laser printer densitometer and microcomputer for estimating molecular sizes of restriction endonuclease digest fragments and application to *Campylobacter jejuni* chromosomal DNA. *Letters in Applied Microbiology* **4**, 5–8.

OWEN, R.J. & BORMAN, P. 1987 A rapid biochemical method for purifying high molecular weight bacterial chromosomal DNA for restriction enzyme analysis. *Nucleic Acids Research* **8**, 3631.

OWEN, R.J. & HERNANDEZ, J. 1990 Genotypic variation in '*Campylobacter upsaliensis*' from blood and faeces of patients in different countries. *FEMS Microbiology Letters* **72**, 5–10.

OWEN, R.J. & PITCHER, D. 1985 Chemical methods for estimating DNA base composition and levels of DNA–DNA hybridization. *Chemical Methods in Bacterial Systematics*, eds Goodfellow, M. and Minnikin, D.E., pp. 67–93. London: Academic Press.

OWEN, R.J., MORGAN, D.D., COSTAS, M. & LASTOVICA, A. 1984 Identification of '*Campylobacter upsaliensis*' and other catalase-negative campylobacters from paediatric blood cultures by numerical analysis of electrophoretic protein patterns. *FEMS Microbiology Letters* **58**, 145–150.

OWEN, R.J., COSTAS, M. & SLOSS, L.L. 1988a Electrophoretic protein typing of *Campylobacter jejuni* subspecies '*doylei*' (nitrate-negative campylobacter-like organisms) from human faeces and gastric mucosa. *European Journal of Epidemiology* **4**, 277–283.

OWEN, R.J., COSTAS, M., SLOSS, L.L. & BOLTON, F.J. 1988b Numerical analysis of electrophoretic protein patterns of *Campylobacter laridis* and allied thermophilic campylobacters from the natural environment. *Journal of Applied Bacteriology* **65**, 69–78.

OWEN, R.J., COSTAS, M. & DAWSON, C. 1989 Application of different chromosomal DNA restriction digest fingerprints to specific and subspecific identification of *Campylobacter* isolates. *Journal of Clinical Microbiology* **27**, 2338–2343.

OWEN, R.J., HERNANDEZ, J. & BOLTON, F. 1990 DNA restriction digest and ribosomal DNA gene patterns of *Campylobacter jejuni*: a comparison with bio, sero, and bacteriophage-types of United Kingdom outbreak strains. *Epidemiology and Infection* **105**, 265–275.

PARK, R.W.A., GRIFFITHS, P.L. & MORENO, G.S. 1991 Sources and survival of campylobacters: relevance to enteritidis and the food industry. *Journal of Applied Bacteriology Symposium Supplement* **20**, 975–1065.

PATTON, C.M., WACHSMUTH, I.K., EVINS, G.M. *et al.* 1991 Evaluation of 10 methods to distinguish epidemic-associated *Campylobacter* strains. *Journal of Clinical Microbiology* **29**, 680–688.

PENNER, J.L. 1988 The genus *Campylobacter*: a decade of progress. *Clinical Microbiology Reviews* **1**, 157–172.

PENNER, J.L. & HENNESSY, J.N. 1980 Passive haemagglutination technique for serotyping *Campylobacter fetus* subsp. *jejuni* on the basis of soluble heat-stable antigens. *Journal of Clinical Microbiology* **12**, 732–737.

PITCHER, D.G., OWEN, R.J., DYAL, P. & BECK, A. 1987 Synthesis of a biotinylated DNA probe to detect ribosomal RNA cistrons in *Providencia stuartii*. *FEMS Microbiology Letters* **48**, 283–287.

PITCHER, D.G., SAUNDERS, N.A. & OWEN, R.J. 1988 Rapid extraction of bacterial genomic DNA with guanidium thiocyanate. *Letters in Applied Microbiology* **8**, 151–156.

ROMANIUK, P.J. & TRUST, T.J. 1987 Identification of *Campylobacter* species by Southern hybridization of genomic DNA using an oligonucleotide probe for 16S rRNA genes. *FEMS Microbiology Letters* **43**, 331–335.

Saiki, R.K., Scharf, S., Faloona, F., Mullis, K.B., Horn, G.T., Erlich, H.A. & Arnheim, N. 1985 Enzymatic amplification of β-globin genomic sequences and restriction site analysis for diagnosis of sickle cell anemia. *Science* **230**, 1350–1354.

Saiki, R.K., Gelfand, D.H., Stoffel, S., Scharf, S.J., Higuchi, R., Horn, G.T., Mullis, K.B. & Erlich, H.A. 1988 Primer-directed enzymatic amplification of DNA with thermostable DNA polymerase. *Science* **239**, 487–494.

Salama, S.M., Bolton, F.J. & Hutchinson, D.N. 1990 Application of a new phage typing scheme to campylobacters isolated during outbreaks. *Journal of Epidemiology and Infection* **104**, 405–411.

Skirrow, M.B. 1987 A demographic survey of campylobacter, salmonella and shigella infections in England. *Epidemiology and Infection* **99**, 647–657.

Skirrow, M.B. 1990 *Campylobacter*, *Helicobacter* and other motile curved Gram-negative rods. *Topley & Wilson's Principles of Bacteriology, Virology and Immunology*, 8th edn, vol. 2, *Systematic Bacteriology*, eds Parker, M.T. & Duerden, B.I., pp. 530–549. London: Edward Arnold.

Skirrow, M.B. & Benjamin, J. 1981 The classification of 'thermophilic' campylobacters and their distribution in man and domestic animals. *Campylobacter Epidemiology, Pathogenesis and Biochemistry*, ed. Newell, D.G., pp. 40–44. London: MTP Press.

Sneath, P.H.A. & Sokal, R.R. 1973 *Numerical Taxonomy*. San Francisco: W.H.Freeman Co.

Tenover, F.C., Williams, S., Gordon, K.P., Harris, N., Norlan, C. & Plorde, J.L. 1984 Utility of plasmid fingerprinting for epidemiological studies of *Campylobacter jejuni* infections. *Journal of Infectious Diseases* **149**, 279.

Tenover, F.C., Knapp, J.S., Patton, C. & Plorde, J.L. 1985 Use of auxotyping for epidemiological studies of *Campylobacter jejuni* and *Campylobacter coli* infections. *Infection and Immunity* **48**, 384–388.

Thompson III, L.M., Smibert, R.M., Johnson, J.L. & Krieg, N.R. 1988 Phylogenetic study of the genus *Campylobacter. International Journal of Systematic Bacteriology* **38**, 190–200.

Vandamme, P., Falsen, E., Rossau, R., Hoste, B., Segers, P., Titgat, R. & DeLey, J. 1991 Revision of *Campylobacter*, *Helicobacter* and *Wolinella* taxonomy: emendation of generic descriptions and proposal of *Arcobacter* gen.nov. *International Journal of Systematic Bacteriology* **41**, 88–103.

Wernars, K., Heuvelman, C.J., Chakraborty, T. & Notermans, S.H.W. 1991 Use of polymerase chain reaction for direct detection of *Listeria monocytogens* in soft cheeses. *Journal of Applied Bacteriology* **70**, 121–126.

Williams, J.G.K., Kubelik, A.R., Livak, K.J. & Tingey, S.V. 1990 DNA polymorphisms amplified by arbitrary primers are useful as genetic markers. *Nucleic Acids Research* **18**, 6531–6535.

Wilson, K. 1987 Preparation of genomic DNA from bacteria. *Current Protocols in Molecular Biology*, Unit 2.4.1. New York: Wiley.

Woese, C.R. 1987 Bacterial evolution. *Microbiology Reviews* **51**, 221–227.

22: Epidemiological Typing of Bacterial Isolates by Random Amplification of Polymorphic DNA

K.Wernars and C.J.Heuvelman
Laboratory for Water and Food Microbiology,
National Institute of Public Health and Environmental Protection,
PO Box 1, NL-3720 BA Bilthoven, The Netherlands

The polymerase chain reaction (PCR) has found increasing application in fundamental and applied research sincc it was described by Saiki *et al.* (1988), who reported the use of a thermostable DNA polymerase for the *in vitro* amplification of DNA fragments. The ability of PCR to generate 10^{10}–10^{12} identical copies of a specific DNA fragment in a short time has improved the detection limits for many microorganisms. In these procedurcs, a fragment of DNA, specific for the organism or group of organisms, is amplified for subsequent detection.

Rather than amplifying a specific predetermined DNA fragment, it is possible to use PCR to generate arbitrary products, which can then be used to obtain the DNA fingerprint of a given isolate. Because of the polymorphism in the DNA of bacterial isolates, differences in the fingerprints can be observed, on the basis of which isolates can be distinguished. This method has been designated random amplification of polymorphic DNA (RAPD; Williams *et al.* 1990). The principle on which the method is based is shown in Fig. 22.1. In contrast to the usual PCR format, the RAPD reaction mixture contains only one synthetic oligonucleotide as primer for the amplification of DNA. This primer has only 10 bases and an arbitrary nucleotide sequence. During the annealing step in the amplification process, the primer attaches to the target DNA at sites with a complementary sequence. This is shown schematically in Fig. 22.1 for corresponding stretches of DNA from two isolates, A and B. In strain A the primer can anneal at two locations, while in strain B only one of the sites is available because of a mutation ('M') at the other site. In this case strain A will give rise to amplification products of a DNA fragment of the designated length, whereas in strain B no such product is formed. When the

New Techniques in
Food and Beverage Microbiology

0-632-03755-5

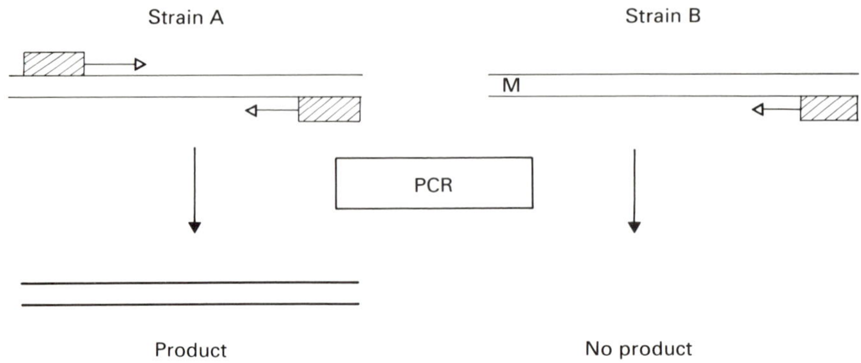

FIG. 22.1.The 10-mer primer molecules are indicated with hatched blocks. DNA synthesis takes place in the direction of the arrows. M = mutation.

amplified reaction mixtures are subjected to electrophoresis on an agarose gel, differences can be observed. Various reports (Mazurier & Wernars 1992; Mazurier *et al.* 1992a, b) have demonstrated, with a simple and efficient experimental procedure, the value of RAPD analysis for studies on the epidemiology of *Listeria monocytogenes* and *Campylobacter jejuni.* This procedure was also found to give satisfactory results for a variety of other bacterial species, including *Bacillus cereus*, *Escherichia coli*, *Plesiomonas shigelloides* and *Staphylococcus* spp. (Wernars *et al.* unpublished).

In this chapter a general protocol for RAPD analysis of bacterial isolates is outlined as a starting point for researchers who wish to apply the technique to microorganisms of their choice. The procedure can be adjusted to meet the specific requirements of the organism under investigation, according to the suggestions in the section 'Optimization and troubleshooting' below.

RAPD Procedure

Bacterial cultures

Bacteria can be cultured either in a liquid medium or, if necessary, on solid media. For the microorganisms mentioned above brain–heart infusion (BHI; Oxoid) has been found to give satisfactory results, both for culture and for RAPD analysis. Cultures in liquid media are incubated at 37 °C to a cell density of 10^8–10^9 colony forming units (cfu)/ml.

Sample preparation

From liquid cultures 0.5–1.5 ml is transferred into a disposable, polypropylene reaction vial (Eppendorf Safe-Lock 2.0 ml, Merck, Darmstadt, Germany) and centrifuged (15 600 *g*, 5 min). The supernatant is then discarded, the cell pellet is resuspended in 1.5 ml of saline (NaCl 9.0 g/l) and the cells are pelleted again by centrifugation. This pellet is resuspended in 0.5 ml sterile distilled water and incubated for 10 min in a boiling water bath. Finally, the boiled lysate is cleared by centrifugation (15 600 *g*, 5 min) and the supernatant is transferred to a clean reaction vial. If necessary, at this stage, the vial can be stored at −20 °C until amplification. The OD_{260nm} is adjusted to 0.15 by dilution with sterile distilled water.

If solid media have been used for culture, cells can be collected by rinsing the surface with saline. The suspension so obtained is treated as described above for the liquid cultures.

RAPD amplification

Several 10-mer oligonucleotides have been found to perform well as amplification primers in the RAPD analysis of a variety of bacterial microorganisms. Primers HLWL74 (5′-ACGTATCTGC-3′), HLWL82 (5′-CGGCCTCTGC-3′) and HLWL85 (5′-ACAACTGCTC-3′) were synthesized on a Gene Assembler (Pharmacia, Milwaukee, USA) by standard phosphoramidate chemistry or they were obtained commercially (10-mer Kit, Operon Technology, Alameda, USA). Primers were dissolved in sterile distilled water and stored at −20 °C. Reaction mixtures (25 μl total volume) contain 0.6 units of *Taq* polymerase (AmpliTaq, Perkin Elmer Cetus, Norwalk, USA), 10 mmol/l Tris (pH 8.3 at 25 °C), 50 mmol/l KCl, 2.0 mmol/l $MgCl_2$, 100 μmol/l of each deoxynucleoside triphosphate (dATP-, dCTP-, dGTP- and dTTP-lithium salts, Boehringer, Mannheim, Germany), 0.3 μmol/l of oligonucleotide primer, 2.5 μl boiled lysate and finally 40 μl of paraffin oil to block evaporation during thermal cycling. The reaction mixture is kept on ice during preparation. Amplification is carried out in a programmable PCR incubator (Thermal Cycler 480, Perkin Elmer Cetus, Norwalk, USA) by repeating the temperature/time sequence: 94 °C for 1 min, 35 °C for 2 min, 72 °C for 2 min, for 45 cycles at the fastest possible transition speed. The first cycle is preceded by an incubation at 94 °C for 3 min to assure complete denaturation of the target DNA, and the final cycle is followed by an incubation of 72 °C for 10 min, to allow the polymerase to extend incomplete products to full length. After amplification, the samples are kept at 5 °C.

Analysis of amplified products

The amplified DNA fragments are resolved by electrophoresis on a 1.6% (w/v) agarose gel (Seakem LE Agarose, FMC BioProducts, Rockland, USA) in TBE buffer (g/l: Tris 10.8, boric acid 5.5, disodium EDTA 0.93), supplemented with ethidium bromide (0.5 μg/ml). After electrophoresis (6 V/cm, 90 min), gels are photographed on a 302-nm UV transilluminator (UV Products Inc., San Gabriel, USA) with a Polaroid camera equipped with a UV filter (Wratten gelatin filter No. 2E, Eastman Kodak Co., Rochester, USA) plus an orange filter (Wratten gelatin filter No. 22, Eastman Kodak Co., Rochester, USA) and Polaroid P/N 665 film. Comparison of the photographed fingerprints is done by eye.

Optimization and Troubleshooting

When RAPD analysis is used as a tool for epidemiological typing of microorganisms, it is essential to obtain reliable and reproducible results. Only then is it possible to compare fingerprints obtained in successive experiments and in different laboratories. Various parameters have been found to have an effect on the appearance and reproducibility of the fingerprints obtained with RAPD. In the following paragraphs the RAPD procedure, as described above, is followed and the most important parameters in each step are discussed.

Bacterial cultures

The action of the *Taq* polymerase enzyme is sensitive to inhibition by compounds present in the reaction mixture. Since inhibitory components from the culture medium can be introduced with the boiled lysate, they can result in poor and irreproducible amplification. The effect of the medium on the action of the polymerase enzyme should, therefore, be examined for interference with the RAPD analysis. We have found that the use of BHI does not have a noticeably inhibitory effect on the amplification. The use of media containing blood should be avoided, because of the strong inhibition produced by haemoglobin. Indeed, culture of bacteria on blood agar plates to obtain cells for RAPD analysis is likely to lead to a complete inhibition of the amplification reaction, because of erythrocytes coharvested with bacteria from the plate.

Sample preparation

The preparation of boiled lysates is simple and has been found to give satisfactory RAPD results for all the bacteria mentioned above. For some bacterial genera the preparation of these lysates can even be omitted. In the case of *Listeria* spp., it was found that good results can be obtained simply by adding whole cells to the RAPD reaction mixture (Mazurier & Wernars 1992; Mazurier *et al.* 1992a). The number of *Listeria* cells added to the reaction mixture is standardized spectrophotometrically (2.5 μl of a suspension with an OD_{600} of 1.5). For *C. jejuni*, however, the addition of whole cells resulted in irreproducible RAPD fingerprints, whereas the use of boiled lysates completely solved this problem.

RAPD amplification

Primers

Although the nucleotide sequence of RAPD primers is arbitrary, the actual sequence of the primer used will determine the appearance of the fingerprint. A satisfactory primer must meet two important conditions: (i) reproducibility of the fingerprints; and (ii) discrimination of different isolates. Although the primers HLWL74, 82 and 85 perform well for the epidemiological typing of the bacterial isolates listed, this is not an absolute guarantee for success with other bacteria, particularly with respect to the level of discrimination obtained. Whether a primer yields satisfactory results must be empirically determined. Guidelines can be given to optimize the design of new 10-mer RAPD primers, with respect to the reproducibility of the fingerprints. First, they should contain at least 50% G and/or C residues. Second, sequences that allow intra- or intermolecular annealing of the primer should be avoided. Finally, primers with G and/or C residues at the two 3′-terminal positions usually perform best. The choice of primers that give satisfactory results is facilitated by the commercial availability of kits containing 20 different 10-mer oligonucleotides.

Magnesium chloride

The concentration of Mg^{2+} in the reaction mixture strongly influences the number of DNA fragments which are amplified. Higher concentrations result in an increase of the number of bands visible in the fingerprint. For each organism the optimal concentration must be experimentally determined in the range 0.5–3.0 mmol/l. Once the optimal concentration of Mg^{2+} ions has been determined, it should be strictly maintained throughout further experiments, because comparison of fingerprints obtained at different concentrations is meaningless.

Target DNA

The concentration of target DNA present in the reaction mixture should be experimentally optimized. For this purpose the values described in the general protocol are a good starting point. Both a very high and a very low concentration will have a negative effect on the final results. A high concentration can lead to a background of numerous faint bands present in the fingerprint, while with a low concentration the fingerprints obtained become irreproducible.

DNA polymerase

We have observed a clear effect of the enzyme preparation used in the RAPD assay. The *Taq* polymerase from some suppliers failed to give reproducible results or yielded no fingerprints at all. Moreover, batch-to-batch differences were observed. Under the experimental conditions described here, the AmpliTaq preparation performed best and no batch-dependent effects were observed.

PCR thermal cycler

The result of the RAPD analysis depends on the sites at which the primer molecules anneal to the target DNA. Therefore, temperature plays an important role, because a decrease in temperature results in a decrease of the annealing stringency. As a result of a lower temperature, both the number and the intensity of bands in the fingerprint will increase. Thus, to obtain reproducible results in the RAPD analysis, it is essential to subject all reaction mixtures to an identical incubation regime. In this respect, lack of temperature uniformity of the PCR machine can be a serious problem, making the appearance of the fingerprint dependent on the position of the reaction mixture in the thermal cycler. Comparison of results obtained with identical reaction mixtures at different positions is a relatively simple way to determine whether a PCR machine is suited for RAPD analysis. As a general rule the variation at 35 °C should not exceed ±2 °C.

Electrophoresis

Since the electrophoretic separation of the amplified DNA fragments finally yields the RAPD fingerprint used for identification and comparison of strains, standardization of the procedure is essential. The shape of the fingerprint is influenced by parameters such as agarose concentration and type, buffer composition, gel temperature, potential difference per centimetre, running time, etc. Only if these are kept constant in subsequent experiments is it possible to compare the results obtained.

Conclusions

RAPD analysis is a powerful new tool for epidemiological studies of micro-organisms. Typical results are illustrated in the publications by Mazurier & Wernars (1992) and Mazurier *et al.* (1992a, b). The relatively simple procedure, outlined in this chapter, should make it possible to apply the technique to many bacterial genera, without any prior information on the DNA sequence. Moreover, highly sophisticated equipment and profound knowledge of molecular biology are not required to produce good results. However, reproducible results are only obtained with great care and adherence to the protocols. Finally, the whole experimental procedure is suited for automation, including the storage and matching of RAPD fingerprints in a database.

References

MAZURIER, S. & WERNARS, K. 1992 Typing of *Listeria* strains by random amplification of polymorphic DNA. *Research in Microbiology* **143**, 499–505.

MAZURIER, S., AUDURIER, A., MARQUET-VAN DER MEE, N., NOTERMANS, S. & WERNARS, K. 1992a A comparative study of randomly amplified polymorphic DNA analysis and conventional phage typing for epidemiological studies of *Listeria monocytogenes* isolates. *Research in Microbiology* **143**, 507–512.

MAZURIER, S., VAN DE GIESSEN, A., HEUVELMAN, K. & WERNARS, K. 1992b RAPD analysis of *Campylobacter* isolates: DNA fingerprinting without the need to purify DNA. *Letters in Applied Microbiology* **14**, 260–262.

SAIKI, R.K., GELFLAND, G.D., STOFFEL, S., SCHARF, S.J., HIGUCHI, R., HORN, G.T., MULLIS, K.B. & EHRLICH, H.A. 1988 Primer-directed enzymatic amplification of DNA with a thermostable DNA polymerase. *Science* **239**, 487–491.

WILLIAMS, J.G.K., KUBELIK, A.R., LIVAK, K.J., RAFALSKI, J.A. & TINGEY, S.V. 1990 DNA polymorphisms amplified by arbitrary primers are useful as genetic markers. *Nucleic Acid Research* **18**, 6531–6535.

Index

THE SOCIETY FOR APPLIED BACTERIOLOGY TECHNICAL SERIES

Book Series Editors: F.A. Skinner (1966–1992)
M. Sussman (1992–)

1 Identification Methods for Microbiologists Part A *1966*
Edited by B.M. Gibbs and F.A. Skinner
Out of print
2 Identification Methods for Microbiologists Part B *1968*
Edited by B.M. Gibbs and D.A. Shapton
Out of print
3 Isolation Methods for Microbiologists *1969*
Edited by D.A. Shapton and G.W. Gould
4 Automation, Mechanization and Data Handling in Microbiology *1970*
Edited by Ann Baillie and R.J. Gilbert
5 Isolation of Anaerobes *1971*
Edited by D.A. Shapton and R.G. Board
6 Safety in Microbiology *1972*
Edited by D.A. Shapton and R.G. Board
7 Sampling – Microbiological Monitoring of Environments *1973*
Edited by R.G. Board and D.W. Lovelock
8 Some Methods for Microbiological Assay *1975*
Edited by R.G. Board and D.W. Lovelock
9 Microbial Aspects of the Deterioration of Materials *1975*
Edited by R.J. Gilbert and D.W. Lovelock
10 Microbial Ultrastructure: The Use of the Electron Microscope *1976*
Edited by R. Fuller and D.W. Lovelock
11 Techniques for the Study of Mixed Populations *1978*
Edited by D.W. Lovelock and R. Davies
12 Plant Pathogens *1979*
Edited by D.W. Lovelock
13 Cold Tolerant Microbes in Spoilage and the Environment *1979*
Edited by A.D. Russell and R. Fuller
14 Identification Methods for Microbiologists (2nd edn) *1979*
Edited by F.A. Skinner and D.W. Lovelock
15 Microbial Growth and Survival in Extremes of Environment *1980*
Edited by G.W. Gould and Janet E.L. Corry
16 Disinfectants: Their Use and Evaluation of Effectiveness *1981*
Edited by C.H. Collins, M.C. Allwood, Sally F. Bloomfield and A. Fox
17 Isolation and Identification Methods for Food Poisoning Organisms *1982*
Edited by Janet E.L. Corry, Diane Roberts and F.A. Skinner
18 Antibiotics: Assessment of Antimicrobial Activity and Resistance *1983*
Edited by A.D. Russell and L.B. Quesnel

19 Microbiological Methods for Environmental Biotechnology *1984*
Edited by J.M. Grainger and J.M. Lynch
20 Chemical Methods in Bacterial Systematics *1985*
Edited by M. Goodfellow and D.E. Minnikin
21 Isolation and Identification of Micro-organisms of Medical and Veterinary Importance *1985*
Edited by C.H. Collins and J.M. Grange
22 Preservatives in the Food, Pharmaceutical and Environmental Industries *1987*
Edited by R.G. Board, M.C. Allwood and J.G. Banks
23 Industrial Microbiological Testing *1987*
Edited by J.W. Hopton and E.C. Hill
24 Immunological Techniques in Microbiology *1987*
Edited by J.M. Grange, A. Fox and N.L. Morgan
25 Rapid Microbiological Methods for Foods, Beverages and Pharmaceuticals *1989*
Edited by C.J. Stannard, S.B. Petitt and F.A. Skinner
26 ATP Luminescence: Rapid Methods in Microbiology *1989*
Edited by P.E. Stanley, B.J. McCarthy and R. Smither
27 Mechanisms of Action of Chemical Biocides *1990*
Edited by S.P. Denyer and W.B. Hugo
28 Genetic Manipulation: Techniques and Applications *1991*
Edited by J.M. Grange, A. Fox and N.L. Morgan
29 Identification Methods in Applied and Environmental Microbiology *1992*
Edited by R.G. Board, Dorothy Jones and F.A. Skinner
30 Microbial Biofilms: Formation and Control *1993*
Edited by S.P. Denyer, S.P. Gorman and M. Sussman
31 New Techniques in Food and Beverage Microbiology *1993*
Edited by R.G. Kroll, A. Gilmour and M. Sussman